Localized Corrosion—
Cause of Metal Failure

A symposium
presented at the
Seventy-fourth Annual Meeting
AMERICAN SOCIETY FOR
TESTING AND MATERIALS
Atlantic City, N. J., 27 June—2 July 1971

ASTM SPECIAL TECHNICAL PUBLICATION 516
Michael Henthorne, symposium chairman

List price $22.50
04–516000–27

75TH ANNIVERSARY ASTM
AMERICAN SOCIETY FOR TESTING AND MATERIALS
1916 Race Street, Philadelphia, Pa. 19103

Library of Congress Catalog Card Number: 72–86243

NOTE

The Society is not responsible, as a body,
for the statements and opinions
advanced in this publication.

Printed in Tallahassee, Fla.
October 1972

Foreword

The Symposium on Localized Corrosion—Cause of Metal Fatigue was presented at the Seventy-fourth Annual Meeting of the American Society for Testing and Materials held in Atlantic City, N. J, 27 June–2 July 1971. The sponsor of this symposium was Committee G-1 on Corrosion of Metals. Michael Henthorne, Carpenter Technology Corp., presided as symposium chairman.

Related
ASTM Publications

Metal Corrosion in the Atmosphere, STP 435 (1968), $27.00 (04–435000–27)

Contents

Introduction

Localized corrosion is a major cause of metal failure in a wide variety of industries. More importantly it often results in unexpected failures because it is difficult to predict and is generally undetectable in its early stages. The total amount of metal attacked may be very small, but by being localized it can have very serious consequences. For example, one relatively small pit in the bottom of a liquid storage vessel will make it useless. Several of the various types of localized attack are not understood, and there is a general lack of standard test methods to evaluate susceptibility.

For these reasons Committee G-1 on Corrosion of Metals sponsored a symposium to review the state of the art for the various types of local attack including exfoliation, intergranular corrosion, crevice corrosion, pitting, and dealloying. These reviews covered: (1) practical aspects—what conditions cause attack, how can it be prevented, etc.; (2) mechanisms; and (3) testing. Papers providing new data on mechanisms and test methods were also presented, and the related topics of galvanic corrosion and hydrogen blistering were covered. Stress-corrosion cracking was not dealt with specifically, but much of the symposium content was relevant to situations which either lead to or exist during stress-corrosion cracking.

The contents will be useful to those who need to prevent, understand, or test for localized corrosion. In almost all cases both the metallurgical and the electrochemical aspects have been dealt with.

Testing methods receive the most emphasis. In particular, the new data and test methods for exfoliation will be of interest to users and producers of aluminum alloys. The papers on pitting, crevice corrosion, intergranular corrosion, dealloying, and galvanic corrosion highlight the mechanistic aspects and the need for better test methods. They will be of special interest to users of stainless steel, nickel base alloys, titanium alloys, and copper alloys.

The practical problems associated with localized corrosion are far from being solved, particularly in the case of pitting and crevice attack. The papers in this volume define the problems and offer some solutions. They will serve as a basis for more work aimed at increasing our understanding of the mechanisms of attack and for the development of standard test methods. Meanwhile they contain much useful information which will enable us to minimize, if not yet completely control, this insidious form of corrosion.

Michael Henthorne
Supervisor, Corrosion Research,
R & D Center, Carpenter
Technology Corp., Reading,
Pa. 19603; symposium chairman.

S. J. Ketcham[1] *and I. S. Shaffer*[1]

Exfoliation Corrosion of Aluminum Alloys*

REFERENCE: Ketcham, S. J. and Shaffer, I. S., **"Exfoliation Corrosion of Aluminum Alloys,"** *Localized Corrosion—Cause of Metal Failure, ASTM STP 516,* American Society for Testing and Materials, 1972, pp. 3–16.

ABSTRACT: This paper is a general review of the subject of exfoliation corrosion of aluminum alloys summarizing both published and unpublished work. Present thinking on mechanisms is presented. Susceptible alloys and the corrosive environments in which exfoliation occurs are discussed, as well as effect of varying degrees of exfoliation on static and dynamic fatigue strength of 7075-T6 and on life of an actual aircraft structure.

Protective coatings and special heat treatments to minimize exfoliation are discussed.

KEY WORDS: corrosion, exfoliation corrosion, aluminum alloys, grain structures, plastic deformation, corrosion tests, salt spray tests, fatigue (materials), accelerated tests

Exfoliation is a term used to describe a type of corrosion in which delamination takes place parallel to the metal surface. When this occurs on the surface, flakes of metal peel or are pushed from the surface due to internal stresses caused by the building up of corrosion products. An example is shown in Fig. 1. This type of attack can also take place below the surface, a condition which is much more insidious, and difficult to detect.

For a metal to exfoliate a highly directional structure of some sort is required. It is most common in rolled or extruded aluminum alloy products in which grains are elongated and flattened. It has been reported also on other alloys such as primary magnesium ingots in chloride solutions [*1*],[2] a magnesium sheet alloy which contained 14 percent lithium and 2 percent aluminum [*2*], and cupro-nickel heater tubes [*3*]. Wrought iron, produced by piling several plates on top of one another, subjected to heat, and passed through a rolling mill, develops zones which behave

[1] Head, Chemical Metallurgy Branch, and metallurgist, respectively, Naval Air Development Center, Warminster, Pa. 18974.

* The opinions expressed are the private ones of the authors and are not to be construed as official or reflecting the views of the Department of the Navy or the Navy Service at large.

[2] The italic numbers in brackets refer to the list of references appended to this paper.

FIG. 1—*Example of severe exfoliation on 7178-T6 aluminum alloy* (×*1*).

differently from one another when subjected to corrosive conditions. Evans considers this zonal corrosion of wrought iron to have much in common with the layer corrosion of light alloys [*4*].

Exfoliation in Aluminum Alloys

Aluminum alloys are most prone to this type of attack. It occurs on 2000 (Al-Cu-Mg), 5000 (Al-Mg) and 7000 (Al-Zn-Cu-Mg) series alloys. It usually is related to susceptibility of these alloys to intergranular attack as a result of heat treatment which has created selective anodic paths at the grain boundaries for electrochemical attack. When intergranular attack develops in a product that has an elongated grain structure parallel to the surface, exfoliation will occur. Such grain structures are found in cold-rolled material and in the unrecrystallized portion of an extrusion. The severity of the exfoliation will depend on the extent to which the grains have been flattened or elongated, as well as to the continuity of the anodic path at the grain boundaries.

In exfoliation susceptible alloys, as corrosion proceeds along multiple narrow paths parallel to the surface, the insoluble products which are formed occupy a larger volume than the metal consumed in producing them. These voluminous corrosion products exert a wedging action which develops lateral tensile forces [*5*]. This results in the splitting, flaking, or delamination of uncorroded layers of metal, as shown in Fig. 2.

The time required for exfoliation corrosion to occur depends greatly upon the thickness of the recrystallized layer normally present on the outside surface of a rolled or extruded product. This outer layer is usually susceptible to intergranular corrosion but, having an equiaxed grain structure, does not exfoliate. In a solid plate or extrusion no exfoliation takes place until the corrosion environment has penetrated to the laminar grain structure of the underlying material which is not recrystallized because it did not receive an equivalent amount of cold work.

Thus, variations in exfoliation susceptibilities and rates of attack can be often attributed to differences in thickness of the recrystallized skin. In service where actual commercial structures or parts have machined surfaces or rivet or fastener holes, the corrosive electrolyte can have easy

FIG. 2—*Example of splitting of layers on 7178-T6 due to buildup of corrosion products (×100 unetched).*

access to the interior directional grain structure, and exfoliation can occur without a long incubation period.

Liddiard et al working with extrusions developed corrosion time curve plots for an aluminum-copper alloy [*6*]. The effective loss in thickness was used as the indication for the degree of exfoliation. By measuring the deflection of specimens under four-point loading before and after exposure in exfoliation tests, the changes in the effective thickness of the extrusion was determined. In contrast to the normal exponential corrosion rates for most aluminum alloys, they found that the corrosion-time curves for specimens that undergo exfoliation are linear. The corrosion-time curves that they developed, for specimens tested with the as-extruded face exposed, showed three distinct portions: the first represents intergranular penetration of the recrystallized skin and is slow; the second represents the rapid undermining of the skin as the attack reaches the highly directional material, and the rate is fast; and the third represents the exfoliation attack on the interior structure which proceeds at a rate intermediate between the first and second. Exposing machined faces which had the recrystallized surface removed produced a curve which after an incubation period showed a single linear rate. Differences in exfoliation resistance on the T/10 plane[3] and the T/2 midplane[3] in a 0.375-in. plate are shown in Fig. 3.

Lifka et al point up another factor which influences susceptibility to exfoliation of sheet and plate material [7]. Grain directionality increases with increased reduction in area of the ingot. Plates over 1 in. in thickness do not show the same degree of flattening at the midplane that thinner plates do and are, therefore, more resistant to exfoliation. However, there is a greater susceptibility to intergranular corrosion because of possible slower cooling on quenching with thick plate. In the thickness range from 0.150 to 0.750 in., the rolled structure is the most prone to exfoliation. With further reduction to thin gage sheet a high degree of recrystallization takes place, and susceptibility to exfoliation decreases.

It has been well established in aluminum alloys that the degree of exfoliation that starts at an edge of a specimen is affected markedly by the method in which the edge was prepared. Where prior plastic deformation at an edge is excessive, exfoliation will be much more severe. The best example of this point is the difference in the degree of exfoliation between a sawed and sheared edge. Since the shearing operation causes severe microstructural deformation it is not surprising that the electrochemical attack along preferred microstructural paths such as grain boundaries would be accelerated. Another prime example of this effect of plastic deformation is in machined holes. When fastener holes are drilled with a

[3] T/10 plane refers to plane exposed when one tenth of the thickness is removed from the plate; T/2 when one half of the thickness is removed.

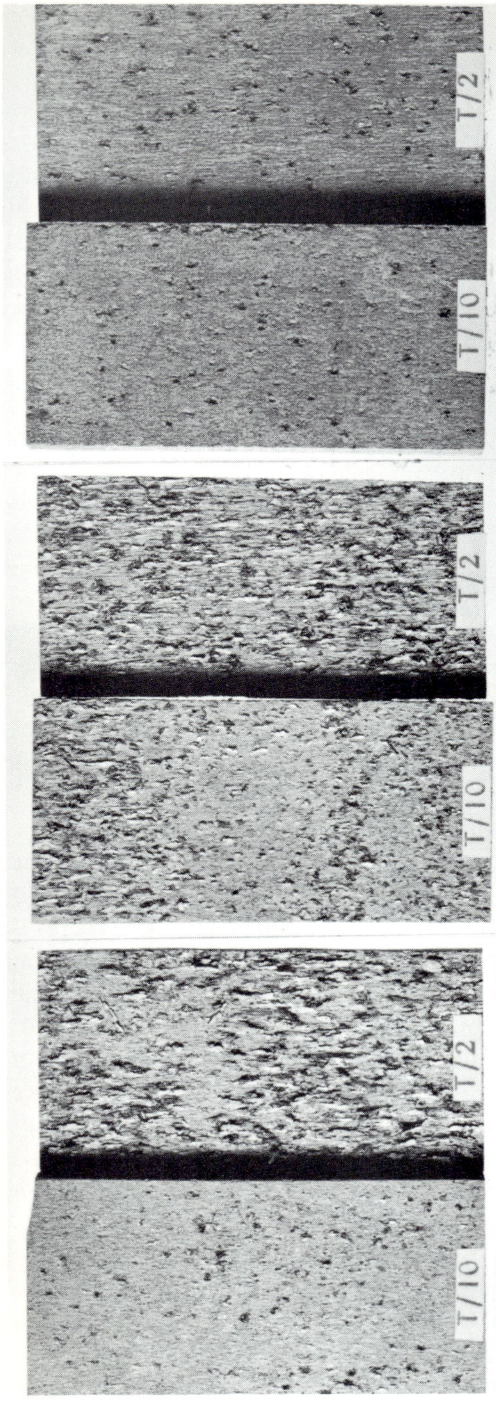

FIG. 3—*Differences in exfoliation resistance on the T/10 plane and the T/2 plane on a 0.375-in. plate of 7178 aluminum. The three panels were aged for different times to produce varying degrees of exfoliation susceptibility* (×0.5).

dull drill bit or when rivets are drilled out of holes, the grain structure distortion that develops can greatly change the exfoliation susceptibility of the end grains in the holes.

Exfoliation has been found in aluminum material which was not susceptible to intergranular corrosion. Godard has referred to cases where it was believed that preferential corrosion paths were provided by insoluble cathodic constituents which formed during solidification in the original cast ingot [*8*]. During subsequent working, these constituents became aligned in stringers unrelated to the final recrystallized grain structure. Electron probe investigations have indicated that the presence of iron and titanium tend to promote exfoliation.

Service Experience

Exfoliation of high-strength aluminum alloys is most prevalent in marine and industrial environments. Pearlstein and Teitell reported severe exfoliation on 2024-T3 exposed to tropical marine environment in Panama [*9*].

It has been stated that exfoliation of light alloys occurs generally in atmospheric exposure, and is not normally encountered under immersion conditions [*6*]. In deep sea exposure, however, at various depths in the ocean, 7002, 7075, and 7178 sheet specimens in the T6 temper exfoliated.

Not long ago, severe exfoliation problems were encountered in the bilge area of service boats constructed of aluminum alloy 5456-H321 [*10*]. Investigation showed that corrosion of the hull plates was due to a high degree of continuity of grain boundary precipitate.

It is in aircraft applications where exfoliation creates the most concern. Several descriptions of service problems have been published [*11–13*]. High-strength aluminum alloys developed during and after World War II, with their significant strength to weight advantages, were at first used conservatively, predominantly in thin gages. The next generation aircraft used extrusions and sculptured plates to produce integrally stiffened parts.

The first problem involving a major structural component occurred about 15 years ago on thick skin material of a seaplane in an area subjected to impingement by the engine exhaust of piston engines. Reinhart has discussed the corrosion of aluminum alloys by exhaust gases [*14*]. Soon other problems developed. One interesting one involved exfoliation in wing fuel tanks of a turboprop plane due to the presence of microorganisms.

Subsequently, the Navy began flying planes that were too large to be stored in the hangar deck of an aircraft carrier and had to be stored on the flight deck exposed to a combination of sea spray and sulfur bearing compounds. Incidence of exfoliation problems, particularly around fastener holes, became much greater. Exfoliation, however, is not restricted to the exterior. Problems have also developed in the interior of wing and fuselage sections where fastener or rivet holes allowed entrance of moisture.

An example of such a problem was in the spar caps of air sea rescue planes. Internal defects were revealed by ultrasonic inspection of the spar caps which were L-shaped 7075-T6 extrusions, bolted and riveted to the wing center section of the aircraft. The defects which were restricted almost exclusively to the skin attachment leg of the spar caps resulted from pitting corrosion that initiated deep within the rivet holes and developed as exfoliation along the highly directional interior grain structure of the extrusions, as shown in Fig. 4.

An investigation was made to determine to what extent this internal exfoliation had compromised the structural integrity of the spar caps, and the results have been published [*15*].

Specimens were taken from locations with varying degrees of surface corrosion and of internal defects as indicated by ultrasonic inspection. Fatigue tests were conducted under constant amplitude and under spectrum loading. Results are presented graphically in Fig. 5.

The results of the constant amplitude fatigue tests indicated that the amount of surface corrosion present had no deleterious effect on fatigue life. Likewise slight laminar attack had only a negligible effect. Three specimens with severe exfoliation in addition to some surface pitting showed an average life reduction of more than 40 percent and a maximum life reduction of almost 60 percent compared to specimens which had suffered no corrosion.

The results of the variable amplitude or spectrum loaded fatigue tests indicated that at the limit load of 30,000 psi the average life was unaffected for the slightly exfoliated material but was reduced more than 50 percent for the severely exfoliated material.

Illustrations of the internal exfoliation which adversely affected the fatigue strength of spar caps are found in Fig. 6. These photomicrographs show that the corroded laminar paths are preferential sites for the initiation of fatigue cracks.

Since the effect of exfoliation on the fatigue life of the spar caps was considerable, it was decided that a full scale wing fatigue test of the aircraft should be conducted, to provide the basis for establishing service life criteria for spar cap replacement or modification.

Subsequently an aircraft of the same model with a total of 7216 flight hours was withdrawn from service and used as the test vehicle. During the fatigue test on the wing 8200 test hours were accumulated before catastrophic failure occurred. Post-failure examination revealed the presence of exfoliation in the wing main beam lower spar cap. At this point the left hand wing was reinforced with steel straps, and testing continued. After another 4000 h, the left wing failed. Post-failure inspection indicated that this left wing had more exfoliation than the right wing which

FIG. 4—*Cross sections through 7075-T6 spar cap rivet holes showing pitting corrosion, and subsequent development into exfoliation* (top) $\times 5$ *and* (bottom) $\times 250$ *unetched.*

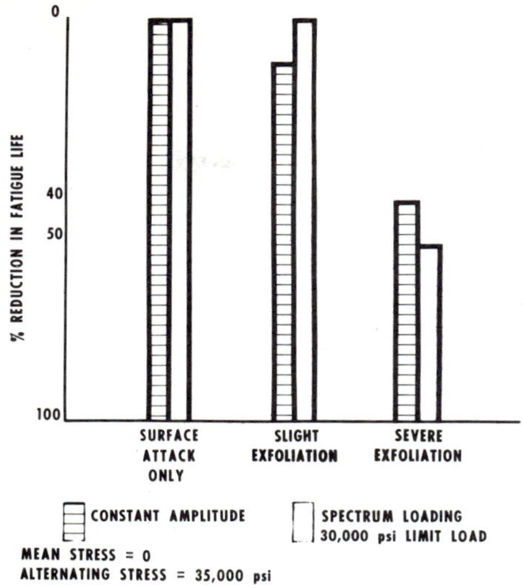

FIG. 5—*Results of fatigue tests on 7075-T6 spar cap specimens.*

failed first. It was concluded that the existence, but not necessarily the amount of exfoliation, is a determining factor in fatigue life. That is, the existence of even a small amount of exfoliation can initiate a fatigue crack [*16*].

Protection Against Exfoliation

Metallic aluminum, zinc, and magnesium coatings (or combinations thereof) will provide protection against exfoliation. Carter et al describe the use of sprayed metal coatings to prevent exfoliation [*17*]. Paint coatings are inadequate to prevent exfoliation around fastener and rivet holes in highly stressed areas because the films are too brittle and crack. On older model military aircraft exfoliation is being minimized by the use of elastomeric polysulfide sealants, particularly over fastener patterns and around faying surfaces [*18*].

The other approach is the use of extended aging to produce exfoliation resistant tempers [*19*]. The effect of these extended or overaging treatments is to lessen the tendency for intergranular corrosion to occur. During extended elevated temperature aging, additional precipitation takes place in the subgrain structures, providing a more equipotential situation which lessens the rate of grain boundary penetration and exfoliation. The effectiveness of these special aging treatments which are used commercially has been demonstrated by laboratory and atmospheric exfoliation tests. The T73 aging temper for 7000 series and the artificially aged tempers of the 2000 series aluminum alloys used to minimize or eliminate susceptibility to stress corrosion cracking also understandably have proven to produce immunity to exfoliation. There is, however, a significant de-

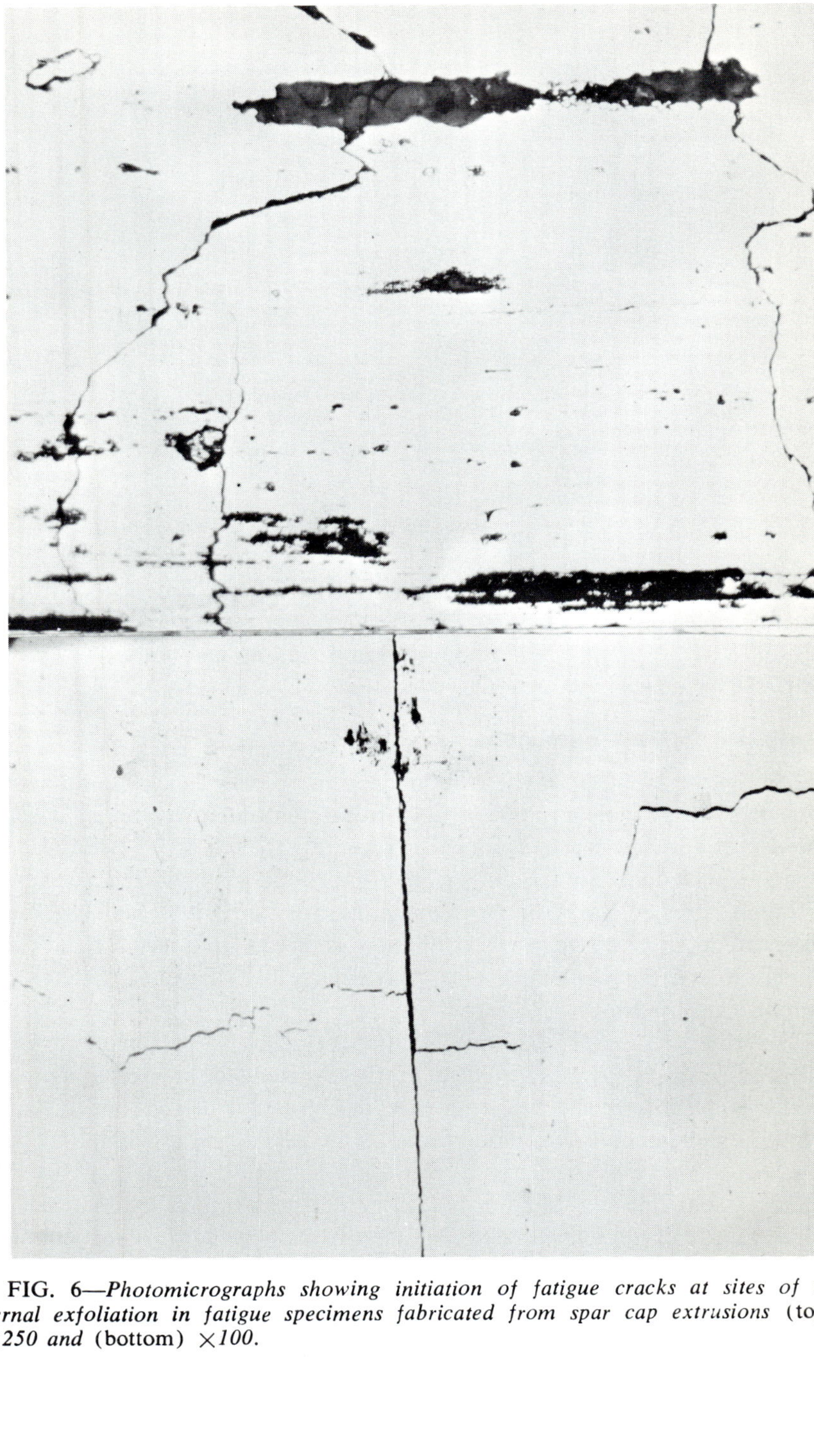

FIG. 6—*Photomicrographs showing initiation of fatigue cracks at sites of internal exfoliation in fatigue specimens fabricated from spar cap extrusions* (top) *×250 and* (bottom) *×100.*

crease in strength (10 to 12 percent) with the T73 temper compared to the older, more commonly used T6 temper. It was found that less overaging was required for exfoliation immunity than for stress corrosion immunity. For this reason a T76 temper having exfoliation immunity and less loss in strength from the T6 temper was developed. Exfoliation resistant tempers (H116 and H117) were developed recently for 5085 and 5456 alloy boat hull plate.

Test Methods

When exfoliation corrosion problems began to crop up in service, users became critical of laboratory tests which had not detected this possibility. The standard 5 percent salt spray test (ASTM B 117–64) produced pitting attack but not exfoliation. A test was developed using 5 percent salt solution acidified to pH3 with acetic acid [*20*]. The test was cyclic, alternately spraying and allowing the specimens to become dry. As more experience was gained with this test, its ability to detect borderline conditions was questionable.

Several researchers experimented with electrochemical techniques. Budd and Booth used a potentiostatic approach [*21*]. In the authors' laboratory a galvanostatic method was tried. Although both appeared promising, the electrochemical techniques did not lend themselves to the multiple testing required for production control of heat treatment.

In the meantime, a cyclic acidified, synthetic sea salt spray test was proposed [*22*] and also a 5 percent salt spray test involving introduction of sulfur dioxide into the cabinet at specified intervals [*23*]. The latter test was developed to simulate the environment of an aircraft carrier which combines sea spray and sulfur from the carrier stack gases.

Results of a comparison of the three salt spray tests and a new immersion test were published which indicated the acidified synthetic sea salt spray was the most discriminating for detecting behavior of exfoliation resistant tempers of 7178 [*23*]. Figure 7 shows a comparison of the exfoliation produced by the acidified sodium chloride and acidified sea salt spray tests on identical specimens of 0.375 in. plate of 7178.

Because of the exfoliation problems on the military boats mentioned earlier, Navy authorities decided to incorporate a test for determining susceptibility to exfoliation into the specification for 5456 alloy. However, there was lack of agreement among producers and users of the exfoliation resistant tempers as to the best test method. This prompted the formation of a task group among Aluminum Association members for 5000 series testing and a joint task group for 7000 series alloys of ASTM G01/05 and the Aluminum Association. Both have been conducting interlaboratory testing programs of several test methods. The object of the ASTM program was to prepare a recommended practice for the method which in the opinion of the participants was the most satisfactory for the 7000

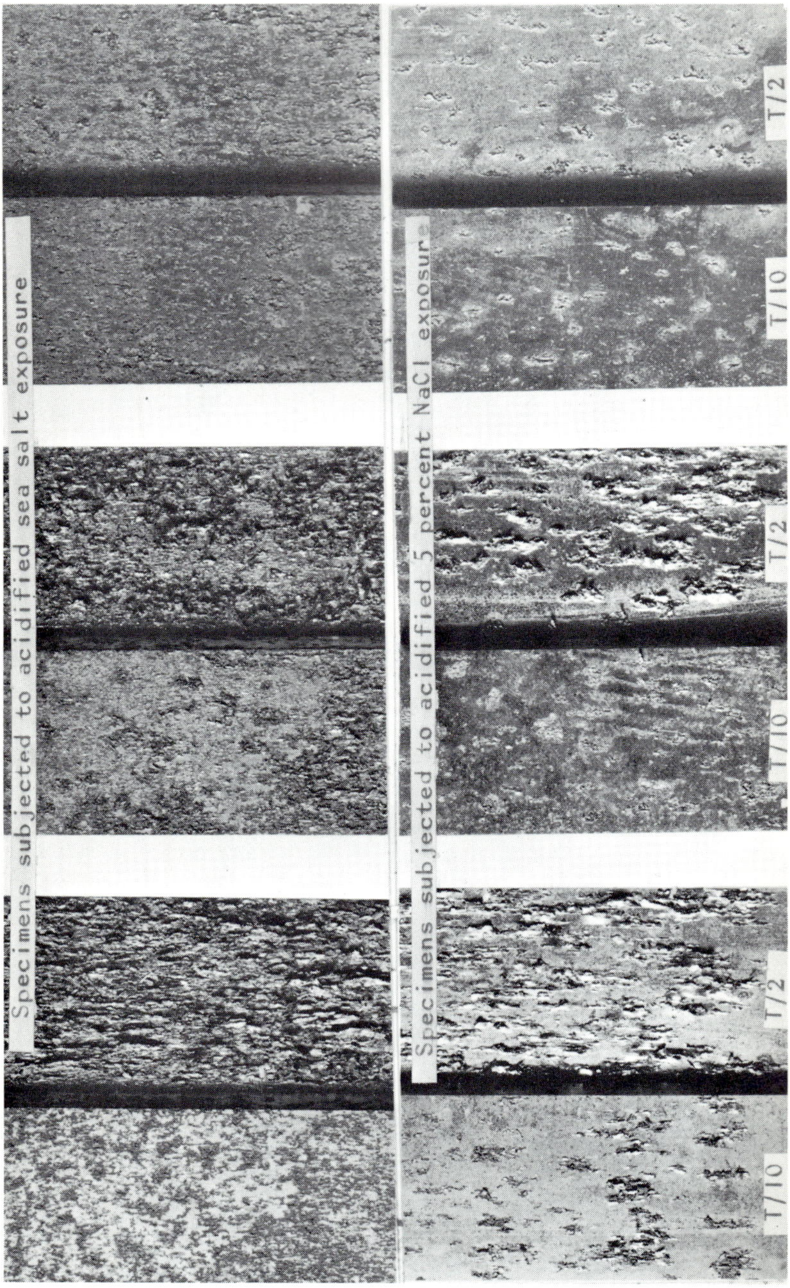

FIG. 7—A comparison of the exfoliation produced by the two salt spray tests on identical specimens of 0.375-in. plate of 7178.

series. A description of the 7000 series program is published elsewhere in this book.

Summary

With the increased use of exfoliation resistant tempers and the improvement in test methods for detecting susceptibility, the incidence of severe exfoliation problems on current and future production of military weapons systems or other equipment fabricated from aluminum alloys should substantially decrease. The exfoliation resistant tempers in the 2000 and 7000 series alloys are, however, more prone to pitting corrosion as a result of the equidistribution of precipitates throughout the matrix. Good protective finishing systems for these alloys, therefore, will still be necessary in marine and industrial environments.

In older equipment where exfoliation prone tempers are still in use, constant vigilance is required so that the inception of exfoliation can be detected and arrested. Ultrasonic inspection equipment of the type used in the spar cap investigation is useful for this purpose. The importance of a continuing program of replacement of structural components fabricated from exfoliation resistant tempers cannot be overemphasized, particularly in those cases where structural integrity under dynamic conditions is required.

References

[*1*] Schiebold, E. and G. Siebel, *Zeitschrift für Physik,* ZEPYA, Vol. 69, 1931, p. 459.
[*2*] Godard, H. P., Jepson, H. P., Bothwell, M. R., and Kane, R. L. in *The Corrosion of Light Metals,* Wiley, New York, 1967, p. 271.
[*3*] Castle, J. E., Harrison, J. T., and Masterson, H. G., *British Corrosion Journal,* BCRJA, Vol. 1, No. 4, Jan. 1966, p. 143.
[*4*] Evans, U. R. in *The Corrosion and Oxidation of Metals,* Edward Arnold Ltd. London, 1961, p. 511.
[*5*] Nielsen, N. A., *Physical Metallurgy of Stress Corrosion Fracture, Metallurgical Society Conferences,* Vol. 4, Interscience, N. Y., 1959, p. 121.
[*6*] Liddiard, E. A. G., Whitaker, J. A., and Farmery, H. K., *Journal,* Institute of Metals, JIMEA, Vol. 89, 1960–1961, p. 377.
[*7*] Lifka, B. W., Sprowls, D. O., and Kaufman, J. G., *Corrosion,* SBIIA, Vol. 23, No. 11, Nov. 1967, p. 335.
[*8*] Godard, H. P., private communication.
[*9*] Pearlstein, F. and Teitell, L., *Proceedings,* 7th Annual National Conference on Environmental Effects on Aircraft and Propulsion Systems, 1967.
[*10*] Sprowls, D. O., Summerson, T. J., Romans, H. B., and Sansonetti, S. J., Aluminum Association Report, issued 1969.
[*11*] Tate, R., *Chemistry and Industry,* CHINA, 27 April 1957, p. 506.
[*12*] Klang, D. R. and Peters, W. L., *Materials Protection,* MAPRA, April 1964, p. 16.
[*13*] Ward, C. B. and Cohen, B., *Light Metal Age,* LMAGA, Vol. 24, Feb. 1966, p. 16.
[*14*] Reinhart, F. M., *Corrosion,* SBIIA, Vol. 10, No. 12, Dec. 1954.
[*15*] Shaffer, I. S., Sebastian, J. C., Rosenfeld, M. S., and Ketcham, S. J., *Journal of Materials,* JMLSA, Vol. 3, No. 2, 1968, p. 400.

[*16*] Berman, L., "Determination of Life Remaining in Model XX Airplane Wing," Naval Air Development Center Report, NADC-ST-7007, June 1970.
[*17*] Carter, V. E. and Campbell, H. S., *Journal,* Institute of Metals, JIMEA, Vol. 89, 1960–1961, p. 472.
[*18*] Miller, R. N., *Material Protection,* Vol. 6, No. 2, 1967.
[*19*] Bell, W. A. and Campbell, H. S., *Journal,* Institute of Metals, JIMEA, Vol. 89, 1960–1961, p. 464.
[*20*] Lifka, B. W. and Sprowls, D. O., *Corrosion,* SBIIA, Vol. 22, No. 1, Jan. 1966, p. 7.
[*21*] Budd, M. K. and Booth, F. F., *Corrosion,* SBIIA, Vol. 18, No. 5, May 1962.
[*22*] Romans, H. B., *Materials Research and Standards,* MTRSA, Nov. 1969.
[*23*] Ketcham, S. J. and Shaffer, I. S., *Proceedings,* 4th International Congress on Metallic Corrosion, 1969.

H. L. Craig, Jr.[1]

Nitric Acid Weight Loss Test for the H116 and H117 Tempers of 5086 and 5456 Aluminum Alloys

REFERENCE: Craig, H. L., Jr., "**Nitric Acid Weight Loss Test for the H116 and H117 Tempers of 5086 and 5456 Aluminum Alloys,**" *Localized Corrosion—Cause of Metal Failure, ASTM STP 516,* American Society for Testing and Materials, 1972, pp. 17–37.

ABSTRACT: This test was developed to indicate the resistance of aluminum-magnesium alloys 5086 and 5456 in the H116 and H117 tempers to exfoliation and intergranular corrosion. The test consists of measuring the weight loss per unit area of a 2 by 0.25 in. by gage specimen immersed in concentrated nitric acid at 30 C (86 F) for 24 h. The acid dissolves a second phase, an aluminum-magnesium intermetallic compound, in preference to solid solution of magnesium in aluminum. When this compound is precipitated in a continuous network along grain boundaries, the effect of the preferential attack is to eat around the grains, causing them to fall away from the specimen. Weight losses for materials proven susceptible to exfoliation or intergranular corrosion amount to the hundreds of milligrams per square inch, while samples of 5086 and 5456 in the H116 and H117 tempers lose only from 10 up to 100 mg/in.^{2}.

Results of tests performed on commercially produced materials are presented, as well as several examples using other tempers and alloys which have shown varied resistance to exfoliation or intergranular corrosion in service.

KEY WORDS: corrosion, aluminum magnesium alloys, exfoliation corrosion, intergranular corrosion, corrosion tests, corrosion resistance, nitric acid, salt spray tests, immersion tests (corrosion), etching

In order to ensure that the Department of Defense has available to it materials with superior resistance to corrosion, members of the Aluminum Association cooperated in an examination of test methods for use in specifying 5086 and 5456 alloys for boat and ship hull construction [*1*].[2] The chemical composition and mechanical properties of these and a closely related alloy, 5083, are given in Table 1.

The present Federal Specification [*2*] calls for a metallographic exami-

[1] Research supervisor, Metallurgical Research Division, Reynolds Metals Co., Richmond, Va. 23218.

[2] The italic numbers in brackets refer to the list of references appended to this paper.

TABLE 1—*Nominal chemical compositions and typical mechanical properties of alloys 5086, 5083, and 5456.*[a]

Alloy-Temper	Element[b] Mg	Mn	Cr	Tension Strength[c] Ultimate	Yield	Elongation, percent in 2 in.
5086-H32 (H116, H117)	4.0	0.45	0.15	42	30	12
5083-H321	4.45	0.6	0.15	46	33	16
5456-H321 (H116, H117)	5.1	0.8	0.12	51	37	16

[a] Reference: Aluminum Standards and Data, 1970–1971, The Aluminum Association, 1, Dec. 1969, Table 1.1, p. 14, and Table 2, pp. 28–29, Federal Specifications QQ-A-00250/19 and 20, have replaced the H32 temper for 5086 and the H321 temper for 5456 with the designations, H116 and H117.

[b] Values in percent—balance, Al plus normal impurities.

[c] Ksi.

tion of each production lot for susceptibility to exfoliation and intergranular corrosion. The criterion is that "the microstructure is predominantly free of a continuous grain boundary network of aluminum-magnesium precipitate." This examination is carried out by etching a polished specimen with 40 percent phosphoric acid solution at 95 F (35 C) for 3 min, and examining the resultant etched surface at 500 magnification.

A reference micrograph is established on a "first-article" basis for sheet and plate which is representative of mill production. The corrosion resistance of this "first-article" reference material is judged by exposing samples for one week to the synthetic seawater acetic acid intermittent salt spray test (SWAAT) [*2,3*].

In an Aluminum Association round robin, the SWAAT test produced exfoliation of samples of 5456-H321 which had experienced exfoliation corrosion in actual service. The same test conditions caused only negligible corrosion of materials believed to have improved resistance to exfoliation, such as 5086 and 5456 in the H116 and H117 tempers [*4*].

This test, it is recognized, does not always cause exfoliation of materials believed to exhibit borderline susceptibility. Most workers ascribe this to the difficulty in conducting a reproducible salt spray test. Because of this, further work is being carried out on immersion tests. This paper presents the results obtained with one of these tests, the nitric acid weight loss test. It also describes the development and reasoning behind the procedure. The second test, called ASSET, is the subject of another report.[3]

Premise of Test Procedure

The nitric acid immersion test is based on the same philosophy as the metallographic examination described in the pertinent Federal Specification [*2*]. In fact, it might be regarded as a quantitative metallographic

[3] See p. 38.

procedure, as it measures a characteristic of the microstructure of the metal. This test is not intended to supplant the metallographic examination of each production lot, but rather is intended to provide a means of quantitative measurement of precipitate continuity. It does not produce the visual appearance of exfoliation in the test specimens. However, the predominant corrosion attack is clearly intergranular as indicated by metallographic examination of the exposed specimens. An example is shown in Fig. 1.

The attack is intergranular because nitric acid dissolves the magnesium-aluminum intermetallic compound while not appreciably attacking solid solution of magnesium in the aluminum matrix. Due to the nature of the precipitation process, it is possible to have a continuous network of the intermetallic compound surrounding each grain, while the grain interior remains relatively free of this precipitate. It is this condition that gave rise to the exfoliation and intergranular corrosion experienced in service by materials in ships hulls [*5*].

Using the SWAAT test described previously and other tests, the aluminum producers developed two tempers for 5086 and 5456 alloys, each of which imparts resistance to exfoliation and intergranular corrosion. In this temper, the alloys do not contain this continuous network of grain boundary precipitate. Consequently, immersion in nitric acid causes little weight loss (see Fig. 2).

On the other hand, in susceptible material, prolonged exposure to nitric acid dissolves the envelope of intermetallic compound precipitated around the grains. Entire grains become dislodged and fall away from the specimen, thereby causing large weight losses, as illustrated in Fig. 1.

This test defines, on a quantitative scale, whether or not precipitation is of such an extent as to cause grain dropping in nitric acid. The problem comes in selecting a pass-or-fail value in relation to the material performance in various environments other than nitric acid. Reliance must be placed, then, on a sampling of materials which have been in service, and which may or may not have suffered exfoliation or intergranular corrosion. Such a selection is presented in this paper.

The decision on the magniture of the weight loss value a corrosion resistant material should exhibit is pending further evaluation within the Aluminum Association.

Test Procedure (summary)

The test is carried out in the following manner: a specimen is machined to size. A convenient size is 2 by 0.25 in. by gage, with the 2 in. dimension in the direction of working. It is convenient to limit the specimen gage to 1 in., and the specimen should be cut to preserve one worked surface.

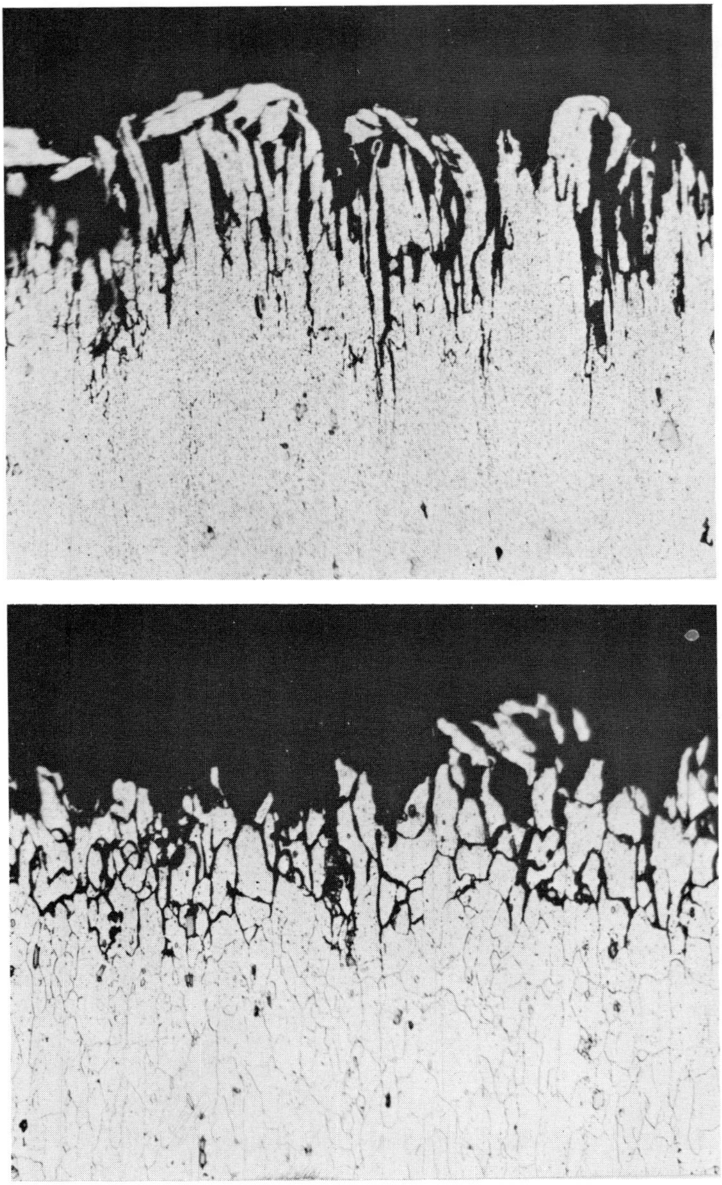

(*top*) Lamellar grain structure, 5456 alloy, nitric acid weight loss, 177 mg/in.^{2}

(*bottom*) Equiaxed grain structure, 5086 alloy, nitric acid weight loss, 129 mg/in.^{2}

FIG. 1—*Intergranular corrosion of test specimens after immersion in nitric acid* (×*250; etch, nitric acid*).

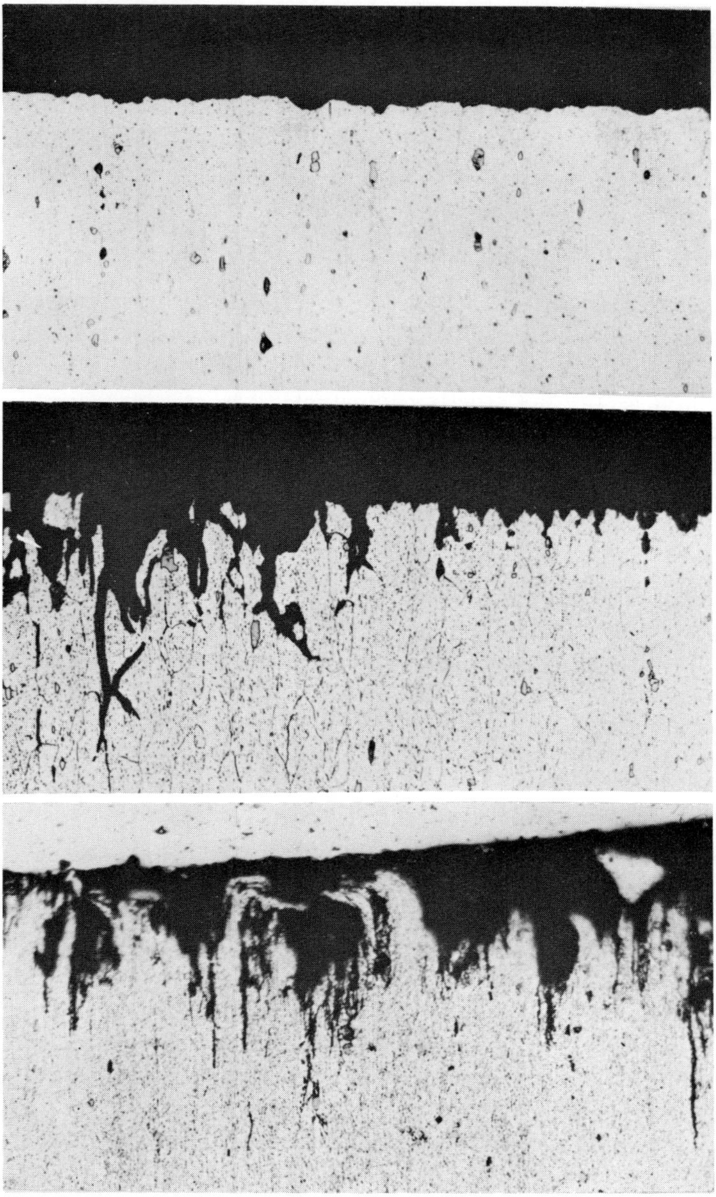

(*top*) 5086 alloy, 8 mg/in.^{2} weight loss in test.
(*middle*) 5086 alloy, 23 mg/in.^{2} weight loss in test, slight intergranular corrosion attack.
(*bottom*) 5456 alloy, 60 mg/in.^{2} weight loss in test, slight intergranular corrosion attack.

FIG. 2—*Corrosion attack on test specimens after immersion in nitric acid* (*×250; etch, nitric acid*).

The specimen is cleaned by caustic etching prior to weighing, and then the weight is determined to the nearest tenth of a milligram. It is then immersed in concentrated nitric acid and held at 30 C (86 F) for 24 h. Upon removing the specimen from the acid, it is rinsed and any loose particles (grains) of metal are brushed off with a stiff bristle brush. The specimen is dried, weighed, and the weight loss per unit area (WLV) is calculated. A detailed procedure is given in the Appendix.

Notes on Test Procedure

It has been observed that the transverse surfaces of the specimen are attacked more heavily than the worked surfaces. This results in greater weight losses per unit area on those surfaces. Therefore, WLV from two different gages are not strictly comparable. This is a second order effect, however, and for purposes of this test may be neglected. It is illustrated in Fig. 3.

The effects of alloy composition, temper, and grain morphology also appear to be of second order in magnitude. Whether these effects are significant should be determined by the person running the test. It should be pointed up that, with a large WLV, the poor reproducibility due to the variation inherent in removing loose grains from the specimen may obscure these effects.

There are undoubtedly environments in which materials with large WLVs can be used with little concern for corrosion. Thus, material acceptability is a joint responsibility of the consumer and the producer. In specifying the H116 and H117 tempers for 5086 and 5456 alloys, the consumer is assured that these alloys are in their most corrosion-resistant tempers.

Test Development

As stated previously, this test is an outgrowth of a metallographic procedure. Two acids are commonly used to reveal the precipitation of the aluminum-magnesium intermetallic: phosphoric acid and nitric acid.

Some data from a comparative study are presented in Table 2; examination of the data illustrates the reasoning for the choice made between the two acids. Condition A represents 5456 alloy which has been heated at a temperature and for a time period which produces an intergranular precipitate around the grains. Condition B, on the other hand, has a structure with all the magnesium in supersaturated solid solution. Concentrated phosphoric acid attacks B to a significant extent whenever A is attacked. Thus, phosphoric acid is not very selective towards the aluminum-magnesium intermetallic compound. Nitric acid, on the other hand, produces only minimal weight losses for B, while registering relatively large weight losses for A.

Another variable was considered, that of specimen size. A comparison

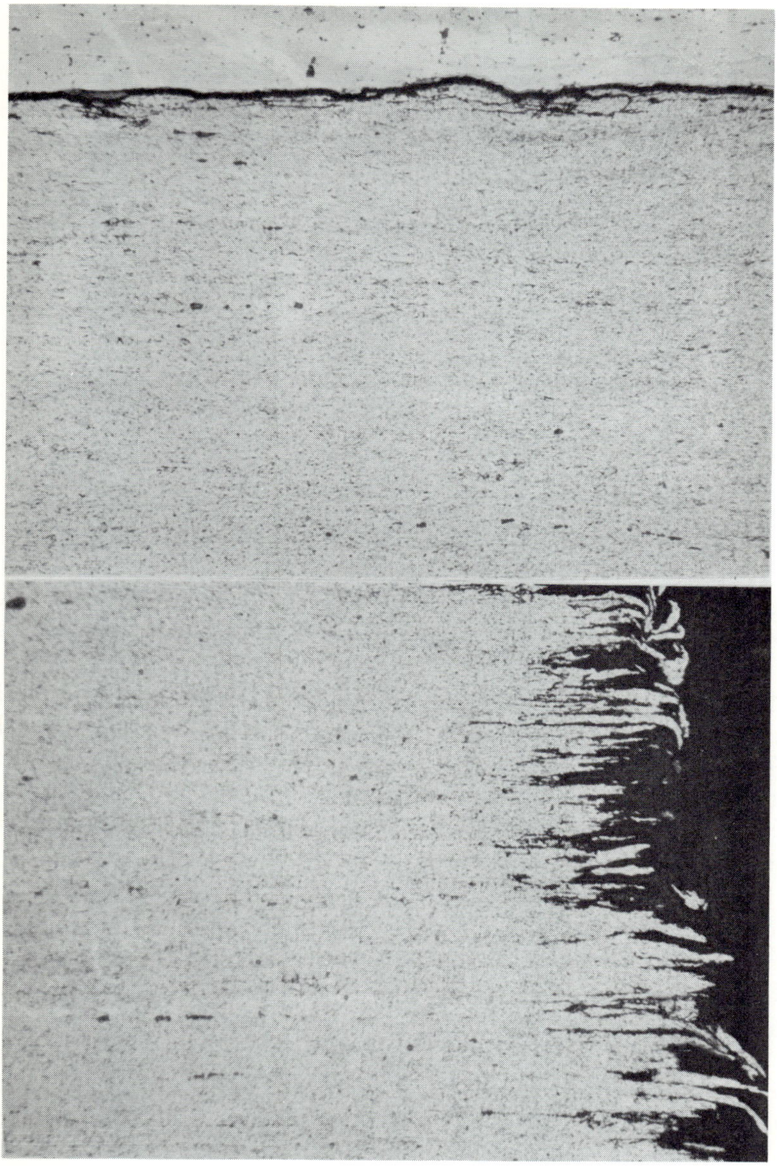

(*top*) Section through surface of specimen lying in the rolling plane.
(*bottom*) Section through surface of specimen lying in a plane transverse to rolling direction.

FIG. 3—*Corrosion attack on worked and transverse surfaces of nitric acid weight loss test specimen, 5456 alloy, 115 mg/in.^{2} weight loss in test* (×*250; etch, nitric acid*).

TABLE 2—*Acid variables affecting the attack on intergranular precipitate in 5456-H321 alloy.*[a]

Time, min	Temperature, deg C	H_3PO_4, 85 weight percent				HNO_3, 70 weight percent			
		15 Volume Percent		Concentrated		15 Volume Percent		Concentrated	
		A	B	A	B	A	B	A	B
15	30	2.0	1.2	...	...	...	...	1.0	1.0
45	30	...	...	16.0	9.1	6.0	0.7	...	...
15	60	...	...	36.0	36.0	16.8	1.2	...	...
45	60	94.0	23.0	...	...	...	...	40.0	2.8

[a] Values in the table represent weight loss per square inch for 5456-H321 alloy, 1.25 in. gage, cut into specimens, 0.5 by 0.5 by 1.25 in.
Condition 'A'—5 specimens—heated one week at 212 F to develop intergranular precipitate.
Condition 'B'—3 specimens—heated 3 h at 800 F, cold-water quenched to minimize intergranular precipitate.

was made between a 2 by 0.25 in. by gage rectangular bar and a 2 by 2 in. by gage square section coupon. Again, two materials were selected for study. One was 0.25-in. gage plate that represented the same condition as B, namely, a structure with most of the magnesium in supersaturated solid solution. To demonstrate the effect on the nitric acid weight loss value, a series of specimens was heated for various time intervals at 100 C (212 F) before determining the weight loss value. This procedure, often called "sensitizing," produces a continuous precipitate of aluminum-magnesium compound around the grains. The second material was 0.188-in. gage sheet that initially represented a structure with both intergranular and intragranular precipitate. Specimens were also subjected to the same heating program as the 0.25-in. gage plate and then tested. The results in Fig. 4 demonstrate three points:

1. The WLVs apparently are not affected greatly by the shape difference.

2. When specimens are heated for prolonged periods at 100 C, (212 F) the WLVs become erratic. This is no doubt due to a redistribution of the intergranular precipitate during this heating period.

3. With the 0.25-in. gage material, a sharp increase in WLV occurs between 10 and 20 h, due to the onset of precipitation around the grains. The weight losses greater than about 50 indicate significant amounts of grain dropping. The 0.188-in. gage material seems to be "stabilized" with respect to the intergranular precipitation, and heating does not increase the weight loss value.

Table 3 shows the effects of time, temperature, and concentration of nitric acid. Again, two materials were selected for the comparison. One, Material C, represents a weight loss value near the 50 mg/in.^{2} value in the normal test procedure, that is, 24 h immersion in concentrated nitric acid at 30 C (86 F). Material D specimens were taken from a sample of 5456-H321 plate which exhibited exfoliation corrosion in service. It was

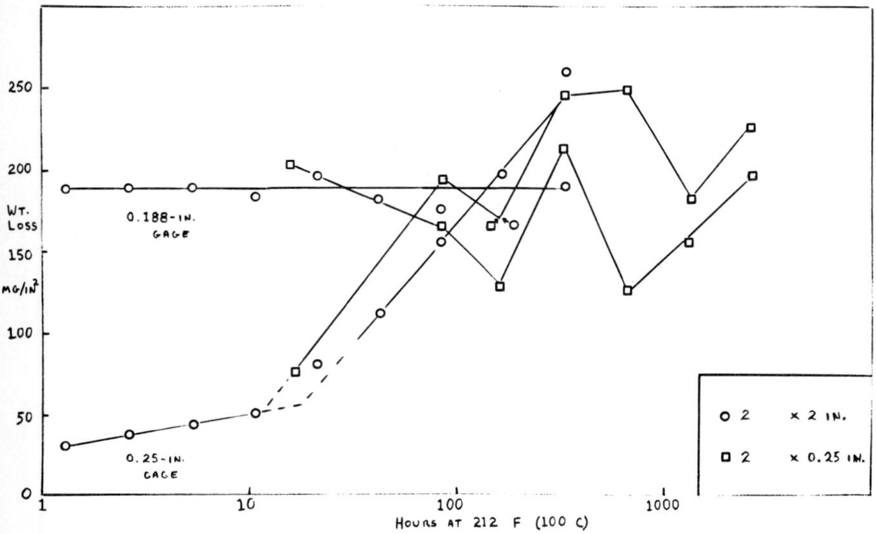

FIG. 4—*5083-H113 sheet and plate comparison of specimen size and its effect on HNO_3 weight loss.*

TABLE 3—*Study of factors affecting the nitric acid weight loss test.*

Factor	Concentration, in volume percent							
	Dilute (25%), c_1			Concentrated (100%), c_2				
MATERIAL C—RESISTANT TO EXFOLIATION								
	t_1	t_2	t_2/t_1	t_1	t_2	t_2/t_1	t_1	t_2
Time, h	6	24	4.0	6	24	4	c_1/c_2	c_1/c_2
Temperature:								
T_1, 30 C	31	234	7.5	4	42	11.7	7.8	5.6
T_2, 40 C	103	510	4.9	16	146	9.1	6.4	3.5
T_2/T_1	3	2.2		4	3.5			
MATERIAL D—SUSCEPTIBLE TO EXFOLIATION								
	t_1	t_2	t_2/t_1	t_1	t_2	t_2/t_1	t_1	t_2
Time, h	6	24	4	6	24	4	c_1/c_2	c_1/c_2
Temperature:								
T_1, 30 C	78	379	4.9	13	108	8.3	6.0	3.5
T_2, 40 C	195	709	3.6	49	281	5.7	4.0	2.5
T_2/T_1	2.5	1.9		3.8	2.6			

D/C

Time, h	6	24	6	24
Temperature:				
T_1, 30 C	2.5	1.6	3.3	2.6
T_2, 40 C	1.9	1.4	3.1	1.9

chosen from the available samples left from the Aluminum Association round robin test program. It had the lowest weight loss value of those samples, 108 mg/in.^{2}. All corroded areas were removed before specimens were taken from the plate.

The ratios of weight loss values were used to assess the effect of each variable. The grand averages are shown in Table 4.

It is expected that the discrimination between Materials C and D would be relatively low, as the samples were selected to represent conditions that were near the high range for resistant materials and on the low side for susceptible materials. Earlier work showed that the effect of immersion time is linear in each of the two branches of the curve for the 0.25-in. gage material shown in Fig. 4. Therefore, if the attack has proceeded far enough to cause grain dropping, the ratio should be 4.0. This condition was met when the smaller weight loss value in a given ratio exceeded 100 mg/in.^{2}. The grand average of 7.0 for this ratio demonstrates that

TABLE 4—*Grand averages of ratios of weight loss values for factors affecting the nitric acid weight loss test.*

Material	D/C	2.3
Time	t_2/t_2	7.0
Concentration	c_1/c_2	4.9
Temperature	T_2/T_1	3.0

many weight loss values obtained in these tests did not reflect the grain dropping condition, since the weight losses are relatively low until grain dropping occurs, and the ratio is therefore greater than 4.0.

The effect of concentration is typical of the behavior of aluminum in nitric acid. With pure aluminum, for equal times and temperatures, a maximum in corrosion rate is observed at a concentration of about 20 volume percent. This rate falls off as the concentration increases to 80 volume percent up to the concentrated acid [*6*]. Under the conditions studied here, the dilute acid was about five times more corrosive than the concentrated acid. This and other work demonstrates that the concentrated acid is the better agent for producing intergranular attack in this weight loss test, as the attack on the grain bodies is minimized.

The usual effect of a 10 C increase in temperature is to double the rate of a chemical reaction. This ratio was approached as a limit in the present test. The same considerations enumerated above for the time effect appear to apply to this factor: once a minimum weight loss value is obtained, the ratio approaches 2.0. This value appears to be between 100 and 200 mg/in.^{2}.

Testing the H116 and H117 Tempers

Another series of tests was performed on samples of 5086 and 5456 alloys in the H116 temper. All these materials passed the Federal Specification, "first-article" qualification test—the SWAAT test for exfoliation and intergranular corrosion. As can be seen from Table 5, the weight loss values ranged from 9 to 80, with most values falling between 10 and 32. There were no apparent trends between alloys, or among gages or laboratories, with the exception that the value for the 2.5-in. gage was high. This specimen, if it had been cut with no more than 1 in. of the transverse surfaces exposed (that it, reduce the gage from 2.5 to 1¼ in. as spelled out in the proposed procedure), would have probably had a much lower weight loss, in the 25 to 30 range.

TABLE 5—*Typical weight loss values for several gages of 5086 and 5456 in the H116 temper.*

Alloy	Gage, in.	Sample	Weight Loss mg/in.^{2}	Remarks[a]
5086	0.188	A	12	specimen size: 2 by 0.250 in. by gage, tested by Laboratory B, unless noted below
		B	11	
	0.250	A	13	
		B	13	
		C	10	Laboratory C
		D	12	Laboratory C
	0.375	A	16	
		B	16	
	1.00	A	15	
		B	15	
	1.5	A	12	specimen size: 2 by 0.25 by 0.75 in.
		B	12	
5456	0.250	A	9	Laboratory B-2[b]
		B	26	Laboratory C
		C	32	Laboratory C
	0.500	A	20	
		B	19	
	0.750	A	16	
		B	16	
		D	11	Laboratory B-2[b]
		E	14	Laboratory B-2[b]
		F	16	Laboratory B-2[b]
	1.00	A	24	
		B	24	
	1.50	A	18	2 by 0.25 by 0.75 in.
		B	18	2 by 0.25 by 0.75 in.
	2.5	A	80	Laboratory C

[a] All materials passed the SWAAT laboratory test for exfoliation corrosion susceptibility.
[b] Laboratory B-2 uses specimens 2 by 2 in. by gage.

The test procedure was given to the other members of the Aluminum Association Committee to study reproducibility among laboratories. The results are shown in Table 6. The samples tested were the same lots that were used in the first Aluminum Association round robin exfoliation testing program [*1*], including three samples which were given treatments to produce conditions of low and medium resistance to exfoliation. An analysis of variance was performed on these results by A. I. Kemppinen, with the following results: there were no significant differences at the 99 percent confidence level due to variations among the laboratories or be-

TABLE 6—*Comparison of nitric acid weight loss data for 5086 and 5456 alloys in the H116 and H117 tempers among laboratories (A–F) participating in the Aluminum Association Round Robin Test Program Two.*

Identification			Weight Loss, mg/in.^{2}								
Alloy	Temper	Gage	A	B		C		D	E	F	Avg
5086	H116	0.250	52	59	59	101	93	61	26	20	59
		1.00	12	14	14	17	...	16	25	24	17
	H117	0.250	8	8	8	10	...	4	3	6	7
		1.00	7	8	8	8	9	2	4	6	7
5456	H116	0.250	14	17	17	21	...	36	8	16	18
		0.750	16	18	18	18	22	10	10	15	16
	H117	0.250	13	16	16	22	...	11	7	17	15
		1.00	9	10	10	10	...	7	6	9	9
5086	[a]	0.250	277	283	269	269	278	315	184	221	262
5456	[b]	0.250	213	219	210	204	242	224	136	155	200
	[a]	0.250	274	310	289	312	...	339	203	283	287

[a] Control—low resistance.
[b] Control—medium resistance.

tween the tempers or the gages in 5086 alloy. Alloys were not compared, nor was the gage tested in 5456 due to the lack of strictly comparable values (that is, the 0.75-in. gage for 5456-H116 rather than a 1.00-in. gage sample). No statistical evaluation of the control samples was made. The grand mean for 5086 alloy was 22 mg/in.^{2} and for 5456 was 16 mg/in.^{2}. The wide spread between these values and those for the low- and medium-resistance samples demonstrates the ability of the test to discriminate among good and poor materials.

Table 7 compares the average weight losses with the individual exfoliation test results. Except for the Q ratings in the ASSET test, there were no indications of susceptibility in any of the three tests for H116 or H117 temper materials. However, in the controls, some question was raised in the ASSET test (one of duplicates in twelve sets) and more in the SWAAT test (four in twelve sets). With the nitric acid WLV, there was good separation of the medium resistance (200) from the two low-resistance values—262 and 287.

Correlation with Service Experience

Since the SWAAT test was correlated with service experience, the nitric acid weight loss test was also used to examine samples removed from actual in-service failures of 5456-H321. The WLVs ranged from 114 to 492 in these specimens, as shown in Table 8. Since there was only slight variation among these samples in their exfoliation behavior, it is difficult to assess the significance of these WLVs in terms of degree of resistance. However, some differences of degree were available in two AA lots—Lot 2, with a weight loss of 180 was marginal. One of three laboratories found slight exfoliation in this Lot. Lot 4, with a weight loss

TABLE 7—*Correlation of nitric acid weight loss with exfoliation and intergranular corrosion behavior of 5086 and 5456 alloys in the H116 and H117 tempers in the Aluminum Association Round Robin Test Program Two (Laboratories A–D).*

Identification			Weight Loss	Exfoliation Tests—Round Two[a]							
			Weight Loss,	ASSET (immersion)				SWAAT (salt spray)			
Alloy	Temper	Gage, in.	mg/in.² Avg of 6 Labs	A	B	C	D	A	B	C	D
5086	H116	0.250	59	Q[d]	P[d]	Q[d]	P[d]	P[d]	N	P[d]	P[d]
		1.00	17	Q[d]	P[d]	Q[d]	N	P[d]	N	N	P[d]
	H117	0.250	7	Q[d]	P[d]	N	N	P[d]	N	P[d]	P[d]
		1.00	7	P[d]	N	N	N	P[d]	P[c]	P[d]	P[d]
5456	H116	0.250	18	Q[d]	P[d]	Q[d]	P[d]	N	N	P[d]	N
		0.750	16	Q[d]	P[d]	Q[d]	P[d]	P[d]	N	N	N
	H117	0.250	15	Q[d]	P[d]	Q[d]	P[d]	P[d]	N	P[d]	N
		1.00	9	P[d]	N	N	N	P[d]	P[c]	P[d]	P[d]
5086	[b]	0.250	262	ES	ES	ES	ES	EM	EM	Q	Q
5456	[c]	0.250	200	EI	EM	EI	EI, N[e]	Q	EM	P	EI
	[b]	0.250	287	ES	ES	ES	ES	ES	EI	ES	ES

[a] Visual code—N = no pitting, exfoliation or appreciable attack.
P = pitting.
Q = questionable exfoliation (pit blister metallographic examination recommended to determine type of attack).
EM = exfoliation, mild.
EI = exfoliation, intermediate.
ES = exfoliation, severe.

[b] Control—low resistance.
[c] Control—medium resistance.
[d] Examined metallographically and passed test on this basis.
[e] Two specimens gave disparate results.

of 215 showed poor behavior. This same material, annealed, gave good results—WLV, 15, and only pitting in the salt spray test.

Table 9 shows data for 5086 alloy similar to that given in Table 8 for 5456 alloy. Since the service failure was attributed to stress-corrosion cracking, a stress-corrosion test was used to evaluate these materials, as well as the exfoliation and intergranular corrosion tests [*7,8*]. The stress relief treatment apparently reduced the susceptibility to stress corrosion cracking to the extent that the material, so treated, passed the test. However, the 425 F (218 C) treatment did not reduce greatly the continuity of the grain boundary precipitate and thereby did not eliminate the concommitant susceptibility to intergranular corrosion. The gap between highly susceptible material (WLV = 204) and that which passes the stress corrosion test (WLV = 193) is narrow, but both materials show moderate to severe intergranular corrosion in both ASSET and SWAAT. A borderline material that was not in service (Sample 2) showed slight pitting with an intergranular component in one of two laboratory tests, and had a WLV of 134. Sample 1 after heating at 450 F (232 C) showed slight to moderate pitting with an intergranular attack on the microscopic scale. Since the material for this test was taken from the original sample which was heavily pitted, this attack may not be characteristic of the 23 WLV but rather represents vestiges of prior structure and its exposure.

TABLE 8—*Correlation of nitric acid weight loss with exfoliation and intergranular corrosion behavior of 5456-H321 alloy plate in service and laboratory tests.*

Identification			Exfoliation or Intergranular Susceptibility				
				SWAAT[a]			
				Surface		Midplane	
Service No.	Boat Frame	Weight Loss,[e] mg/in.^{2}	In Service	Visual	Micro	Visual	Micro
2	4P	201	pitting and	B	P	C	I
	7S	139	intergranular	B	I	C	I
	. . .[d]	492	corrosion	B	I	B	I
3	10S	129	severe	C	Exf	B	Exf
	10P	138	exfoliation	C	Exf	B	Exf
	5P	200	corrosion	C	Exf	C	Exf
4	7S	204		D	Exf	C	Exf
	7P	210		D	Exf	C	Exf
	10P	217		D	Exf	C	Exf
5	5S	127		D	Exf	B	Exf
	5S	123		D	Exf	D	Exf
	9S	138		B	Exf	A	Exf
	10P	114		C	Exf	A	Exf
	. . .[d]	299		D	Exf	D	Exf
AA lot	2[b]	180	. . .	A	P + SI	A	P
				(Reynolds shows slight Exf)			
AA lot	4	215	. . .	C	Exf	C	Exf
	4A[c]	15	. . .	A	P	A	P

[a] Results of AA round robin test performed by Alcoa, Kaiser, and Reynolds, taken from Ref 1.
Code—Visual rating of degree: A = no blistering or delamination.
B = slight blistering or "lipping" of pits.
C = large blisters or localized delamination.
D = severe delamination.
Microscopic examination of type of attack: P = pitting.
P + SI = pitting plus slight intergranular.
I = intergranular.
Exf = exfoliation.
[b] 0.188-in. gage.
[c] 5456–0 annealed plate.
[d] Reported by T. J. Summerson, Kaiser Aluminum.
[e] Nitric acid weight losses—performed by Reynolds Metals Co.

The conclusion from this series is that the nitric acid weight loss test is as effective in evaluating 5086 alloy service failures as it is with 5456 alloy samples.

Long-Term Behavior

To show the broader application of the nitric acid test in the examination of service behavior, a series of samples that have been collected over the past fourteen years is presented in Table 10. The samples are discussed briefly next.

Peraluman 50 (*al-5 percent magnesium, 0.3 percent manganese*)

Three samples were removed from a ferryboat which sails the Caribbean. One sample (No. 1) was badly corroded, showing heavy exfoliation

TABLE 9—*Correlation of nitric acid weight loss with stress-corrosion cracking and intergranular corrosion behavior of 5086-H32 alloy sheet in service and laboratory tests.*

Identification	Weight Loss, mg/in.^{2}	Stress-Corrosion Cracking [*8–9*] and Intergranular Susceptibility			
		Service	Laboratory Test[c]		
			SCC	ASSET	SWAAT
5086-H32 sheet, as produced, 0.125-in gage:					
Sample 1[a]	204	failed by intergranular corrosion and stress-corrosion cracking	failed	P/IG-MOD-SEV	P/IG-SEV
Sample 2[b]	134	not in service	passed	N	P/IG-SL
Above material stress relieved 8 h at:					
425 F					
Sample 1[a]	193	...	passed	P/IG-MOD-SEV	P/IG-SEV
Sample 2[b]	10	...	passed	N	N
450 F					
Sample 1[a]	23	...	passed	P/IG-MOD	P/IG-SL
Sample 2[b]	7	...	passed	N	N

[a] Taken from amphibious vehicle—welded ramp: chemical analysis, 4.36Mg.
[b] Commercial source: chemical analysis, 3.72Mg.
[c] Rating system—Visual/Microscopic;
P = pitting
IG = intergranular
N = no intergranular corrosion
SEV = severe
MOD = moderate
SL = slight

and corrosion product buildup. Metallographic examination revealed severe intergranular corrosion. Etching a polished surface developed the expected continuous grain boundary network. The WLVs were very high, especially in comparison with those of the two control samples, which had seen ostensibly the same service but had not corroded. It does not seem reasonable to ascribe the difference in behavior to the magnesium content alone. The hardness and gages were comparable for three materials, so it is assumed the temper and production practices were the same. Thus, on this evidence, it was judged that Sample 1 had probably been exposed to elevated temperature aging somewhere in its history, but that other two samples had not. It was suggested that the materials should be replaced with a more resistant alloy and temper.

5083–0

These samples were removed from a full sized experimental bridge section, which had suffered corrosion attack in the industrial marine atmosphere of a large East Coast city while undergoing fatigue testing. The corrosion product formed on inner surfaces where condensation occurred but where rain could not wash off contaminants. The WLVs were higher for specimens with heat affected zones caused by welding,

TABLE 10—*Use of nitric acid weight loss to investigate precipitation phenomena in aluminum-magnesium alloys.*

Identification	Weight Loss, mg/in.[2]	Description
Peraluman 50Al (5Mg, 0.3Mn) 0.125-gage sheet:		
Sample 1—5.16Mg	275	exfoliation and severe intergranular
	443	corrosion in service
Sample 2—4.71Mg	157	not corroded in service
	144	
Sample 3—4.83Mg	168	not corroded in service
	155	
5083–0, 0.125 gage sheet as produced:		
Sample 1	22	pitting plus slight intergranular corrosion in service
Sample 2	20	
Sample 3	41	Samples 3 & 4 contained weld beads and heat affected zones.
Sample 4	27	
Above, annealed, cold-water quenched:		
Sample 1	8	passed stress-corrosion test [8]
Sample 2	8	
Above, heated 7 days:		
212 F (100 C)	156	passed stress-corrosion test
221 days 212 F (100 C)	380	failed stress-corrosion test
5083-H112 extrusions:		
Sample 1	9	as produced
Sample 2	8	annealed and cold-water quenched
Sample 3	22	production material exposed to sun (Phoenix, Arizona) 21 April–Sept. 27
5456-H112 extrusions:		
Sample 1	10	as produced
Sample 2	7	annealed and cold-water quenched.
Sample 3	61	exposed to sun (Phoenix, Ariz.) 21 April–27 Sept.
5456-HX extrusions:		experimental temper:
Sample 1	90	as produced
Sample 2	8	annealed and cold-water quenched
Sample 3	85	exposed to sun as above
N6-¼H:		
Al5, Mg 0.3Mn	184	deck plating in naval patrol boat failed by cracking and intergranular corrosion in service
5083-H113 welded plate:		evaluation of weight loss samples:
As welded		
Sample 1—0.388 in.	17	no corrosion attack in weld, heat affected zone, or parent metal
Sample 2—0.600 in.	13	
Heated 1 week 212 F (100 C)		
Sample 1	114	intergranular attack in heat affected zone; slight undercutting of welds at rolled surface-weld interface
Sample 2	108	
5083-H113 plate 0.3125 in.		
Car A		
Sample 1 (side of car)	43	samples removed from aluminum hopper car body (Alumina Service); no susceptibility to intergranular or exfoliation attack in service; samples also passed a stress-corrosion cracking test
Sample 2 (end of car)	34	

Continued

TABLE 10—(*Concluded*)

Identification	Weight Loss, mg/in.[2]	Description
Car B		
Sample 1 (side of car)	27	(Petroleum Coke Service); slight pitting;
Sample 2	36	maximum depth 8 mils; no evidence of intergranular or exfoliation attack in service
5083-H113 plate 0.3125 in.:		plates, 4 by 4 by 0.3125 in. were removed
Car A (few months in service)	65	from the sides of coke cars; samples were removed and tests performed: no failures
Car B (three years service)	58	in the stress corrosion test; overall eval-
Car C (three years service)	63	uation of panels was: superficial discoloration and slight attack in service
(Average of two samples for each car)		
5083-H113 plate 0.500 in.		panels, 4 by 12 by 0.500 in. were installed in steel cars used to haul petroleum coke—90 to 450 F
As produced (control)	14	slight pitting and traces of intergranular attack in the nitric acid weight loss test
Car A (55 days)	37	in the nitric acid weight loss test, sample
Car B (45 days)	41	showed more intergranular attack than control—more grain boundary precipitate than control
Car C (57 days)	30	passed the stress corrosion test
Cars A,B,C (185 days)	. . .	failed the stress corrosion test
As produced, then heated at 212 F (100 C) for one month	103	failed the stress corrosion test; one of three specimens failed at atmospheric exposure stress corrosion test at Kure Beach, N. C. [*8*]; samples of this material showed severe exfoliation in the SWAAT test while the as produced condition showed no exfoliation

but the atmospheric corrosion attack was no heavier in these areas than on the rest of the surface. After 10-years' exposure at an industrial atmospheric test site, the corrosion did not progress beyond the initial slight pitting and intergranular attack. The WLVs indicate this material to be corrosion resistant, and the subsequent performance bears it out. Note that after prolonged elevated temperature aging, the WLV increases, but susceptibility to stress corrosion cracking does not occur until a value of 380 is reached.

5083, 5456 Extrusions

These samples were tested by the plant laboratory, located at Phoenix. As can be seen from a comparison of the data, the heating which occurs during the summer caused increases in the WLV, from 9 to 22 for 5083-H112, and from 10 to 61 for 5456-H112. When the latter alloy was produced with a small amount of randomly distributed precipitate in an experimental temper, the condition was considered stable towards the heating, since a continuous network did not occur and the weight loss value did not increase.

N6–1/4H (5 percent magnesium)

A sample removed from a German-built patrol craft, operated in the North Sea, showed a weight loss of 184, in a plate that had failed by cracking and intergranular corrosion (test by a foreign affiliate laboratory).

5083-H113 Weldments[4]

This test was conducted to see if weld zones might become sensitized to corrosion if exposed to elevated temperatures, such as the defrosting procedures used with railroad coal cars. Even prolonged heating (one week or 168 h) at 100 C (212 F) did not produce a severe weight loss, and the extent of intergranular attack by the nitric acid was judged superficial.

5083-H113 Plate[4]*—Railroad Car Bodies*

Several sets of samples were either removed from aluminum car bodies, or aluminum test plates were carried in steel car bodies. The purpose was to determine if the hot cargo caused heating of the metal to the extent of producing extensive precipitation within the alloy. Since 5083 has some fraction of magnesium in a supersaturated solid solution, this precipitation is always a possibility.

The first set was removed from a car used in alumina service. It showed WLVs of 34 to 43, which were typical of that gage and temper as produced in the late fifties. The cars had been in service over a year at the time of the sampling. Since then, they have been in service over ten years. No complaints or problems have developed in the corrosion behavior of this material in this application.

The second and third sets were removed from cars in petroleum coke service. The cargo is intrinsically more corrosive than alumina. The WLVs again were moderate, the corrosion attack due to the coke liquors was slight, and there was no intergranular component to the pitting. Some of these cars had been in service three years. Comparison of WLVs from these cars with one from a car in service for only a few months indicated that no precipitation was occurring.

The last set is a series of plates that initially exhibited a very low WLV, 14. In the field test, after a relatively short period, the WLVs increased by a factor of two or more. The samples passed the stress-corrosion test.

However, samples from the same cars, exposed for 185 days to hot cargo (the actual test period was longer, as deadhead time was not counted) did fail the stress corrosion test. WLVs were not obtained on these specific samples, but heating the as-produced plates for one month

[4] This temper has been superseded by the H321 temper.

at 100 C (212 F) produced a WLV of 103, and specimens failed the stress corrosion test.

This sample is one of a series that is included in a long range study of 5083 alloy. When precipitation is produced by heating at 100 C (212 F) for one month, the metal also shows susceptibility to stress corrosion in atmospheric tests. It also shows severe exfoliation in the SWAAT test. The as-produced material, to date (eleven years) is immune to all these problems, in immersion and atmospheric testing in Reynolds Metals Company's comprehensive, multienvironment program [*8*].

Discussion of Examples (long-term behavior)

The precipitation of the aluminum-magnesium intermetallic compound is a complex metallurgical process. Susceptibility to exfoliation or intergranular corrosion depends not only on the site and nature of the precipitate but also on the grain structure. Exfoliation can occur only when the grains are elongated. Equiaxed grains produce intergranular corrosion and deep pitting. Whether or not a material does corrode also depends on its environment. The results from the previous quoted examples attempt to put some of this behavior in its proper perspective. The following factors are judged to be important:

1. The magnesium content of the alloy.
2. The degree of cold work.
3. The fabrication history.
4. The environment, including exposure to elevated temperatures for prolonged periods.

There is only a slight difference in the magnesium concentration among 5456, 5083, and 5086 alloys. Yet, again and again, that slight difference seems to be critical, especially under otherwise equivalent conditions of cold work, exposure to elevated temperatures, and environment.

Cold work is an accelerator of precipitation. This, then, makes it a negative factor in producing a corrosion resistant material. It also leads to precautionary measures when a structure is fabricated. Our best judgment at this time is that a structure with controlled precipitation, such as produced in the H116 temper, is superior in its behavior to a structure with all or most of the magnesium in supersaturated solid solution. Again, this difference can be measured with the nitric acid weight loss test.

In addition to cold work, structures may be "warm-formed" or otherwise exposed to elevated temperatures, as in a welding process. The nitric acid test is a convenient measure of the effect of these processes.

When it comes to judging the severity of environments, it can only be suggested that, first, concentrated nitric acid is representative of a real life environment for many aluminum structures, used in a chemical plant. Secondly, nitric acid is a "most severe" environment compared to immersion in seawater, exposure to marine or industrial atmospheres, or other chemical solutions in which aluminum-magnesium alloys may be expected to offer excellent service. The nitric acid weight loss test can be used to detect subtle changes in the metal structure and does so prior to the onset of susceptibility as measured by other, more conventional laboratory tests.

Despite the complex behavior and multifactored situation described above, the nitric acid weight loss test provides a quantitative scale covering all degrees of behavior. This scale may be used for "pass-or-fail" judgments based on the expectation of service performance, or it may be used to subdivide materials into pass, fail, and "borderline" cases.

I conclude that 5086 and 5456 alloys in the H116 and H117 tempers with weight loss values less than 50 are resistant to exfoliation and intergranular corrosion, while those materials with weight losses greater than 50 to 100 are suspect. The degree of susceptibility increases with the amount of the weight loss, and materials with values of 200, 300, or 400 should be increasingly susceptible to exfoliation or intergranular corrosion.

Acknowledgments

The permission of Reynolds Metals Company to publish these procedures and results is gratefully acknowledged.

The development of this procedure was, in a large part, due to the careful work of the late H. B. Romans. The assistance of L. E. Householder, is acknowledged, in developing the procedure for plant use, and encouraging its use in plant laboratories.

The contributions of J. S. Prestley and A. I. Kemppinen are also acknowledged.

I appreciate the contribution of the various members of the AA Committee and especially acknowledge those who granted their permission for me to include their test results. The Committee members are. S. G. Epstein, The Aluminum Association, secretary; P. W. Jeffrey, Alcan Research and Development, Ltd,; F. E. Loftin, Reynolds Metals Co.; J. F. Pashak, Dow Chemical Co.; B. R. Pathak, Alcan Research and Development, Ltd.; W. C. Rotsell, Harvey Aluminum, S. J. Sansonetti, Reynolds Metals Co., chairman; M. B. Shumaker, Alcoa Research Laboratories; D. O. Sprowls, Alcoa Research Laboratories; and T. J. Summerson, Kaiser Aluminum & Chemical Corp.

APPENDIX

Nitric Acid Weight Loss Procedure

1. Prepare specimens measuring 2.000 by 0.250 in. by gage with the 2 in. dimension parallel to the rolling direction. If the gage exceeds 1 in. reduce to ½ thickness while retaining one original surface. Machine all sawn surfaces.
2. Smooth all edges with a fine file or fine emery cloth.

3. Measure all three dimensions to the nearest 0.001 in. with a micrometer. Calculate the total surface area.

4. Immerse the specimen in 5 percent by weight sodium hydroxide (NaOH) solution at 180 F (82 C) for 1 min followed by a distilled water rinse, a 30 s immersion in concentrated, reagent grade nitric acid, and a distilled water rinse.

5. Allow the specimen to air dry completely. Do not wipe dry with a rag or paper towel. From this point on in the procedure, handle the specimens with tongs or tweezers.

6. Weigh the specimens to the nearest 0.1 mg.

7. Fully immerse each specimen in 80 ml of concentrated nitric acid,[5] contained in a 100-ml beaker.[6] It is suggested that beakers without spouts be used with watchglasses as covers. Place the beaker in a water bath controlled to 30 $\pm$ 0.1 C (86 F).

8. After 24 h, remove the specimens, rinse in distilled water and brush with a stiff nylon bristle toothbrush to remove loosely adhering particles, then allow to dry.

9. Weigh the specimens and determine the weight losses.

10. Determine the weight loss per unit area and express it in terms of mg/in.^{2}.

References

[*1*] Sprowls, D. O., Summerson, T. J., Romans, H. B., and Sansonetti, S. J., "The Aluminum Association Tentative Exfoliation Test for Al-Mg Alloys for Boat and Ship Hull Construction," The Aluminum Association, New York, 1969.

[*2*] Federal Specification QQ-A-00250/19 and QQ-A-00250/20—5086 and 5456-H116 and H117 Sheet and Plate.

[*3*] Romans, H. B., *Materials Research and Standards,* MTRSA, Vol. 9, No. 11, 1969, pp. 31–34.

[*4*] Brooks, C. L., *Naval Engineers Journal,* NVEJA, Aug. 1969, pp. 29–32.

[*5*] Craig, H. L., Jr., and Loftin, F. E., *Proceedings,* 26th Annual Conference, National Association of Corrosion Engineers, Houston, Tex., 1970.

[*6*] "Aluminum with Food and Chemicals," 2nd ed., The Aluminum Association, Feb. 1969, p. 59.

[*7*] Craig, H. L., Jr., and Romans, H. B. in *Stress Corrosion Testing, ASTM STP 425,* American Society for Testing and Materials, 1969, pp. 51–65.

[*8*] Romans, H. B. and Craig, H. L., Jr., in *Metal Corrosion in the Atmosphere, ASTM STP 435,* American Society for Testing and Materials, 1968, pp. 61–82.

[5] Reagent Grade, 70 percent by weight. Fresh acid should be used for each test.
[6] Beaker, Berzelius (Corning 1040).

D. O. Sprowls,[1] *J. D. Walsh,*[1] *and M. B. Shumaker*[1]

Simplified Exfoliation Testing of Aluminum Alloys

REFERENCE: Sprowls, D. O., Walsh, J. D., and Shumaker, M. B., **"Simplified Exfoliation Testing of Aluminum Alloys,"** *Localized Corrosion—Cause of Metal Failure, ASTM STP 516,* American Society for Testing and Materials, 1972, pp. 38–65.

ABSTRACT: A program was undertaken to develop accelerated tests to determine susceptibility to exfoliation in aluminum alloys. Experimentation was concentrated on total immersion procedures to develop a simple, easily controlled test. Chloride solutions did not cause exfoliation during immersion; however, formulations of chloride-nitrate solutions were found that produced severe exfoliation of highly susceptible Al-Mg-Mn alloy specimens in only one or two days. Investigation of the most effective chloride-nitrate ratios, concentrations, solution pH, and temperature led to the development of a new, simple, and rapid exfoliation test for Al-Mg-Mn alloys. Additional investigation resulted in the development of similar test media for Al-Cu-Mg, Al-Zn-Mg, and Al-Zn-Mg-Cu alloys, and showed that the optimum test conditions differ for the separate alloy families.

Test data are presented for alloys 2024, 5086, 5456, 7005, 7075, and 7178 in metallurgical conditions that provide comparisons of high and low resistance to exfoliation. Comparisons with exposures to natural environments also are included.

KEY WORDS: corrosion, exfoliation corrosion, seawater corrosion, corrosion tests, salt spray tests, immersion tests (corrosion), aluminum alloys

Exfoliation corrosion, sometimes referred to as layer, stratified, or lamellar corrosion, is a form of corrosion resulting from a relatively rapid lateral attack along electrochemically anodic strata parallel to the metal surface. In the more familiar occurrences of exfoliation corrosion, attack progresses along grain boundaries. Because it occurs most readily in alloys and tempers that have a relatively low resistance to stress-corrosion cracking in the short-transverse direction, exfoliation sometimes is regarded as a special form of intergranular attack or of stress-corrosion cracking. Such a limited perspective of exfoliation, however, is an oversimplification. Experience with corrosion testing a wide variety of alloys

[1] Section head, technician, and senior research engineer, respectively, Chemical Metallurgy Division, Alcoa Research Laboratories, New Kensington, Pa. 15068.

and tempers at Alcoa Research Laboratories has shown that exfoliation also can result from *intra*granular corrosion, and that not all products that are susceptible to *inter*granular corrosion or stress-corrosion cracking will develop exfoliation. Thus, the requisite of a highly directional microstructure is more essential than a specific mode of corrosion. Another requisite is a specific type of corrosive environment.

In recent years the development of exfoliation-resistant alloys and tempers [*1–5*][2] has brought into focus a need for improved accelerated corrosion tests. There are increasing numbers of U. S. Government specifications for aluminum alloy products that include exfoliation test requirements [*6,7*]. A suitable accelerated exfoliation test must not only be rapid, but it must also enable the investigator to distinguish between alloys and tempers of differing susceptibility to exfoliation and to rate them in the same order as service environments. Liddiard et al [*8*] in 1960, reported that "no satisfactory accelerated test for layer corrosion has been found." Liddiard and other investigators [*9*] at that time used intermittent spray exposures to neutral or acidified sodium chloride (NaCl) solution. The main disadvantage of such tests is that they require considerable time, usually several months. Although improved tests have been developed in recent years by Budd and Booth [*10*], Lifka and Sprowls [*11*], Romans [*12*], and Ketcham [*13*], these tests require sophisticated apparatus, and the test results can be subject to considerable variability (Fig. 1).

Liddiard et al [*8*], after experimenting with aluminum-copper (Al-Cu) alloys immersed in solutions of 3 percent NaCl, 6 percent NaCl + 0.11 percent hydrogen peroxide (H_2O_2), 0.5 *N* ammonium chloride (NH_4Cl) and 3 percent NaCl buffered at pH 11, concluded that layer corrosion does not occur under total immersion conditions because corrosion products are formed away from the site of attack, and there is always considerable attack on the grains themselves. Indeed, this was the opinion at Alcoa Research Laboratories at the time the improved cyclic acetic acid acidified salt spray test was developed [*11*]. However, a renewed investigation of immersion tests prompted by U. S. Navy interest in an accelerated exfoliation test for screening aluminum alloy boat hull plate led to the development of some very effective and reproducible test procedures. The purpose of this paper is to present results of this investigation and to compare test results from immersion tests with tests in service environments.

Procedure and Results

General Test Conditions

Many years of testing experience and consultation on customer service problems have shown that exfoliation corrosion is most likely to occur in

[2] The italic numbers in brackets refer to the list of references appended to this paper.

Tank A (133 tests): (*a*) 0%, (*b*) 14%, (*c*) 18%, (*d*) 68%.
Tank B (98 tests): (*a*) 6%, (*b*) 10%, (*c*) 29%, (*d*) 56%.

FIG. 1—*This photograph illustrates the range in severity of exfoliation produced in weekly tests of specially sensitized (boiling water quenched ½-in.-thick plate) 2024-T42 control specimens by two similar spray tanks* [11]. *The numbers indicate the percentage occurrences of each degree of severity.*

certain types of environments. More aggressive environments involve chloride ions from sources such as seawater combined with alternate wetting and drying. Other accelerating factors are acidity, elevated temperatures, and galvanic stimulation. Tests by spray techniques generally incorporate some or all of the aforesaid conditions, and similar conditions were investigated in the evaluation of corrodents for immersion testing.

Test specimens 1½ by 3 in. were sawed from rolled plate with the 1½ in. dimension parallel to the rolling direction. Edges of the specimens were dressed by machining or sanding, after which the specimens were etched 1 min in 5 percent by weight sodium hydroxide (NaOH) solution at 82 C (180 F), rinsed in water, desmutted 30 s in concentrated nitric acid (HNO_3) at 27 C (80 F), rinsed in distilled water, and air dried.

The experimental solutions were made with reagent grade chemicals and distilled water. Tests were performed in glass beakers with a volume of solution sufficient to provide a minimum volume-to-metal surface ratio of 40 ml/in.^{2}. Solution temperatures were maintained at the indicated temperature within ±1 C.

Aluminum-Magnesium (Al-Mg) (5XXX) Alloys

Initial emphasis was on 5XXX alloys. Four items of plate were selected to provide a variation in microstructure and susceptibility to exfoliation based on prior performance in service and in spray tests. The microstructures and a description of the spray test performances are shown in Fig. 2.

Exploratory Tests of Various Corrodents—Acetic acid acidified chloride solutions of the following formulations have been shown to develop exfoliation of susceptible materials in seven days of intermittent spray exposure at 49 C (120 F).

1. Five percent NaCl solution buffered to pH 3 with acetic acid [*11*]
2. Synthetic seawater, ASTM D 1141–52 (without heavy metal additions) plus 10 ml of glacial acetic acid per liter of solution [*12*].

Immersion in these solutions at 49 C (120 F) for seven days did not develop even a trace of exfoliation of low-resistance material. Exploratory immersion tests, therefore, were extended to a variety of acidified chloride, chloride-acetate, chloride-nitrate, and miscellaneous solutions. A number of these corrodents and visual ratings of the exfoliation effects are listed in Table 1. It should be noted that in these exploratory tests the objective was to find a solution that would develop exfoliation in susceptible material within a 5-day period at temperatures not exceeding 82 C (180 F). The most promising corrodents found were chloride-nitrate solutions with an appropriate combination of chloride and nitrate ions. It is known that exfoliation will occur in certain of the other solutions with longer exposure periods, and it is likely that under suitable conditions exfoliation might also occur in other types.

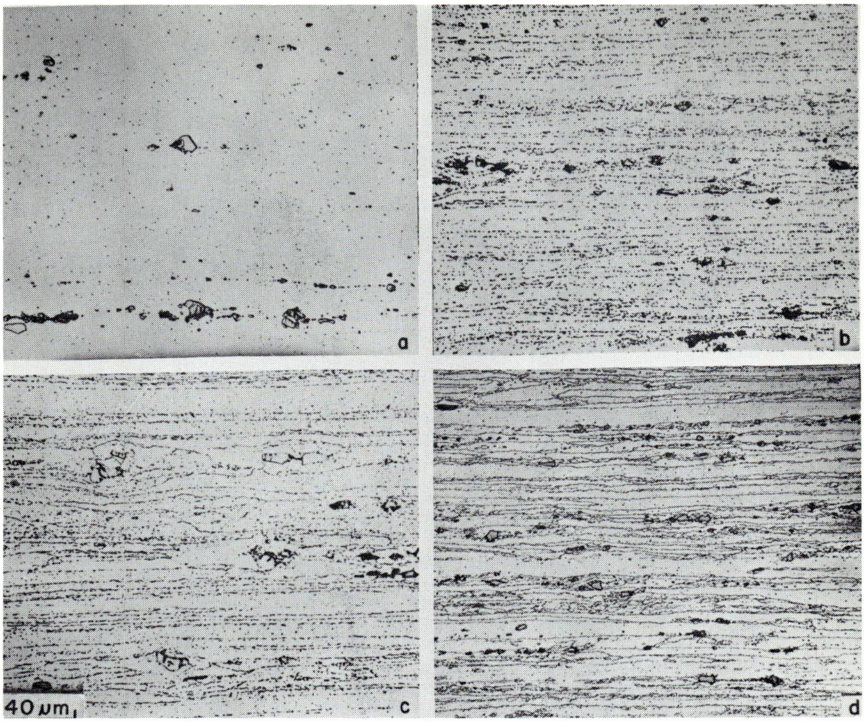

(*a*) 5456-H117 . . . . P
(*b*) 5086-H32 . . . . P
(*c*) 5456 control . . . . Q
(*d*) 5456 control . . . . EM

FIG. 2—*Microstructures (phosphoric acid etch) of four items of 0.250 or 0.375-in.-thick hot-rolled plate of Al-Mg alloys with differing resistance to exfoliation rated as above by an acidified cyclic salt spray test* [5,11]. (*For explanation of exfoliation rating refer to footnote* a *of Table 1.*

Preliminary tests of chloride-nitrate solutions were performed with a solution containing 4 *M* NaCl + 0.5 *M* potassium nitrate (KNO_3) + 0.1 *M* HNO_3 (pH 0.4) developed by Zaretskii [*15*] for stress corrosion testing Al-Cu alloys, and adapted for exfoliation testing by Alcan Research and Development, Ltd.[3] This solution was ineffective at room temperature, but at 66 C (150 F) produced mild exfoliation of the 5456 control specimen, Fig 2*d,* and rapid general corrosion and thinning of the other test materials identified in Fig. 2. Extensive testing of additional chloride-nitrate solutions, including variations in solution pH and temperature, led to a solution containing 1 *M* NH_4Cl + 0.25 *M* ammonium nitrate (NH_4NO_3) + HNO_3 at pH 3 that at a temperature of 66 C (150 F)

[3] Jeffrey, P. W., personal communication, 22 Oct. 1968.

TABLE 1—*Exploratory continuous immersion exfoliation tests of Al-Mg alloys in various corrodents.*

Corrodent	pH	Temperature, deg F	Exposure Time, days	Visual Examination[a,b] 5456 Control Specimens: Low Resistance, Fig. 2d	Visual Examination[a,b] 5456 Control Specimens: Medium Resistance, Fig. 2c	Visual Examination[a,b] 5086-H32, Fig. 2b
Chloride solutions:						
5% NaCl	3	120	7	N	N	N
1.0 *N* $AlCl_3$	1.0	70	7	P	P	P
	2.7	70	7	P	P	P
1.0 *N* NH_4Cl	1.0	70	7	P	P	P
	5.65	70	7	P	P	P
1.0 *N* NH_4Cl + 1 g/liter $CuCl_2$	3.55	70	7	P	P	P
0.5 *N* $AlCl_3$ + 0.5 *N* $FeCl_3$	1.5	70	7	P	P	P
0.5 *N* NH_4Cl + 0.5 *N* $FeCl_3$	1.7	70	7	P	P	P
Chloride-acetate solutions:						
5% NaCl + 0.1 *M* CH_3COOH	3	120	7	N	N	N
3.5% NaCl + 0.18 *M* CH_3COOH	2.6	150	5	N	N	N
4.2% NaCl + 0.18 *M* CH_3COOH	2.8	80	14	N	N	N
4.2% NaCl + 0.18 *M* CH_3COONa + 0.05 *M* $KH_8C_4H_4O_4$	4.2	150	5	N	N	N
4.2% NaCl + 0.05 *M* CH_3COONa + 0.05 *M* $KH_8C_4H_4O_4$	4.2	150	5	N	N	N
4.2% NaCl + 1.0 *M* CH_3COONa + 0.05 *M* $KH_8C_4H_4O_4$	4.0	150	5	N	N	N
1 *M* NaCl + 0.1 *M* CH_3COONa + 0.1 *M* CH_3COOH [*14*]	4.3	150	5	N	N	N
4.2% synthetic sea salt + 0.18 *M* CH_3COOH	2.8	80	10	P	P	N
		120	7	N	N	N
		150	3	N	N	N
4.2% synthetic sea salt + 0.18 *M* CH_3COOH + 1 ppm $CuCl_2$		150	5	N	N	N
4.2% synthetic sea salt + 0.50 *M* CH_3COOH	2.5	150	3	N	N	N
4.2% synthetic sea salt + 1.0 *M* CH_3COOH	2.3	150	3	N	N	N
1 *M* NH_4Cl + 0.25 *M* CH_3COONH_4	4	150	5	N	N	N
Chloride-nitrate solutions:						
4 *N* NaCl + 0.5 *N* KNO_3 + 0.1 *N* HNO_3 [*15*]	0.4	70	4	P	N	N
		150	3	EM	P	P
2 *N* NaCl + 0.25 *N* KNO_3 + 0.1 *N* HNO_3	0.9	150	3	EM	P	P

(*Continued*)

TABLE 1—(*Concluded*)

Corrodent	pH	Temperature, deg F	Exposure Time, days	Visual Examination[a,b]		
				5456 Control Specimens		
				Low Resistance, Fig. 2*d*	Medium Resistance, Fig. 2*c*	5086-H32, Fig. 2*b*
1 *M* $NaCl$ + 0.25 *M* $NaNO_3$	3	150	5	EM	N	N
		150	5	EM	EM	Q
1 *M* NH_4Cl + 0.25 *M* NH_4NO_3	2	120	4	N	N	N
		150	4	EI	N	N
	3	80	10	N	N	N
		120	4	N	N	N
		150	4	ES	Q	N
	5	120	4	N	N	N
		150	4	N	N	N
2 *M* NH_4Cl + 0.25 *M* NH_4NO_3	3	150	5	EI	N	N
4 *M* NH_4Cl + 0.25 *M* NH_4NO_3	3	150	5	Q	P	P
1 *M* NH_4Cl + 1 *M* NH_4NO_3	3	150	5	N	N	N
0.25 *M* NH_4Cl + 1 *M* NH_4NO_3	3	150	5	N	N	N
Miscellaneous solutions:						
5% $NaCl$ + 0.05 *N* H_2SO_4	1.1	120	21	N	N	N
0.5 *M* $NaCl$ + 0.25 *M* Na_2SO_4	3	150	4	N	N	N
1 *M* Na_2SO_4 + 0.5 *M* $NaNO_3$	2	150	5	N	N	N
2 *M* $NaHSO_4$	0.6	150	2	N	N	N
1 *M* NH_4Cl + 0.25 *M* $NH_4H_2PO_4$	4	150	5	N	N	N
1 *M* NH_4Cl + 0.25 *M* $(NH_4)_2HPO_4$	7	150	5	N	N	N

[a] Code for visual examination:
N = no appreciable attack, surface may be etched or discolored.
P = pitting.
Q = questionable exfoliation (pit blisters).
EM = exfoliation, mild.
EI = exfoliation, intermediate.
ES = exfoliation, severe.

[b] In no case was there any indication of exfoliation of test coupons of **5456-H117** (Fig. 2*a*).

caused very little general corrosion and produced severe localized exfoliation of the lower resistance 5456 control material. Although the solution did not actually produce exfoliation of the less susceptible 5456 control specimen, Fig. 2*c*, it appeared most worthy of further evaluation as a corrodent for exfoliation testing of Al-Mg alloys by immersion.

Optimization of Chloride-Nitrate Solution Test—Experiments with intermittent immersion methods to get oxygen to the surface of the test specimen to simulate the intermittent spray effect were unsuccessful. On the other hand, aeration of the solution with fine bubbles of compressed

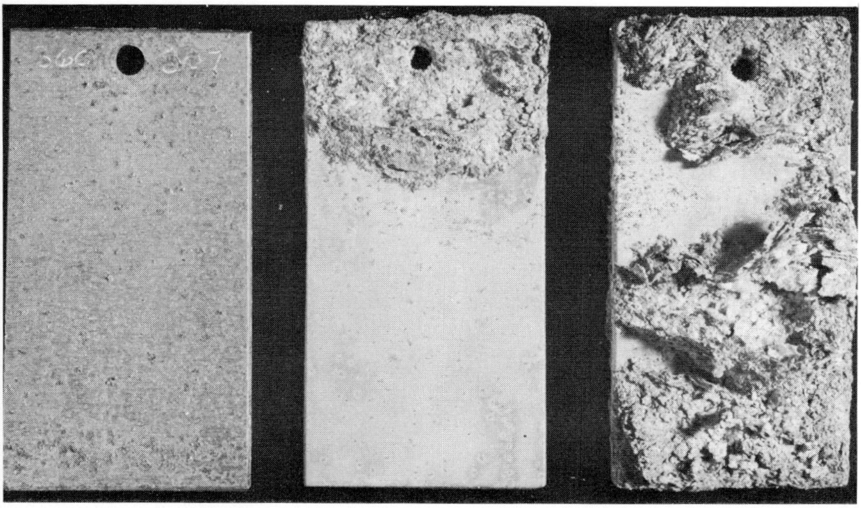

(*a*) Intermittent immersion (10-min immersion; 50-min drying).
(*b*) Continuous immersion, quiescent.
(*c*) Continuous immersion, aerated.

FIG. 3—*Photograph (approximately ×1) of specimens of 5456 control, Fig.* 2d, *exposed five days by the methods noted above to a solution containing 1* M *NH_4Cl + 0.25* M *NH_4NO_3 + HNO_3 to pH 3 at 66 C (150 F) (pH adjusted daily).*

air caused a striking increase in the exfoliation of the lower resistance control specimen (Fig. 3). However, oxygen saturation of the chloride-nitrate solution with H_2O_2 instead of bubbling with air was more effective and decidedly easier to control.

The addition of magnesium chelating compounds, such as potassium or ammonium tartrate or tartaric acid, further increased the effectiveness of the test solution with less susceptible plate samples (Fig. 4).

The immersion test solution (*e*) described in Fig. 4 was compared with the specification salt spray test [*6*] in a round robin program by an Aluminum Association Task Group to develop a standard exfoliation test for aluminum-magnesium (Al-Mg) alloys. Tests performed in six different

(*a*) 1 *M* NH_4Cl + 0.25 *M* NH_4NO_3 + HNO_3 to pH 3.
(*b*) 1 *M* NH_4Cl + 0.25 *M* NH_4NO_3 + 10 ml/liter H_2O_2 + HNO_3 to pH 3.
(*c*) 1 *M* NH_4Cl + 0.25 *M* NH_4NO_3 + 0.005 *M* $K_2C_4H_4O_6$ + 10 ml/liter H_2O_2 + HNO_3 to pH 4.5.
(*d*) 1 *M* NH_4Cl + 0.25 *M* NH_4NO_3 + 0.01 *M* $K_2C_4H_4O_6$ + 10 ml/liter H_2O_2 + HNO_3 to pH 4.5.
(*e*) 1 *M* NH_4Cl + 0.25 *M* NH_4NO_3 + 0.01 *M* $K_2C_4H_4O_6$ + 10 ml/liter H_2O_2, pH 5.4.
(*f*) 1 *M* NH_4Cl + 0.01 *M* $K_2C_4H_4O_6$ + 10 ml/liter H_2O_2 + HNO_3 to pH 4.5.
Solution (*d*) included in Interim Federal Specifications QQ-A-00250/19, 20, 31 Dec. 1968.

FIG. 4—*Photograph (approximately* ×*0.5) of specimens of 5456 control, Fig.* 2c, *exposed three days at 66 C (150 F) in solutions varied as noted above to evaluate the relative effectiveness of specific components. (Solution changed daily).*

laboratories demonstrated the superior reproducibility of the immersion test procedure. The immersion test not only effectively produced exfoliation of susceptible materials but also caused objectionable pitting of some specimens not susceptible to exfoliation. Subsequent studies of the test variants showed that the pitting attack could be minimized by modifications such as: reduction of chloride:nitrate ratio (Fig. 5), increasing the temperature (Fig. 6), and shortening the exposure. Certain of these modifications were incorporated into two similar test procedures (Fig. 7). The procedure referred to subsequently in this article as the "ASSET" test is listed next:

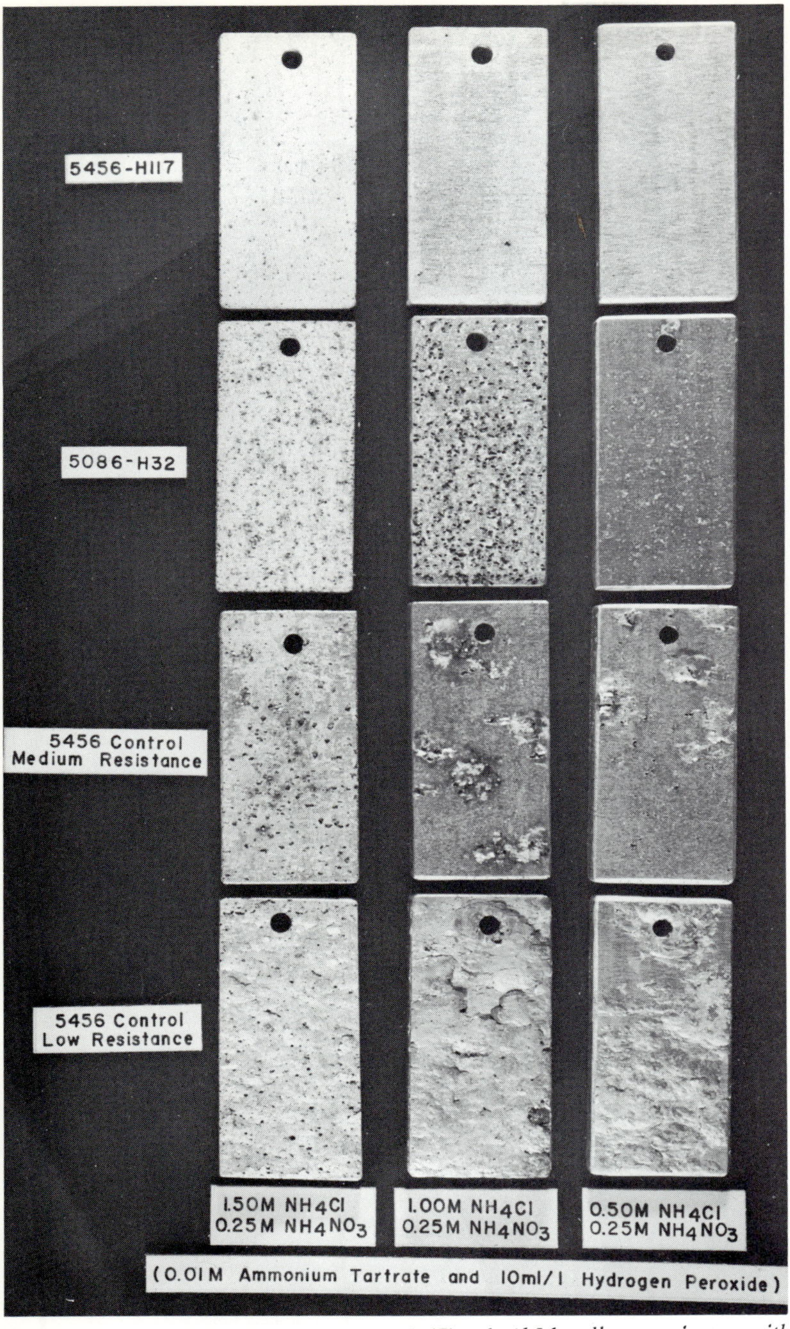

FIG. 5—*Photograph (approximately ×0.67) of Al-Mg alloy specimens with differing resistance to exfoliation exposed two days at 66 C (150 F) to solutions of varied chloride-nitrate ratio.*

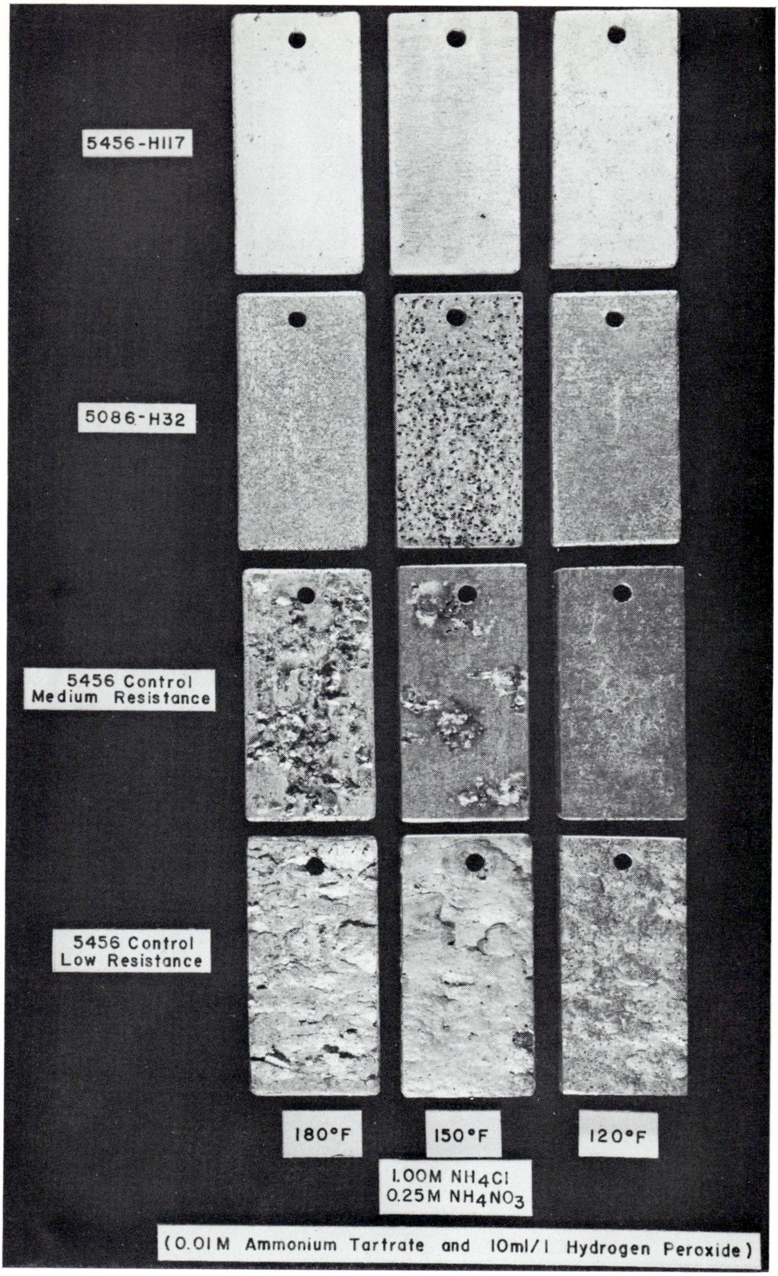

FIG. 6—*Photograph (approximately ×0.67) of Al-Mg alloy specimens with differing resistance to exfoliation exposed two days at temperatures of 49, 66, and 82 C (120, 150, and 180 F).*

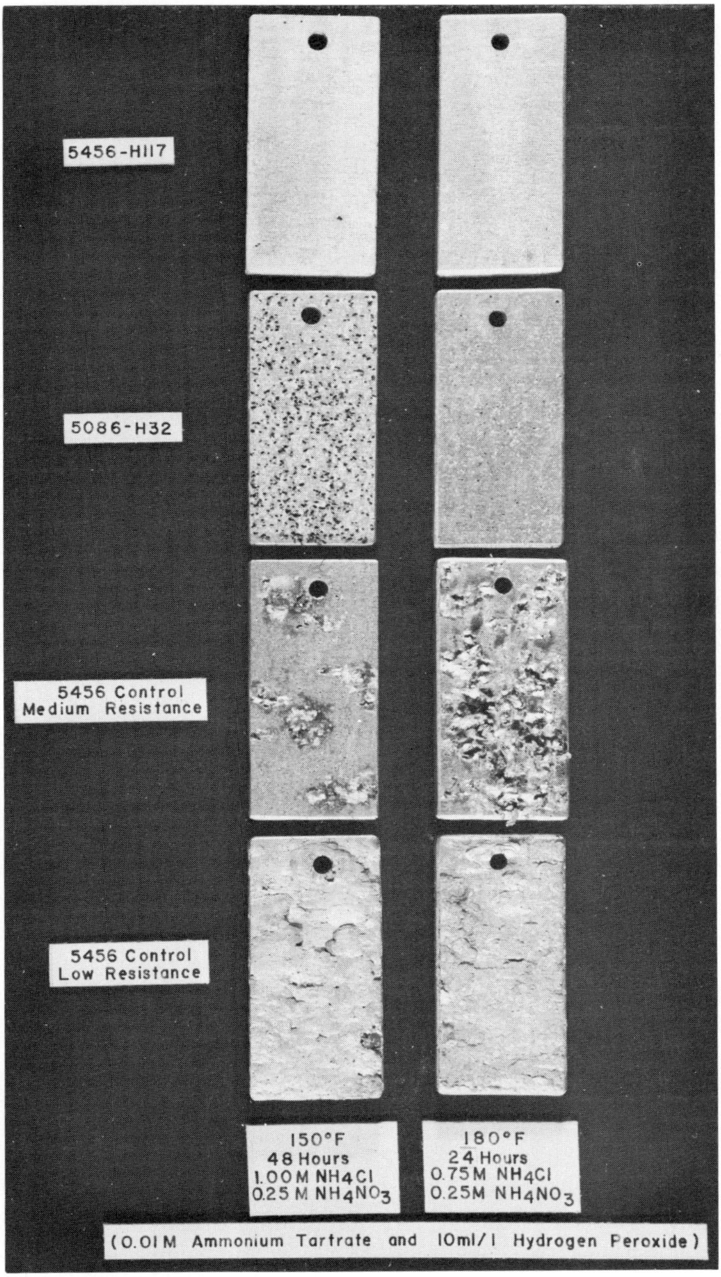

FIG. 7—*Photograph (approximately ×0.67) of Al-Mg alloy specimens with differing resistance to exfoliation exposed to two slightly different versions of the ASSET immersion test.*

Solution: 0.75 *M* ammonium chloride.
0.25 *M* ammonium nitrate.
0.01 *M* ammonium tartrate.
10 ml/liter hydrogen peroxide (30 percent H_2O_2).
Temperature: 80 C (180 F).
Volume-to-metal surface ratio: 40 ml/in.^{2}, min.
Duration: 24 h.

The other procedure (Appendix A) has been similarly designated the "ASSET" test by the Aluminum Association Task Group which recently recommended it as the preferred method for exfoliation corrosion testing of 5086 and 5456 alloys.

Tests on Aluminum-Zinc-Magnesium-Copper (Al-Zn-Mg-Cu) (7XXX) Alloys—Preliminary testing of 7178 alloy test specimens by the ASSET procedure showed that it was too severe for these less corrosion resistant alloys. Additional studies, therefore, were initiated to modify the test for this alloy system. Specimens were obtained from a sample of ⅜-in.-thick 7178 alloy plate that was solution heat treated, quenched in cold water, and artificially aged to provide a low, medium, and high resistance to exfoliation.

Lowering the temperature of the ASSET solution to 27 C (80 F) materially reduced the severity of the corrosion. Other experiments showed that the ammonium tartrate addition was not essential. With the reduced test temperature and deletion of the ammonium tartrate, tests were performed to verify the optimum ratio and concentration of the ammonium salts.

Variations in chloride : nitrate molar ratios ranging from 10 : 1 to 2 : 1 (the nitrate concentration was fixed at 0.25 *M*) showed that the most favorable results were obtained with ratios of 4 or 5 : 1. Higher ratios reduced the exfoliation of the low-resistance material and caused some blistering and slivering of the high-resistance specimen. Lower ratios reduced and even eliminated exfoliation of the low-resistance plate. Lowering the concentration of the ammonium chloride and ammonium nitrate salts, while retaining the 4 : 1 ratio, reduced the aggressiveness of the solution but did not affect the exfoliation of the susceptible specimens (Fig. 8).

A solution of this type (Test A) and the Zaretskii solution mentioned above (Test B) were tested by a task group formed under ASTM Subcommittee G01.05.02 to evaluate improved exfoliation tests for Al-Zn-Mg-Cu alloys. At the conclusion of the round robin test program performed in eight different laboratories, it was the consensus of the task group[4] that the results of the immersion tests agreed with the results of

[4] The results of this task group study are given in more detail by S. J. Ketcham and P. W. Jeffrey in another paper in this symposium.

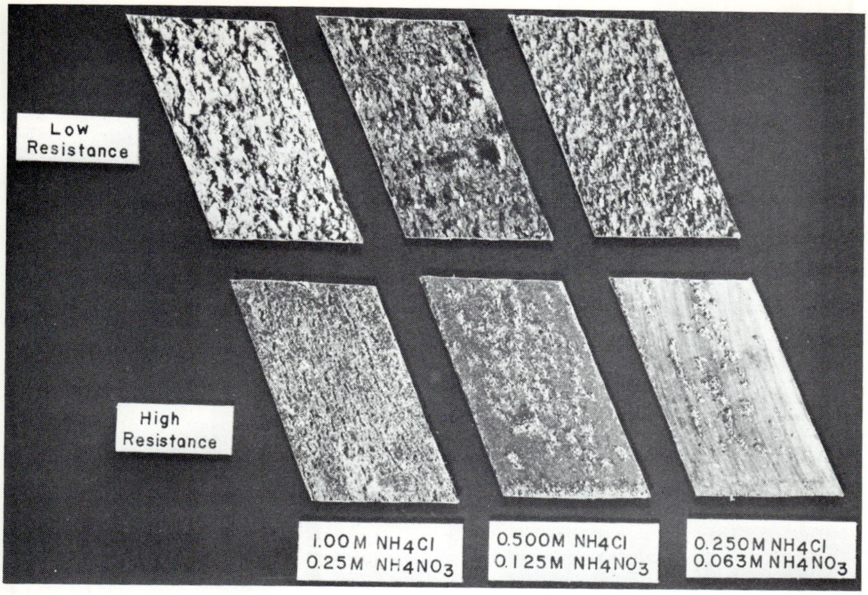

FIG. 8—*Photograph (approximately ×1) of midplane (T/2) of 7178 alloy exposed 16 h at 27 C (80 F) to varied concentrations (4 : 1 chloride : nitrate ratio) of the ASSET solution modified for exfoliation of Al-Zn-Mg-Cu alloys.*

the salt spray test (Fig. 9), and the Zaretskii solution (Test B) gave the most reproducible test results. The name "EXCO" was adopted by the task group for this test procedure (Appendix B).

Experiments with variations in the EXCO test yielded the following additional observations:

1. Substitution of 0.5 *M* sodium nitrate ($NaNO_3$) for 0.5 *M* KNO_3 did not affect the performance.

2. Reduction of the concentration of the salts by 50 percent also did not affect the performance so long as the 8 : 1 ratio of NaCl to KNO_3 (or $NaNO_3$) was maintained.

Aluminum-Zinc-Magnesium (Al-Zn-Mg) (7XXX) Alloys

These alloys have a very good resistance to corrosion, and exfoliation corrosion has not been reported to be a problem with the artificially aged tempers. Exfoliation will occur in F and W temper material, and it has been shown to occur in the heat affected zone of weldments exposed in natural environments such as seawater. Weldments of X7005 alloy sheet which had exhibited these exfoliation characteristics in tests conducted by intermittent immersion in a 3.5 percent NaCl solution were selected to determine whether such exfoliation susceptibility could be detected by immersion test procedures.

Exposure for a period of 24 h by the EXCO procedure produced

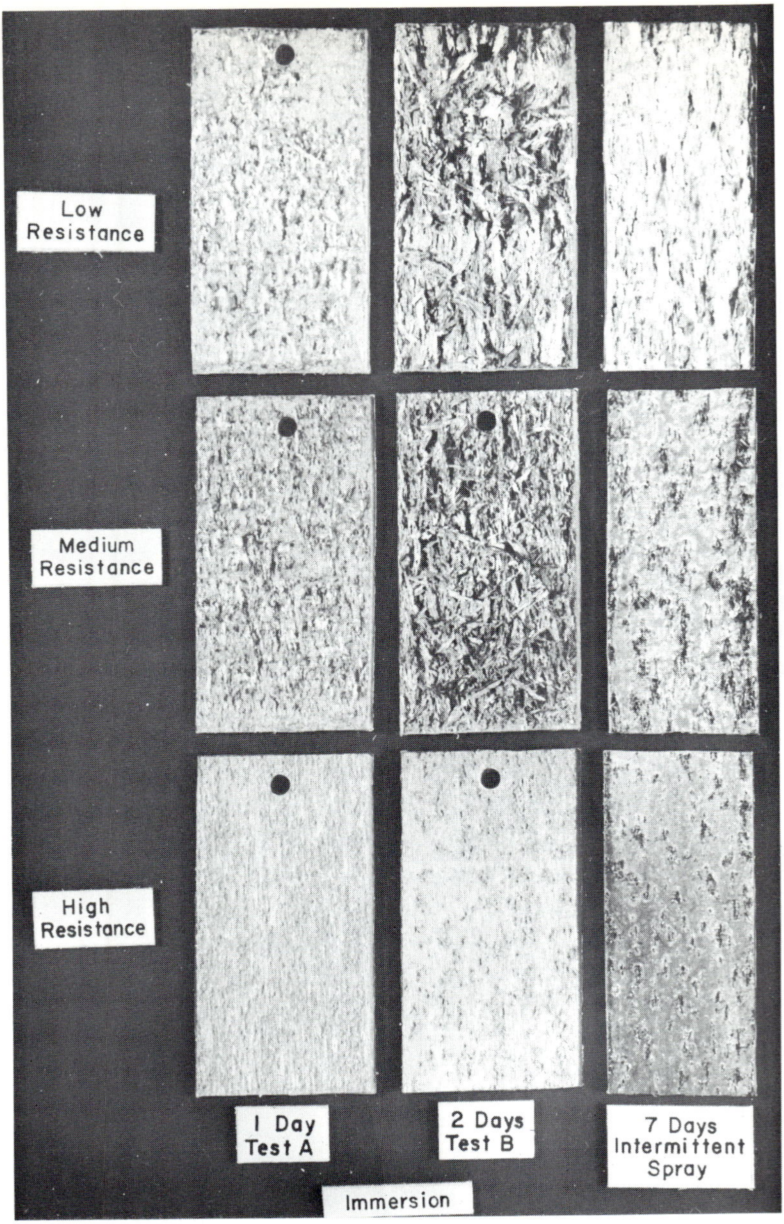

Test A: 0.50 *M* ammonium chloride.
0.125 *M* ammonium nitrate.
10 ml/liter hydrogen peroxide (30% H_2O_2).
Test B: 4.0 *M* sodium chloride.
(EXCO)
0.5 *M* potassium nitrate.
0.1 *M* nitric acid.

FIG. 9—*Photograph (approximately* ×*0.4) of midplane specimens (T/2) of 7178 alloy exposed to two different immersion tests and the acidified salt spray exfoliation test (QQ-A-00250/24, 25)* [11].

exfoliation in the susceptible specimens and provided a clear delineation between susceptible and resistant specimens. Figure 10 shows the appearance of welded specimens after exposure to the EXCO test, the 3.5 percent NaCl alternate immersion test, and the acidified salt spray exfoliation test. It is evident that the immersion exfoliation test provides a better correlation with the salt water exposure than the acidified spray test.

Aluminum-Copper (Al-Cu) (2XXX) Type Alloys

Material in the naturally aged tempers, T3 and T4, of these alloys have shown considerable susceptibility to exfoliation in the atmosphere, and this susceptibility is revealed readily by acidified salt spray tests. In fact, slowly quenched 2024-T42 control panels are employed as controls for the spray test cabinets. However, as illustrated in Fig. 1, appreciable variability has been encountered in the spray tests, even with such highly susceptible material. Therefore, tests were conducted with these materials to determine whether exfoliation could be developed by immersion test procedures.

Room-temperature exposure in both the ASSET solution for 5XXX alloys, and the EXCO solution for 7XXX alloys effectively developed exfoliation in the 2024-T42 control specimens. The ASSET solution developed exfoliation in less time than the EXCO solution (one day versus four days).

Comparison with Natural Environments

When using highly accelerated corrosion tests, one is naturally concerned that the test conditions may be too drastic and produce unrealistic corrosion effects. Thus, it is essential that performance of known materials in the accelerated test be compared with that of the same material in service or in a service type of environment. The following section describes several examples illustrating a good relationship between results of immersion type exfoliation tests and exposures to service environments. Exfoliation was produced in the immersion tests duplicating that which occurred in actual service or in service type environments. And of equal significance, alloys and tempers that did not exfoliate in the service environments likewise did not exfoliate in the immersion tests.

5086-H32 and 5456-H321 Weldments in Seawater

A weldment of 5456-H321 that developed exfoliation of the parent plate during a 2-year period of immersion in seawater at Miami, Fla., is shown in Fig 11. The microstructure of the plate is also shown, and the marked outlining of the grain boundaries with precipitated Al-Mg constituent indicates that exfoliation could be expected in some environments (compare with Fig. 2*c*). The test coupons exposed to the immersion tests indicated an intermediate susceptibility. A specimen of welded

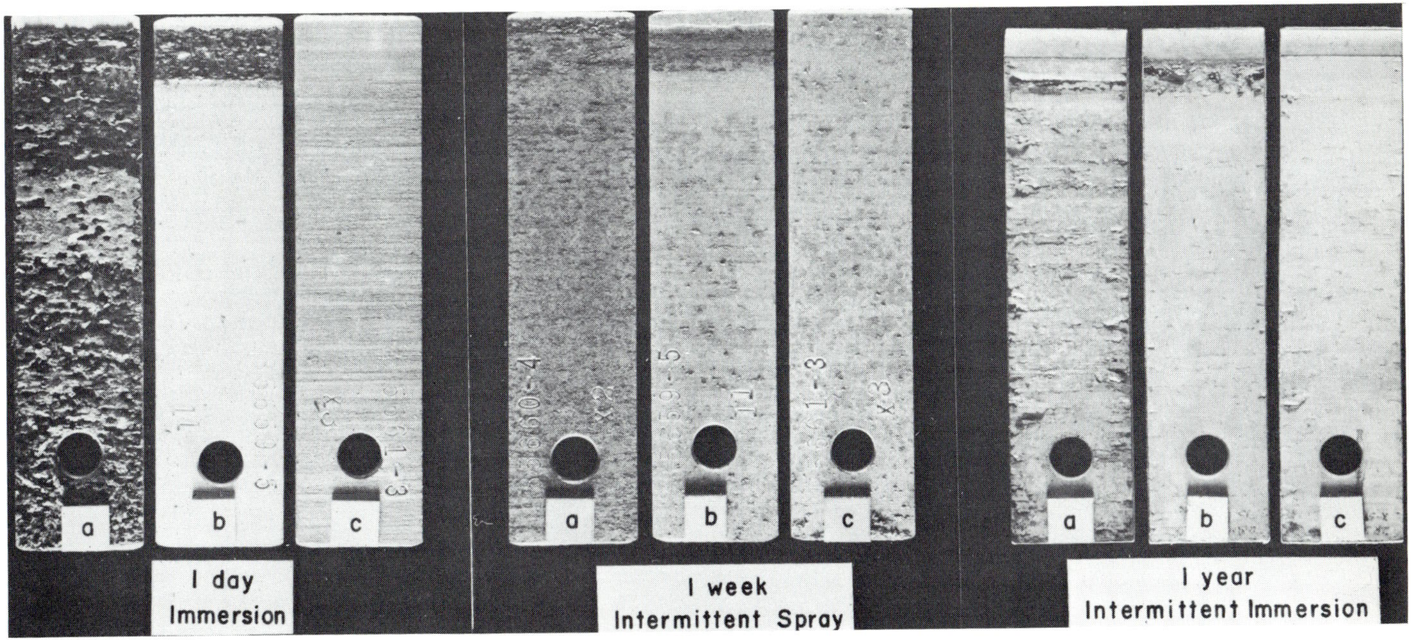

(*a*) W temper, no postweld aging.
(*b*) T63 temper, no postweld aging.
(*c*) W temper, postweld aged by T63 practice.

FIG. 10—*Photograph (approximately ×1) of welded X7005 alloy sheet, 0.125 in. thick, in the conditions noted above exposed to the EXCO immersion test, the acidified salt spray exfoliation test* [11], *and the 3.5% NaCl alternate immersion test. A portion of the weld bead is shown at the upper edge of each specimen.*

FIG. 11—*Correlation of immersion test and seawater exposure on 5456-H321 welded specimen. Inset illustrates the microstructure of the plate and the exfoliation attack which was similar for the accelerated tests and the service environment.*

5086-H32 immersed for a 4-year period in seawater without exfoliating in shown in Fig. 12. The test coupons exposed to the immersion test likewise did not exfoliate, although the coupon immersed in the specification test solution developed marked pitting.

5456-H321 Alloy Boats in Lake Maracaibo

Severe corrosion of the exterior surfaces of 5456-H321 hull plates of two crew boats occurred during 2-years' service in the brackish water of Lake Maracaibo, Venezuela, in the absence of normal protective systems. Figure 13 shows the severe exfoliation observed on Boat A, and Fig. 14 shows the deep localized pitting observed on Boat B. A comparison of microstructures readily indicates that the continuity of precipitate in the plate from Boat A was much greater than in the plate from Boat B and that some difference in the type of corrosion could be anticipated. The immersion test coupons showed a performance identical with that observed in service, as illustrated in Figs. 13 and 14. Both forms of corrosion could have been prevented on the crew boats by the proper use of established protective systems.

Exfoliation Resistant Tempers of 5456 Alloy [6]

Rolled plate in exfoliation resistant strain hardened tempers of 5456 and similar alloys can be produced by proprietary techniques [*16*] that

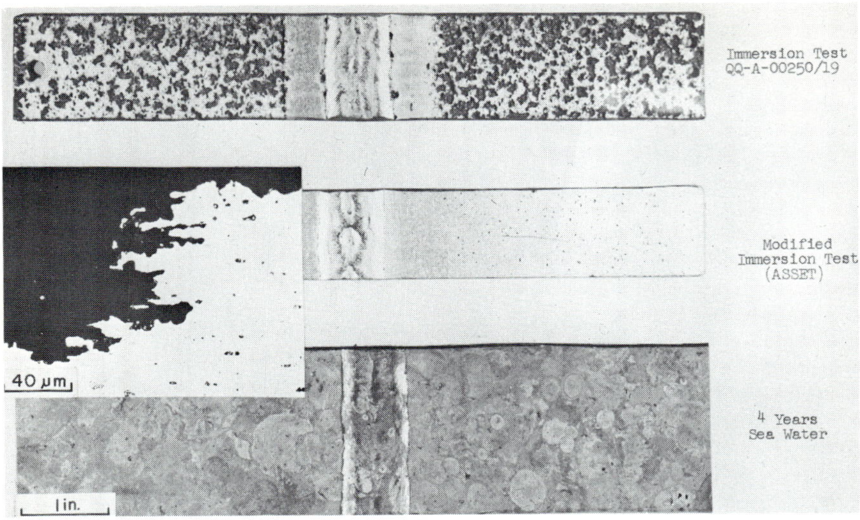

FIG. 12—*Correlation of immersion test and seawater exposure on 5086-H32 welded specimen. Inset illustrates the microstructure of the plate and the pitting attack which was similar for the accelerated tests and the service environment.*

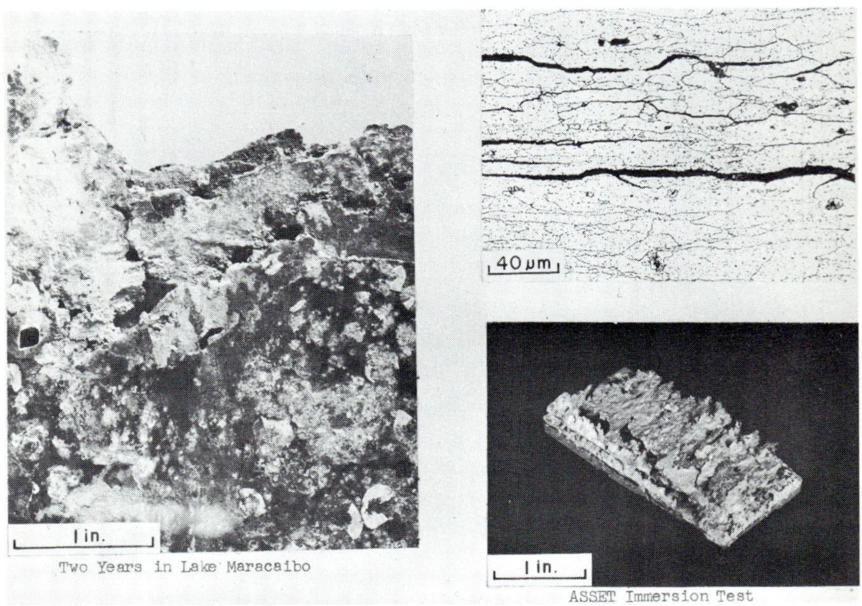

FIG. 13—*Correlation of immersion test and service on a 5456-H321 welded specimen of Crew Boat A in Lake Maracaibo. Photomicrograph illustrates the microstructure of the plate and the exfoliation attack which was similar for both the accelerated test and the service environment.*

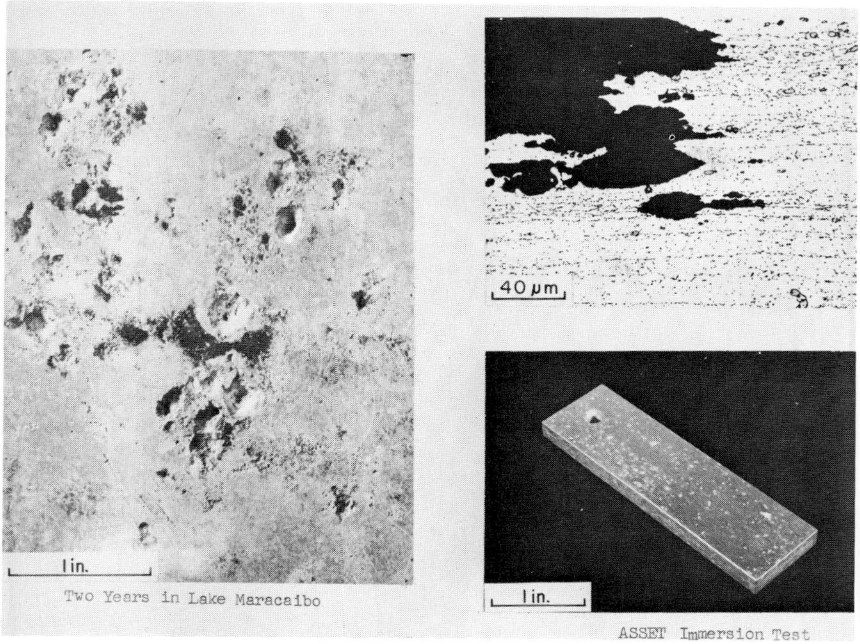

FIG. 14—*Correlation of immersion test and service on a 5456-H321 welded specimen of Crew Boat B in Lake Maracaibo. Photomicrograph illustrates the microstructure of the plate and the pitting attack which was similar for both the accelerated test and the service environment.*

achieve either of two metallurgical conditions: (*a*) homogeneous Al-Mg solid solution or (*b*) homogeneous Al-Mg solid solution plus random dispersion of Mg_2Al_3 precipitate. Intermediate conditions that have heterogenous solid solution or selective precipitation along deformation bands and grain boundaries or both may have some degree of susceptibility to intergranular corrosion or exfoliation. The greatest tendency for exfoliation is associated with a high degree of continuity of Mg_2Al_3 precipitate along deformation bands and grain boundaries. Examples of these metallurgical conditions, together with the results of accelerated exfoliation tests, are shown in Fig. 15. Neither plates in the H117 nor in the H343 temper showed any tendency for exfoliation in the ASSET immersion test or in the salt spray test.

2024-T4 and T6 Extrusions in Seacoast Atmosphere

Because it is difficult to quench thick sections of 2024 alloy extrusions rapidly enough to keep all of the copper in solid solution, it is not uncommon to observe exfoliation with the naturally aged tempers, T3 and T4, after several years' exposure in the atmosphere. Figure 16 shows the exfoliation that occurred with a 2024-T4 extrusion during a 4-year

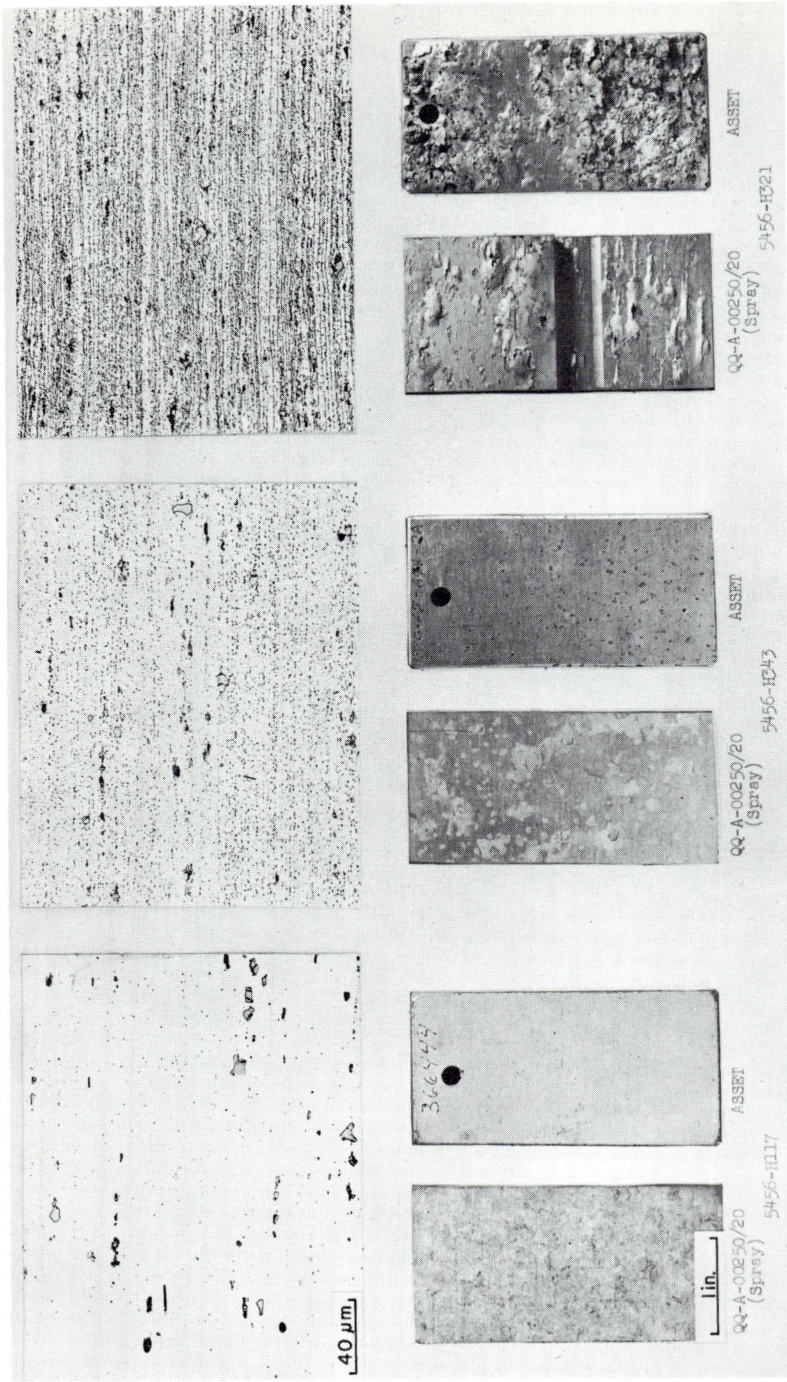

FIG. 15—*Comparison of attack after exposure to specification acidified salt spray test and ASSET immersion test.*

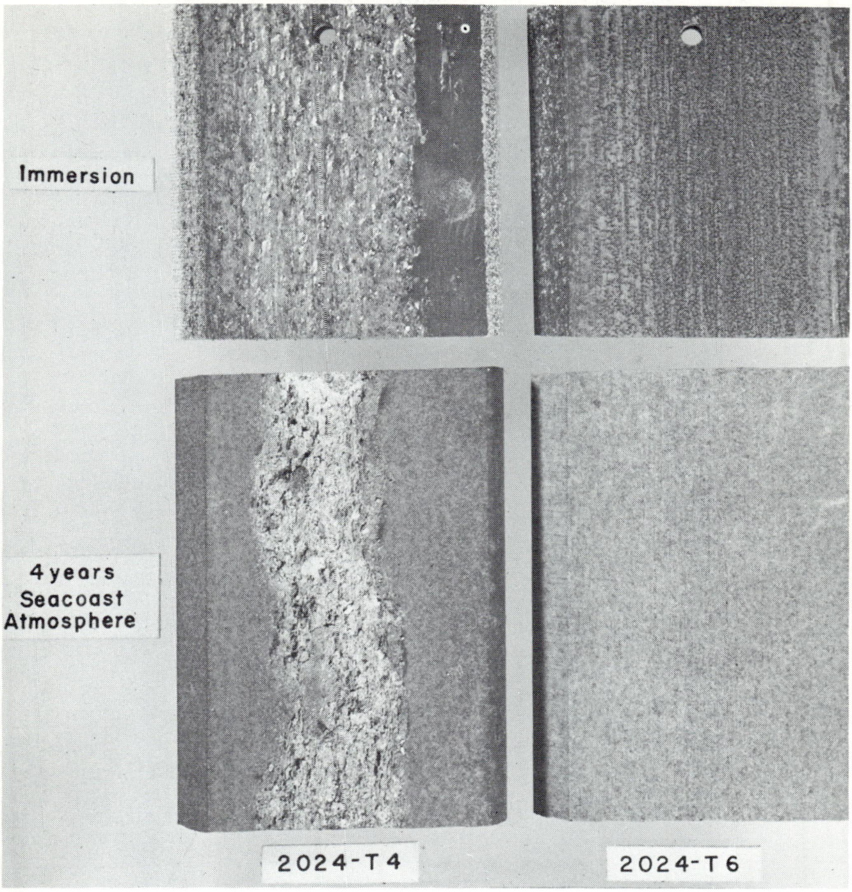

FIG. 16—*Correlation of immersion test and seacoast atmospheric exposure.*

exposure to the seacoast atmosphere at Point Judith, R. I. This problem can be corrected by employing the artificially aged T6 and T8 tempers. The ASSET immersion test, requiring but a 24-h exposure at room temperature, produced exfoliation in the T4 temper and pitting in the T6 temper specimens. Therefore, the immersion test could prove very useful both to determine whether an Al-Cu alloy is susceptible to exfoliation, and, if so, whether the corrective treatment to be employed is valid.

7075-T6 and T76 Extrusions in Seacoast Atmosphere

The resistance to exfoliation of high-strength heat-treatable aluminum alloy products is principally a function of variations in grain shape, quench rate, and the artificial aging treatment. A markedly directional grain structure and a slow quench rate are generally adverse in the case of material in the T6 temper, and exfoliation may occur in these products within a year in a seacoast atmosphere. However, exfoliation attack can be virtually eliminated by the use of the resistant T76 temper. Figure 17 illustrates the exfoliation that occurred in a 7075-T6510 extruded section during a 1-year exposure. It also demonstrates the excellent resistance afforded by a 7075-T76510 section during a 3-year exposure to the seacoast atmosphere at Point Judith, R. I. The EXCO immersion test effectively demonstrated similar corrosion characteristics in these T6510 and T76510 temper sections and required an exposure of only 48 h duration.

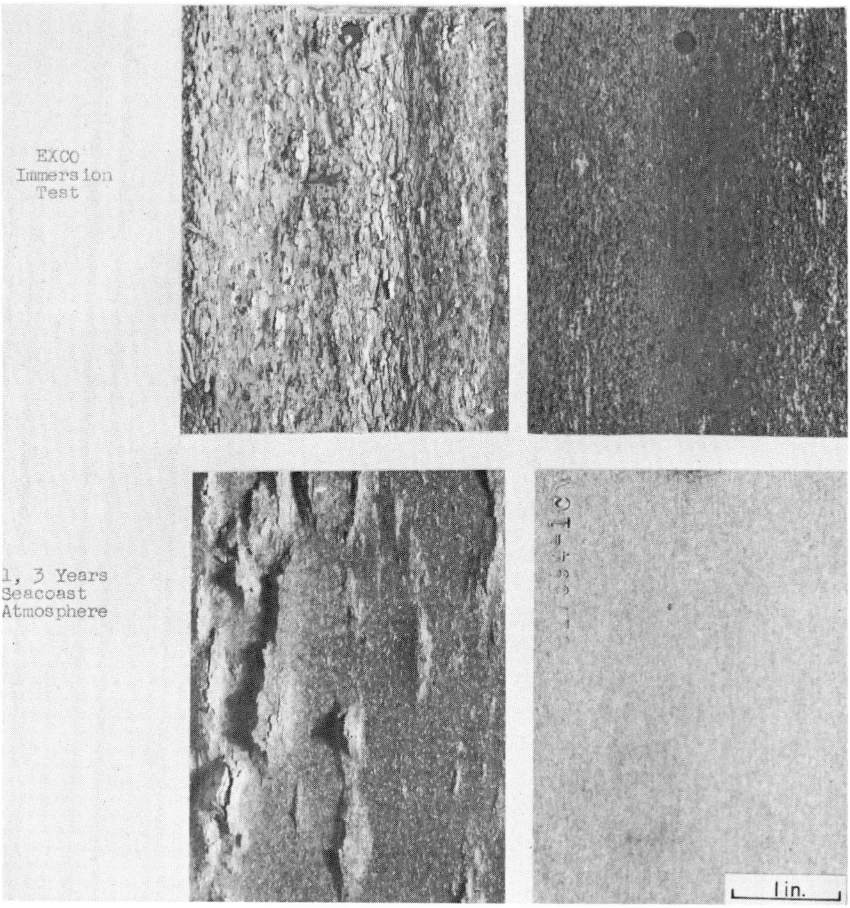

FIG. 17—*Correlation of immersion test and seacoast atmosphere. Midplane of extruded wing planks (0.7 in. thick).*

Discussion

With the optimum immersion test conditions for an alloy type, specimens of most products develop easily distinguishable forms of corrosion such as exfoliation, pitting, or etching. Exfoliation may initiate as blisters that progress into lamellar corrosion, and, some specimens that approach borderline susceptibility to exfoliation may develop an "in-between" form of corrosion. The term "pit-blisters" has been used to describe this form of corrosion (Figs. 12 and 18), where the attack has the appearance of small blisters that have opened into pits with residual undercutting around the edges. Although, generally speaking, pit-blisters can be due to

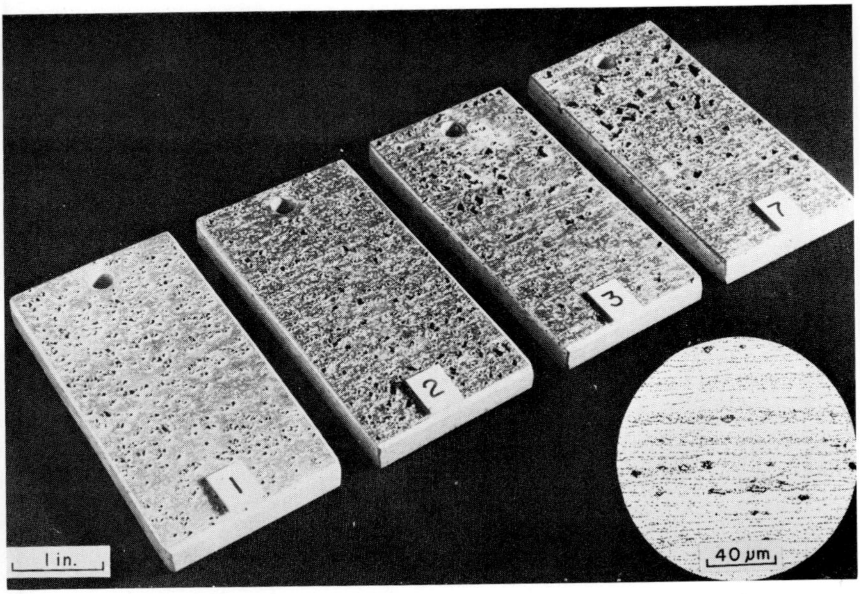

FIG. 18—*Visual appearance of specimens of 5086-H32 alloy, Fig. 2b, after 1, 2, 3, and 7-days' exposure to specification immersion test.*

more than one microscopic type of attack, their occurrence in this very aggressive test denotes negligible susceptibility to exfoliation. One way to gain assurance of this is to extend the exposure period and watch for a change in the form of corrosion. This procedure is illustrated in the following example for 5XXX alloys, and the same procedure is applicable to other alloys. As indicated in Fig. 18 the pit blisters present after the first day of exposure became larger with extended exposure but did not progress into exfoliation. Compare this with the specimen in Fig. 19 where the pit blisters present after the first day grew with extended exposure into larger blisters and distinct exfoliation. (The difference in performance of the specimens illustrated in Figs. 18 and 19 is consistent

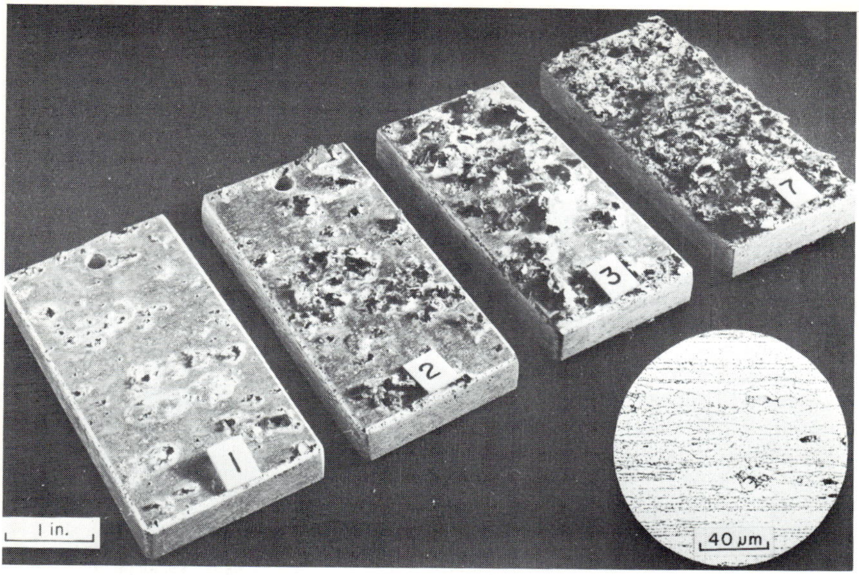

FIG. 19—*Visual appearance of specimens of 5456 control, Fig. 2c, after 1, 2, 3, and 7-days' exposure to specification immersion test.*

with the greater tendency for the precipitated Al-Mg constituent to outline the elongated grain structure in the 5456 alloy specimens shown in Fig. 19). To determine the path of the corrosion and its mechanism, metallographic examination of the corroded specimens is required. In these examples, the 5086-H32 alloy specimens revealed pitting and undermining pitting, whereas the 5456-H321 alloy specimens showed a combination of undermining pitting and intergranular corrosion (Fig. 20). It is noteworthy that the intermittent salt spray test (QQ-A-00250/20) [7] of the latter specimens did not indicate exfoliation susceptibility.

It is evident that the test methods just described involve aggressive corrodents, and it is believed that results of these tests can be used to predict the serviceability of aluminum alloy products in very corrosive marine environments. In relatively mild marine environments, and in many other types of service, the corrosion effects shown in these tests may never occur. There is always a considerable risk associated with the use of highly accelerated corrosion tests such as these, and the results of such tests must be interpreted with caution. While the comparisons with service experience just reported are very encouraging, still more comparisons are desirable to relate the test results with specific environments.

It is important that the optimum formulation of test solution and choice of test duration be used for a given type of alloy. Although further experimentation with these test procedures may lead to additional refine-

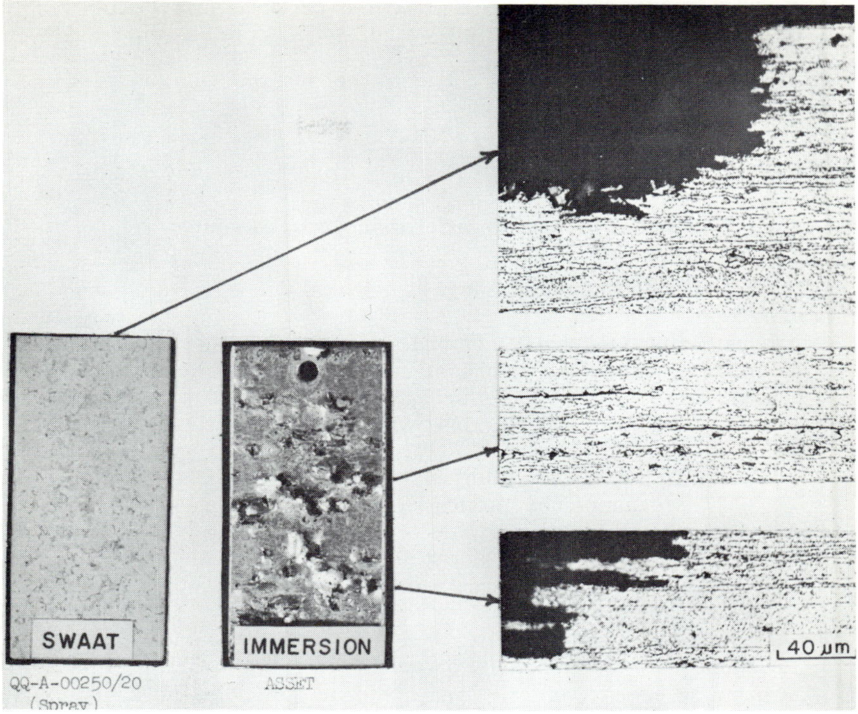

FIG. 20—*Comparison of attack after exposure of specimens of 5456 control, Fig. 2c, to specification acidified salt spray test and immersion test.*

ments, they, too, should be compared with exposures in service environments.

Advantages of an Immersion Test Method

Continuous immersion tests offer several readily apparent advantages over other acceptable spray or electrochemical type corrosion tests. Results may be determined rapidly (one to four days) and visually, without the aid of a magnifying glass or microscope. Test apparatus is simple and economical, and the test can be performed by a laboratory technician. Because of the simple technique a continuous immersion test has a high inherent reproducibility. This has been demonstrated by round robin test programs with two of the methods just described. Tests can be performed with equal ease on test coupons or upon small fabricated parts and structures. Although no special preparation of the surface of the test specimen is required, a light caustic etch is preferred. Finally, and most important, the preceding advantages can be realized with continuous immersion test methods because it has been possible to demonstrate very encouraging comparisons with exfoliation experience in service or with tests in service environments. Additional comparisons of these test methods with service experience should be obtained.

APPENDIX

A. ASSET Continuous Immersion Exfoliation Test Procedure[5]

Application

1. 5086 and 5456 aluminum alloys.

Procedure

1. Specimen size: Optional, a nominal size of 1.5 by 4 in. is recommended.
2. Degrease specimen and etch 1 min in 5 percent by weight sodium hydroxide at 82 C (180 F), rinse in water, desmut 30 s in concentrated nitric acid at 27 C (80 F), rinse with distilled water, and air dry.
3. Solution:

    1.0 *M* ammonium chloride.
    0.25 *M* ammonium nitrate.
    0.01 *M* ammonium tartrate.
    10 ml/l hydrogen peroxide (30 percent H_2O_2).

    Use reagent grade chemicals and distilled water.
    pH as made is about 5.3.
    Operate test at 66 ± 1 C (150 ± 2 F).
4. Recommended minimum solution volume to metal area is 40 ml/in.^{2}.
5. Suspend specimens vertically and completely immerse in solution. Use suitable inert container (that is, glass, etc.).
6. Normal test duration is 24 h.
7. Upon test completion, soak specimens in concentrated nitric acid until clean and rinse in water.

B. EXCO Continuous Immersion Exfoliation Test Procedure[6]

Application

1. Al-Zn-Mg-Cu (1.2 to 2.8 percent copper).

Procedure

1. Specimen size: Optional, a nominal size of 2 × 4 in. is recommended.
2. Degrease specimen with suitable solvent; etching or other pretreatment is not necessary but will not interfere.
3. Solution:

    4.0 *M* sodium chloride.
    0.5 *M* potassium nitrate.
    0.1 *M* nitric acid.

    Use reagent grade chemicals and distilled water.
    pH as made is 0.4.
    Operate test at room temperature, that is, 25 ± 3 C (77 ± 5 F)
4. Recommended minimum solution volume to metal area is 50 ml/in.^{2}.

[5] Procedure recommended by joint Aluminum Association–American Society for Testing and Materials Task Group for Exfoliation Testing 7000 Series Copper Containing Aluminum Alloys.

[6] Procedure recommended by Aluminum Association Task Group on Exfoliation Corrosion and Stress Corrosion of 5000 Series Aluminum Alloys.

5. Immerse specimens in the solution. The horizontal orientation of the test surface is preferred to prevent loss of corrosion products.
6. Normal test duration is 48 h. Inspection of specimens (without cleaning) at intermediate periods should be performed to detect early development of exfoliation that might be obscured with continued exposure.
7. Upon test completion, soak specimens in concentrated nitric acid until clean and rinse in water.

References

[1] Lifka, B. W., Sprowls, D. O., and Kaufman, J. G., *Corrosion,* SBIIA, Vol. 23, No. 11, 1967, p. 335.
[2] Rotsell, W. C. and Long, J. R., *Metal Progress,* MEPOA, Vol. 91, No. 5, 1967, p. 106.
[3] Vaccari, J. A., assoc. ed., *Materials Engineering,* MAENB, Vol. 71, No. 6, June 1970, p. 22.
[4] Luhan, J. V. and Summerson, T. J., *Metals Engineering Quarterly,* MENQA, Vol. 10, No. 4, 1970, p. 35.
[5] Brooks, C. L., *Naval Engineers Journal,* NVEJA, Vol. 82, No. 4, 1970, p. 29.
[6] Interim Federal Specification QQ-A-00250/19, 20 (NAVY-Ships), 31 Dec. 1968.
[7] Federal Specifications QQ-A-00200/14, 15 (ASG), 12 Aug. 1969 and QQ-A-00250/24, 25 (ASG), 8 Oct. 1969.
[8] Liddiard, E. A. G., Whittaker, J. A., and Farmery, H. K., *Journal, of the Institute of Metals,* JIMEA, Vol. 89, 1960–1961, p. 377.
[9] Booth, F. F. and Godard, H. P., *Proceedings,* First International Congress on Metallic Corrosion, 1961, p. 703, Butterworth, London.
[10] Budd, M. K. and Booth, F. F., *Corrosion,* SBIIA, Vol. 18, No. 5, 1962, p. 197t.
[11] Lifka, B. W. and Sprowls, D. O., *Corrosion,* SBIIA, Vol. 22, No. 1, 1966, p. 7.
[12] Romans, H. B., *Materials Research and Standards,* MTRSA, Vol. 9, No. 11, 1969, p. 31.
[13] Ketcham, S. J. and Shaffer, I. S., *Proceedings,* Fourth International Corrosion Congress, 1969, to be published by National Association of Corrosion Engineers.
[14] Sigalovskoya, T. M. and Zaretskii, E. M., *Zashchita Metallov,* ZAMEA, Vol. 3, No. 6, 1967, p. 730.
[15] Zaretskii, E. M. and Kireeva, A. F., *Zavodskaya Laboratoriya,* ZVDLA, Vol. 29, No. 9, 1963, p. 1098.
[16] Anderson, W. A., U. S. Patent 3,232,796, 1 Feb. 1966.

Michael Henthorne[1]

Intergranular Corrosion in Iron and Nickel Base Alloys

REFERENCE: Henthorne, Michael, **"Intergranular Corrosion in Iron and Nickel Base Alloys,"** *Localized Corrosion—Cause of Metal Failure, ASTM STP 516,* American Society for Testing and Materials, 1972, pp. 66–119.

ABSTRACT: Intergranular corrosion in stainless steels (austenitic, ferritic, and two phase) and nickel base corrosion resistant alloys (nickel-chromium-iron, nickel-chromium-molybdenum, and nickel-molybdenum) is reviewed. Practical aspects discussed include thermal treatments producing susceptibility, media that cause attack, influence of alloy composition, and prevention methods. The currently most plausible mechanisms are also outlined. Test methods, particularly the standard methods for quality control, are dealt with as well as techniques used to simulate the thermal effects of welding. The relationships between intergranular attack and other forms of localized corrosion are noted.

KEY WORDS: corrosion, intergranular corrosion, stainless steels, nickel alloys, tests, evaluation, grain boundaries, precipitation (chemistry), stress-corrosion cracking, pitting

Grain boundaries are physically distinct regions separating different or differently oriented grains [*1*].[2] The boundaries are favored sites for precipitation and segregation which makes them chemically as well as physically different. These differences may result in intergranular attack as shown in Fig. 1.

Most metal alloys are susceptible to intergranular corrosion when exposed to specific corrodents. This paper is concerned only with aqueous media and certain iron and nickel base alloys used primarily for their corrosion resistance. In these alloys the attack is usually dependent upon prior thermal history and is a result of chemical rather than just physical differences at the boundaries. The aims are to review:

1. practical information—metallurgical treatments that produce susceptibility, the types of media that corrode susceptible material, and how one can avoid the problem.

[1] Supervisor, Corrosion Research, Carpenter Technology Corp., Reading, Pa. 19603.

[2] The italic numbers in brackets refer to the list of references appended to this paper.

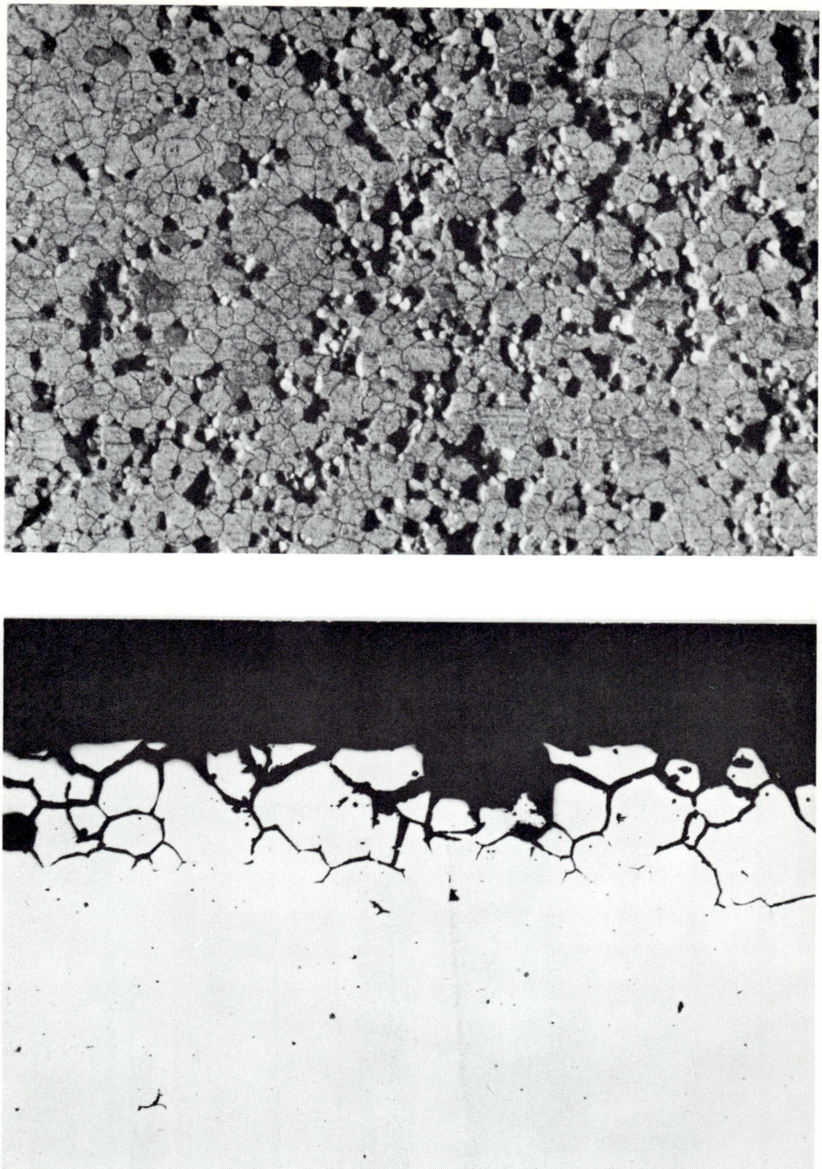

(*top*) Plan view of surface showing some grains dislodged (×50).
(*bottom*) Polished section perpendicular to surface (×100).
FIG. 1—*Intergranular corrosion.*

2. mechanisms—the currently most plausible explanations for intergranular attack.

3. testing—available test procedures, their limitations, advantages, etc.

The alloys discussed are all used commercially and are those for which intergranular corrosion susceptibility can be a limitation to their optimum use. The subjects covered in turn are: (a) austenitic stainless steels, (b) ferritic stainless steels, (c) austenitic-ferritic stainless steels, (d) high-nickel alloys, (e) testing, and (f) relationship to other forms of corrosion.

Austenitic Stainless Steels

This family of alloys has received more attention from researchers of intergranular corrosion than any other material. Most of the work has been with the basic 18Cr-9Ni analysis—AISI Type 304.

Conditions Leading to Attack

The great majority of intergranular corrosion problems in austenitic stainless steels are related to metallurgical changes occurring in the range of about 800 to 1500 F (427 to 816 C). The temperatures most likely to cause problems are usually close to the middle of this range but are also dependent upon the time of exposure. For example, short times are more likely to produce susceptibility at temperatures in the upper part of the range, whereas long exposures (many thousands of hours) may be required at the lower end. This is shown in Fig. 2. For a typical Type 304 analysis an exposure of several minutes at a temperature such as 1250 F (677 C) is usually adequate to cause noticeable attack in a severe intergranular corrosion test. We shall see later that susceptibility is attributed primarily to the precipitation of chromium rich carbides at the grain boundaries although other grain boundary phases and segregates play a role in some cases.

Austenitic stainless steels can be attacked intergranularly in a few

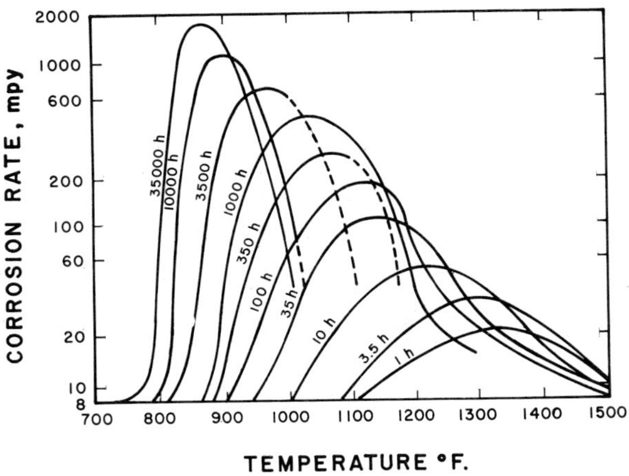

FIG. 2—*Effect of sensitizing time and temperature on corrosion of Type 347 stainless steel in boiling 65 percent* HNO_3 [2].

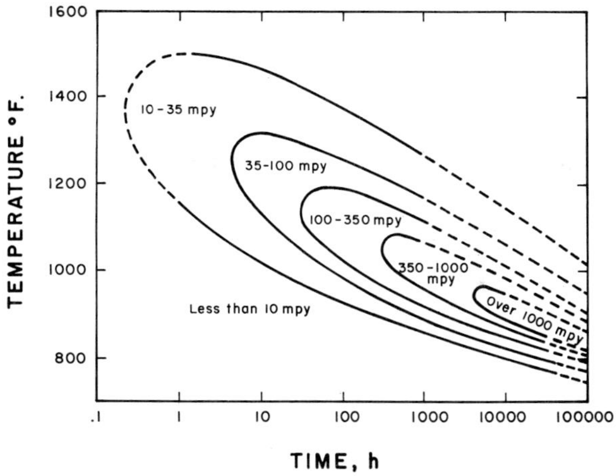

FIG. 3—*Time-temperature-sensitization curves for Type 347 stainless steel in boiling 65 percent* HNO_3 [2].

specific media even when they are in the solution annealed condition. This is a rarity and will be discussed later.

The term sensitization is commonly used to describe heat treatments that may make the alloy susceptible to intergranular attack. The word is used in this context in this paper and does not necessarily imply susceptibility to attack. Sensitization data usually are plotted as a function of the sensitizing temperature and time as shown in Fig. 3. These curves are called TTS, C, or Rollason curves.

It is customary to plot corrosion rate contours like those in Fig. 3, but sometimes the onset of metallurgical changes known to relate to the problem are determined (for example, by metallography, X-ray diffraction) and plotted as shown in Fig. 4. There is a relationship between intergranular corrosion and $M_{23}C_6$ precipitation in the grain boundaries, although of course the position of the intergranular attack curve will depend upon the severity of the corrosion test used to evaluate susceptibility. At the higher temperatures in Fig. 4 there is grain boundary precipitation but no intergranular corrosion. The reason for this will become apparent when the mechanisms of attack are discussed.

In practice, susceptibility to intergranular attack will result usually from one of three situations:

1. Slow cooling from an annealing treatment (or from casting or forging). This is obviously more of a factor in large product sizes where the rate of cooling through the sensitization range is slower.

2. Stress relieving for several hours in the sensitization range, for example, 1100 F (593 C).

3. During welding.

Welding is probably the most common cause of intergranular corro-

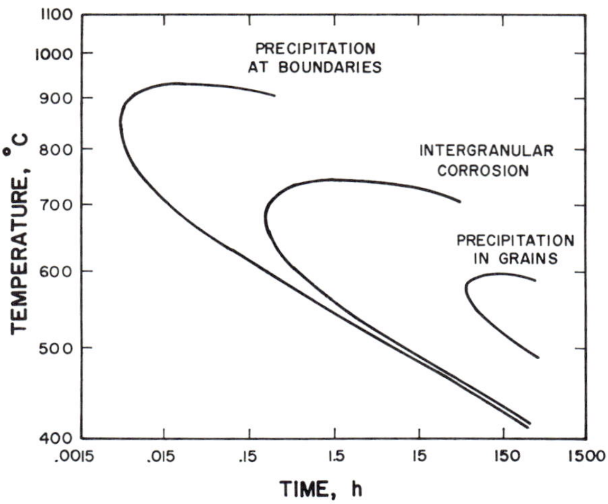

FIG. 4—*Influence of time and temperature on* $M_{23}C_6$ *precipitation and intergranular corrosion in Type 304 stainless steel* [3].

sion. Parent metal on both sides of a weld is heated in the sensitization range for several seconds or minutes during welding, and this can result in susceptibility as shown in Fig. 5. This type of attack is sometimes called "weld decay" or "low-temperature heat affected zone corrosion." It will be shown later that the high-temperature heat affected zones of a weldment can be attacked in some of the modified austenitic stainless steels.

It is very important to realize that material heated in the sensitiza-

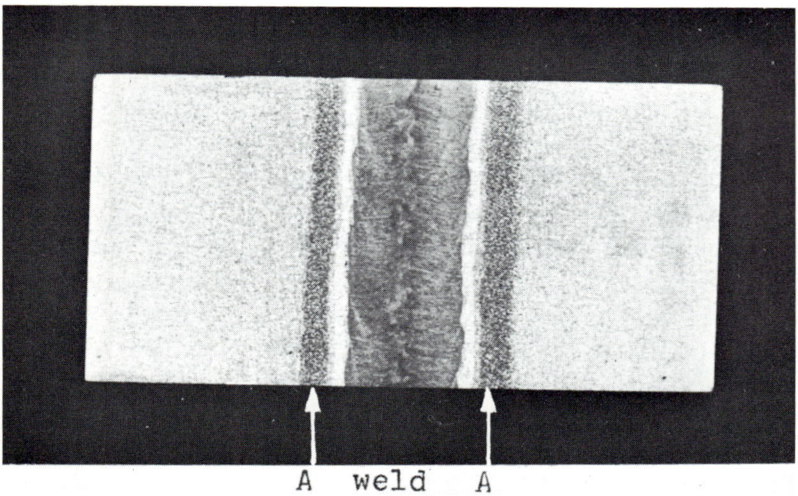

FIG. 5—*Preferential attack in low-temperature heat affected zones* (A) *of Type 302 stainless steel welded strip* (×2).

tion range may not show intergranular corrosion in service. This is true even if the alloy is attacked in a laboratory intergranular corrosion test. Whether or not the problem will occur depends upon the specific service environment. A lack of information as to which media cause attack and which do not was highlighted by Warren [*4*] who compiled a list of media reported to have caused problems. Since then the results of comprehensive test programs to study this aspect have been reported by Auld [*5*] and the High Alloys Committee of the Welding Research Council [*6*]. These programs covered a variety of chemical process and naturally occurring corrodents, and listings of those media which did and did not cause attack in several different welded and sensitized stainless steels were obtained.

It is difficult to summarize the findings of Warren's survey and the field test data because the ability of a corrodent to cause problems is dependent upon the specific alloy, its degree of sensitization, and corrodent variables—concentration, temperature, impurity contents, etc. Most service experiences of intergranular corrosion have been in oxidizing, inorganic or organic acids, but there are exceptions (for example, attack in seawater) and by no means all acid solutions cause problems.

Mechanism of Attack in Sensitized Austenitic Stainless

Intergranular corrosion susceptibility in these steels increases as their carbon content increases [*7*]. Whereas the carbon (typically 0.07 percent by weight) is soluble during annealing (1950 F (1066 C)) it has very restricted solubility at lower temperatures, for example, less than 0.01 percent by weight at ambient temperature. Rapid cooling from the annealing temperature results in supersaturation of carbon in solution. Subsequent holding in the sensitization range enables this carbon to precipitate out as chromium rich carbides. For example, $(Cr,Fe)_{23}C_6$ type carbides have been detected [*8,9*] with chromium contents as high as 70 percent by weight [*10*]. There is little doubt that intergranular corrosion is related to these carbides.

Relatively early in the history of these alloys a plausible theory for their intergranular corrosion was suggested by Strauss et al [*11*] and Bain et al [*7*]. They proposed that chromium depletion occurs adjacent to the high-chromium grain boundary carbides. These low-chromium areas have lower corrosion resistance in certain media and so are attacked preferentially. A modification was proposed by Hatwell [*12*] who suggested that chromium depletion in the boundary itself (between carbide dendrites) was the cause of attack, at least in the early stages of sensitization.

Other roles have been assigned to the grain boundary carbides. For example there is the strain theory of Kinzel [*9*] which attributes attack to corrosion of the distorted lattice adjacent to the precipitate. This theory is difficult to prove or disprove, but the fact that corrosion rates of the bulk material do not increase significantly with applied strain is inconsistent with it unless other arguments (for example, electrochemical effects) are invoked.

Also there is the electrochemical theory promoted by Stickler and Vinckier [*13,14*] and suggested by Kinzel [9] and Stefanides [*15*]. Intergranular attack is attributed to galvanic corrosion between the more noble carbides and the adjacent matrix. There is some evidence for the more noble character of chromium rich carbides [*9,14*], but this is disputed by Osozawa et al [*16*]. However, if one combines the chromium depletion and galvanic corrosion concepts there is feasibly quite a large potential difference between the carbide and the "anodic" depleted zone. In fact, the grain itself could serve as a cathode to a depleted zone, independent of the potential of the carbide.

Bäumel et al [*17*] argue that a galvanic mechanism would result in much more widespread corrosion than just the area immediately adjacent to the carbide. Again this objection would be invalid if one considered a mechanism dependent upon depletion and galvanic corrosion with the depleted layer being the anode.

Stickler and Vinckier [*14*] calculate from their experimental data that the width of metal attacked is much greater than could be attributed to corrosion of chromium depleted material. They use this as evidence against the chromium depletion theory but do not acknowledge the possibility that once corrosion starts in depleted material it could spread to "good" material, that is, the alloy content of the steel might be sufficient to maintain the passivity of an air formed protective but not enough to passivate a more actively corroding surface.

The three main theories are demonstrated in Fig. 6*a,b,c,* and the the combined "depletion-electrochemical" concept is shown in Fig. 6*d*. Bäumel et al [*17*] and Wilson [*18*] have reviewed the depletion, strain, and electrochemical intergranular corrosion models and concluded that the chromium depletion theory is the most plausible. One obvious advantage of this concept is that it is simple and can be understood by users of stainless steel who must learn how to avoid the problem. It can be related readily to the remedial measures discussed later, and providing a simple understanding of the problem probably helps in their successful application. The main disadvantage of the chromium depletion theory has been the lack of direct evidence for the depleted zone. Some have detected a depleted layer [*19,20*], but the data are not conclusive.

In addition to the discussed arguments in favor of chromium depletion it is interesting to note the work of Gellings and deJongh [*21*]. They argue that if chromium depletion is a viable theory there should be preferential elevated temperature oxidation in susceptible material. Their

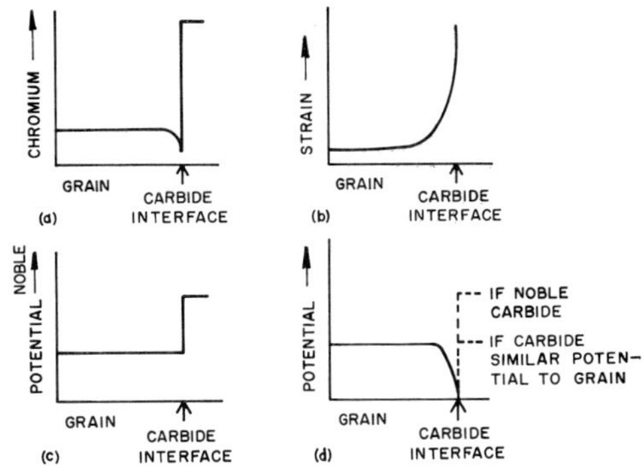

(*a*) Chromium depletion.
(*b*) Strain.
(*c*) Electrochemical—more noble carbide.
(*d*) Electrochemical—more active chromium depleted region.

FIG. 6—*Schematic representation of intergranular corrosion mechanisms for sensitized stainless steels.*

oxidation tests showed this to be the case. It is also worth noting that in recent years there have been attempts by Stawström and Hillert [*22*] and Tedmon et al [*23*] to calculate the chromium content in and adjacent to precipitated grain boundaries, and their findings in general agree with experimental observations and the chromium depletion model. They also indicate a very narrow depleted zone which is consistent with the practical problems experienced in detecting it.

While discussing the mechanism of attack, it is also important to note the influence of: (1) precipitate morphology and (2) solution chemistry. Lack of attention to these variables is probably the main reason for apparently conflicting evidence and theories.

It is not always necessary to have a continuous grain boundary precipitate to get intergranular corrosion because chromium diffusion (and depletion) can occur within the boundary itself. However, having a continuous grain boundary precipitate will generally increase the chances of rapid attack. Precipitate morphology is very much a function of prior thermal and mechanical history as well as sensitizing temperature. Its importance has been recognized by several workers [*8,9,12,13, 24,25*] and should be considered in any intergranular corrosion evaluation. This subject will be dealt with again when nonsensitized stainless steels are discussed.

The chemistry of the corrodent is important because the same mechanism of attack may not apply to all situations. For example, it will be shown later that very strongly oxidizing solutions can attack annealed material which does not contain grain boundary carbides. The utilization of data from one specific solution to refute or support a theory therefore, can be misleading.

A consideration of the information available at this time strongly suggests that in the regular 18Cr-9Ni austenitic stainless steels, chromium depletion is a prerequisite for intergranular attack in most media. The electrochemical theory may work together with chromium depletion as noted previously. It will be shown later that for very strongly oxidizing solutions, molybdenum containing alloys, and some of the high-nickel alloys, chromium depletion adjacent to carbides is not the only cause of intergranular attack.

Influence of Alloy Composition

There are many different types of austenitic stainless steel. Major and minor alloying elements are added to or removed from the typical 18Cr-9Ni analysis to provide specific benefits in corrosion resistance, mechanical properties, or ease of fabrication. These changes can significantly influence intergranular corrosion susceptibility [*26–57*].

When considering composition effects it is important to separate those due to the element itself from those occurring as a result of structure changes. The following discussion assumes a completely austenitic structure. It will be shown later that the presence of a second phase can have a marked influence, and therefore some of the elements dealt with could have additional effects by promoting or removing ferrite. It should be noted also that the effect of composition can vary with the environment and the mode of sensitization (for example, welding compared to stress relieving).

Elements which can be detrimental to intergranular corrosion resistance are carbon, molybdenum, silicon, and nickel. The role of carbon is relatively straightforward, since one of the major causes (in some alloys the only cause) of susceptibility is the precipitation of chromium carbides [*7,30,31*]. Molybdenum, at the 2 to 3 percent level present in AISI Type 316 and its modifications, slightly decreases carbide precipitation but can increase susceptibility in certain media primarily because it promotes the formation of deleterious intermetallic compounds such as σ-phase [*4,31–34*]. These compounds can apparently cause susceptibility to attack even when present in amounts too small to detect metallographically [*25*].

Silicon (1 to 2 percent) has been shown to be detrimental [*26,35,36*] and particularly so in molybdenum containing steels because it appears to accelerate the precipitation of deleterious intermetallic compounds [*29*]. The role of this element in intergranular corrosion will be discussed again in connection with the attack of annealed material. Increases in nickel content can be deleterious to intergranular corrosion because they reduce the solubility of carbon in austenite and so promote chromium carbide formation. Figure 7 demonstrates the deleterious effects of carbon and nickel (and the beneficial effect of chromium).

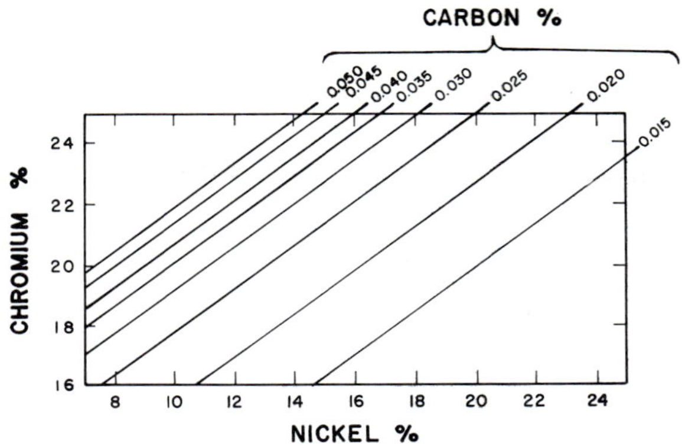

FIG. 7—*Effect of stainless steel chromium and nickel content on the tolerable carbon content to avoid intergranular attack in the* $CuSO_4$-H_2SO_4 *test after 650 C, 1 h* [3].

It will be shown later that higher levels of nickel than shown in Fig. 7 can be increasingly deleterious.

Elements which form stable carbides (for example, columbium [*30,37*] and titanium [*37–40*]) are generally beneficial because they tie up carbon which could otherwise form chromium carbide. Note that titanium can be deleterious in certain strongly oxidizing media such as nitric acid (HNO_3). The reason for this is dissolution of titanium carbide (TiC) [*41,42*]. Chromium can be beneficial as shown in Fig. 7, but in the amounts normally present (up to about 25 percent) the effect is not generally sufficient to prevent attack.

Boron is an element which in amounts from less than 0.001 up to 0.01 percent has been reported to be both deleterious and beneficial depending upon the boron content within this range, the base composition, the heat treatment, and unfortunately the investigator [*43–47*]. The boron atom is intermediate in size between the interstitial element carbon and the substitutional elements iron, chromium, and nickel. Therefore, it would be expected to play a role in grain boundary areas where there is considerable misfit and has been reported to influence the rate of formation of grain boundary carbides and intermetallics [*43,48*]. Although the role of boron is believed mostly due to its effect on chromium carbide, σ-phase, etc., it should be noted also that chromium borides have been detected [*48,49*]. The effect of this element is in need of further investigation with careful consideration of base alloy content, boron level, thermal and mechanical processing, precipitate morphology, etc.

Nitrogen additions (typically 0.15 percent or more) are made to stainless steels to increase their strength or permit savings in nickel (nitrogen is also an austenite former). Manganese is also added as a substitute for nickel and to increase the solid solution solubility of the nitrogen. High-nitrogen Cr-Ni, Cr-Mn-Ni (AISI 200 Series), and Cr-Mn stainless steels are commercially available so it is pertinent to consider their intergranular corrosion resistance. They can be described best as approaching or equalling the resistance of the regular nitrogen (0.03 percent) AISI 300 Series grades [*30,50–54*], although there have been claims that they are slightly inferior [*55–57*]. It should be noted that in one of the latter cases [*56*] the conclusions are questionable because of carbon differences.

Probably the most significant difference between low- and high-nitrogen stainless steels is that the latter cannot be effectively stabilized with elements such as titanium or columbium. This is because these elements when combined with nitrogen reduce both its effectiveness in solid solution and their ability to tie up carbon.

Avoiding Intergranular Corrosion in Austenitic Stainless Steels

Intergranular corrosion is avoided generally by using annealed material which has been cooled rapidly through the sensitization range to avoid a continuous grain boundary network of susceptible material.

Other commonly used methods of avoiding the problem are to use low-carbon grades such as Type 304L ($<$0.03 C) or stabilized grades such as Type 321 (titanium added) or Type 347 (columbium added). Neither of these techniques make an alloy immune to intergranular corrosion, but generally they do make it sufficiently resistant to be used in the welded or stress relieved conditions. The theory behind these approaches is quite simple. The low-carbon grade removes the source of the problem, that is, it limits the amount of chromium carbide that can be formed. The stabilized grades take the different approach of putting in a strong carbide former to tie up carbon before the chromium gets a chance to do so.

In welding applications, the stabilized grade can be at a disadvantage (compared to the low-carbon alloys), because the titanium or columbium carbide goes into solid solution at the high temperatures associated with welding. This problem is evident immediately adjacent to the actual weld metal and is sometimes described as knifeline attack. It normally shows up in stabilized material which has been multipass welded or welded and then stress relieved in the sensitization range.

Knifeline corrosion should not be confused with fissure attack which can occur in titanium stabilized alloys (for example, Type 321) exposed to strongly oxidizing media such as boiling HNO_3 [*39*]. This problem is related very probably to the fact that near the fusion zone in a weld a continuous grain boundary network of TiC forms, and it is soluble in HNO_3 [*41,42*].

The three techniques used to avoid intergranular corrosion in almost all cases are either annealed material, low-carbon alloys, or stabilized grades. There are, however, other factors which influence intergranular corrosion, and these have been reported as prevention methods. They include stabilize annealing, cold working, and grain size control.

Stabilize annealing (for example 1600 F (871 C)) has been suggested as a means of improving the stabilized grades and those containing molybdenum [*25*]. In stabilized material the intent is to optimize the formation of the stabilizing carbide. However, it is possible that the 1600 F (871 C) temperature is too low because there are reports of resistance deteriorating when this anneal is used [*5,6*]. In nonstabilized grades the function of the treatment is probably to form "harmless" chromium carbides. At this temperature they are more likely to be discrete, and chromium has had an opportunity to diffuse back into depleted regions. Again, however, there is evidence that such treatments can be harmful [*6*].

Stabilization treatments very probably can be beneficial by forming random precipitates without chromium depletion. Unfortunately, however, the morphology of grain boundary precipitates is dependent upon prior thermal and mechanical processing. This and the fact that deleterious intermetallic compounds can form in some of the alloys during the anneal is probably why there is conflicting evidence regarding this "prevention method." In any event, stabilize annealing can be hardly considered as a reliable technique for improving resistance unless prior evaluation of similarly processed lots of the material involved has shown it to be so.

The beneficial effects of cold work prior to sensitizing were noted relatively early in the history of the stainless steels by Bain and coworkers [7]. They attributed the benefits to the introduction of carbide nucleation sites within the cold-worked matrix so minimizing continuous grain boundary networks.

More recently Tedmon et al [*58*] confirmed this as shown in Fig. 8. Their specimens were wires, and after the corrosion exposures they used tension tests to evaluate the degree of intergranular attack. Using thin-film electron microscopy, they observed increased matrix precipitate in the cold-worked alloy and also saw some recrystallization in material that was cold reduced 55 percent and held at 1112 F (600 C) for 24 h. They noted that this recrystallization would have the beneficial effect of removing any continuous grain boundary network that did form.

Although there is apparent agreement in the literature as to the effect of cold work, it seems likely that in practice it is not quite so straightforward. For example, cold work in 18Cr-8Ni stainless steels can lead to transformation to martensite which then may be preferentially attacked. Also it seems likely that small amounts of cold work (the

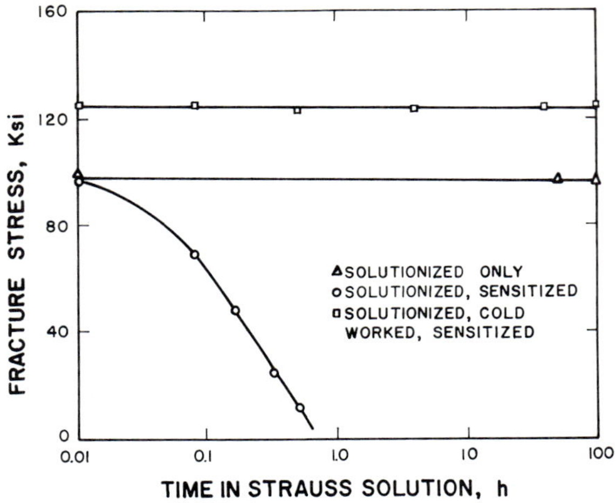

FIG. 8—*Influence of cold work on intergranular attack of 18Cr-8Ni-0.1C steel. A reduction in the fracture stress signifies intergranular attack* [58].

previous discussion has concerned 50 percent or more) could be deleterious in inducing grain boundary precipitation, especially in grades susceptible to σ-phase formation. Yet another area in which cold work is potentially deleterious is in increasing end-grain attack (particularly in boiling HNO_3).

The influence of grain size has been studied in some detail [*25,59–62*], but there is disagreement as to its effect. It is difficult to study because thermal and mechanical-thermal treatments used to vary grain size can cause grain boundary chemical and structural changes which also influence corrosion. For example, high-temperature anneals used to produce large grain size will increase the susceptibility of stabilized grades by virtue of dissolving the stabilizing carbide. On the other hand these high-temperature treatments can be beneficial by completely removing deleterious compounds such as σ-phase and residual grain boundary cold-work effects which might accelerate precipitation.

Following the early work of Newell [*59*], which showed increasing susceptibility with increasing grain size, the most active investigator has been Čihal [*60,61*]. Using calculated and experimental values he generally agrees with Newell but does acknowledge more of the exceptions discussed next. Čihal explains the grain size effect on the basis that in small grain size material there is more grain surface area and therefore less chance of a continuous network of precipitate. Whether this argument is significant for the range of grain size in commercial materials is questionable.

In an effort to minimize thermal effects in producing varying grain size, Levin and Kaloshina [*62*] gave their small and large grain material the same final anneal. For both Types 304 and 321 they found no effect of grain size in the copper-acidified copper sulfate test.

Likewise, Streicher [*25*] did not duplicate the work of Newell and Čihal. Using both weight loss and electrical resistance techniques he concluded that the effect of grain size depends upon the method of measurement and the testing solution. He also notes that small grains are removed more readily and that small grains are surrounded most frequently by ditching in the oxalic acid etch test (see later for discussion of this test). It should be noted also that once a small grain has been removed, attack in HNO_3 is accelerated due to the buildup of hexavalent chromium in the crevice.

In conclusion there is conflicting evidence on the effect of grain size. The weight of available experimental evidence indicates a slightly greater susceptibility for large grain material, but to a large extent this is most likely a result of factors leading to it rather than an inherent effect of grain size itself.

Intergranular Corrosion in Nonsensitized Austenitic Stainless Steels

Annealed (that is, nonsensitized) austenitic stainless steels can be susceptible to intergranular attack in strongly oxidizing media [*25,63–80*]. The solution most commonly used to study this phenomenon is hot nitric acid plus sodium dichromate, for example, boiling 5 N $HNO_3 + 4$ g/liter Cr^{+6}. The attack is attributed to the presence of hexavalent chromium and increases with the amount of this ion as shown in Fig. 9. The data shown are reported to represent intergranular attack. Other ions (Fe^{+3}, Ce^{+4},Mn^{+7}) also have been reported to accelerate attack [*25,64*].

The problem received renewed interest in the 1960s as a result of intergranular failures (probably stress assisted) in boiling water nuclear reactor fuel cladding. Investigations in France and the United States duplicated each other to some extent, and, in addition, researchers on both sides of the Atlantic have each published similar data in a variety of different journals.

The accelerating role of oxidizing cations such as Cr^{+6} is attributed to their depolarizing the cathodic reaction which in turn raises the corrosion potential and anodic reaction rate [*25,64*]. There are also indications that some anodic depolarization occurs [*25*]. Although a rise in the anodic reaction rate is consistent with the observed higher corrosion rates, it does not explain why grain boundaries are attacked preferentially. To explain the latter there are two main theories—(1) that grain boundaries are attacked because they are high-energy regions [*25,64, 70,71*] or (2) that elements segregated at grain boundaries result in a chemical difference between the boundary and the matrix, and this serves as the driving force for corrosion [*74–80*]. There have been attempts to

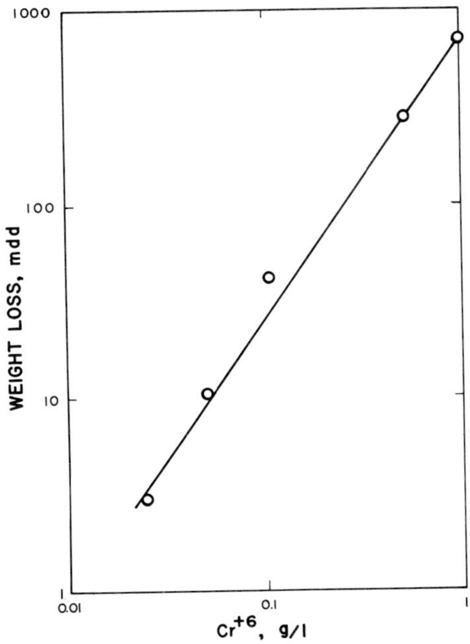

FIG. 9—*Increase in intergranular corrosion rate of annealed Type 304 stainless steel in 5* N *HNO_3 (80 h at 108 C) as a result of hexavalent chromium additions* [68].

combine the two theories [*72*] on the basis that the high-energy regions are those most likely to show segregation.

Chaudron [*65*] has shown that high-purity steels are not susceptible to attack. This supports the segregation theories as opposed to those dependent solely on grain boundary energy. Armijo [*79*] confirmed Chaudron's findings and studied the influence of specific elements in more detail. Using high-purity metals, the effects of carbon, nitrogen, manganese, phosphorus, and sulfur were evaluated. The only elements found to have a significant effect on annealed intergranular corrosion resistance were phosphorus and silicon. Intergranular corrosion rates in a nitric acid-sodium dichromate mixture increased with increasing phosphorus content (with significantly increased attack at the 0.09 percent level), whereas the role of silicon is more complicated as shown in Fig. 10. Similar effects for silicon in commercial purity alloys were also reported by Coriou et al [*66,67,69*].

Efforts to explain the effect of silicon have been numerous, but there is disagreement as to its precise role. There is considerable evidence [*80,82,83*] showing silicon to be present in surface oxide films on the higher silicon alloys. Quite logically this led Armijo and Wilde [*80*] to propose that high-silicon alloys are resistant due to polarization of the cathodic reaction by the silicon rich film. They support this idea with

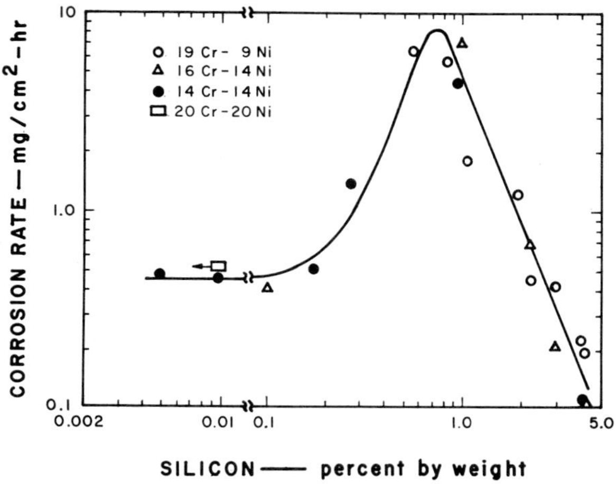

FIG. 10—*Variation of corrosion rate with alloy silicon content. (boiling nitric acid-dichromate solution). Peak attributed to intergranular attack* [80].

electrochemical measurements showing increased polarization as silicon increased from 1 to 4 percent. Armijo and Wilde attribute the higher intergranular corrosion rates at intermediate silicon levels to silicon segregation at the grain boundaries and increased attack of these due to galvanic corrosion.

Scharfstein [*28*] invokes film effects to explain the role of silicon at all levels. He proposes that intermediate levels of silicon (1 to 2 percent) are sufficient to maintain passive films on the grains but not the grain boundaries. This accelerates the attack of the latter. At higher levels (4 percent) there is sufficient silicon to promote a protective film over both, whereas at very low levels ($<$0.1 percent) there is no film effect of silicon on either grain or grain boundary.

Desestret et al [*72*] relate the high intergranular corrosion rates of 1 percent silicon heats to segregation and grain boundary energy. These amounts of silicon create grain boundaries with higher energy than lower or higher levels as shown in Fig. 11. They claim that this results in maximum segregation. The reduced attack in the higher silicon alloys is attributed to a lessening of the chemical concentration gradient between boundary and matrix. Streicher [*73*] had shown previously that high-silicon alloys were much more resistant to chemical etching and thermal etching. Desestret et al [*72*] disagree with Armijo and Wilde [*80*]. The former claim that silicon influences anodic but not cathodic polarization, while the latter show the opposite. Until these differences are resolved the exact role of silicon is difficult to assess.

Vermilyea and Tedmon [*81*] suggest that the role of grain boundary segregation may be to increase the electronic conductivity of the oxide

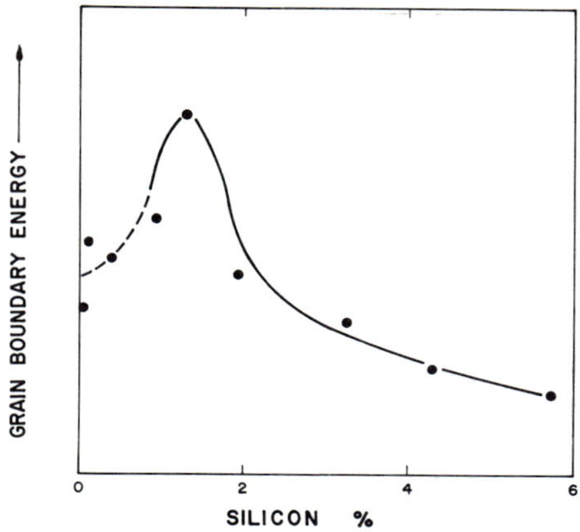

FIG. 11—*Influence of silicon content on grain boundary energy of 16Cr-14Ni alloys* [71].

at the boundary. This increases attack because it accelerates the oxidation of Cr^{+3} in the oxide film to Cr^{+6} which is soluble.

The proponents of the solute segregation theory for attack in annealed material—Aust, Armijo, Westbrook, and co-workers [*74–80*]—support their arguments with heat-treatment studies which are summarized in Fig. 12. For Type 304, quenching from about 1050 C (1922 F) produces no grain boundary carbides but solute segregation as depicted in Fig. 12*a*. If the steel is held in the sensitization range, say 600 C (1112 F) then continuous grain boundary precipitation can occur as shown in Fig. 12*b*. Material treated at slightly higher temperatures

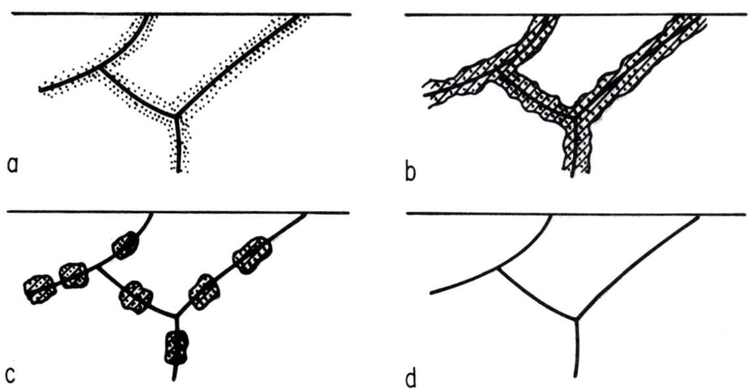

(*a*) solution treated at 1050 and 1300 C, solute segregation.
(*b*) sensitized at 500 to 700 C, continuous carbides or second phase.
(*c*) Stabilized at 800 to 900 C, discontinuous carbides incorporating segregant.
(*d*) High-purity alloy, "clean" grain boundaries.

FIG. 12—*Schematic illustration of local changes in microstructure and composition near grain boundaries* [74].

such as 800 to 900 C (1472 to 1652 F) contains isolated grain boundary carbides which may also incorporate solute impurities from adjacent grain boundary regions as in Fig. 12*c*. High-purity alloys contain neither precipitate nor segregate (Fig. 12*d*).

In summary, annealed austenitic steels can be susceptible to attack in specific highly oxidizing media. Grain boundary segregation of elements such as silicon is very probably the cause, but the exact mechanism by which the segregate increases corrosion is not yet clear. Figures 12 *c* and *d* suggest two methods of avoiding intergranular corrosion in annealed material, that is, use high-purity alloys or stabilize annealing treatments. It should be noted that the latter reduces rather than eliminates attack and will depend upon alloy composition and prior processing history. High-silicon alloys are an alternative solution to this specific problem although they may be prone to fabrication difficulties. Again, it should be emphasized that annealed stainless steels are resistant to intergranular attack in almost all situations, and the discussion previously relates to just a few specific, highly oxidizing media.

Ferritic Stainless Steels

Ferritic stainless steels, like the austenitic grades, can be susceptible to intergranular corrosion which is largely dependent upon prior thermal history.

The ferritic alloys (Fe-Cr and Fe-Cr-Mo) are commercially attractive because of their low raw materials cost and their resistance to stress-corrosion cracking. On the negative side they have generally poorer formability than austenitics, can be susceptible to "885 F embrittlement," and are more prone to brittle failure and intergranular corrosion (particularly in the welded condition). There are several commonly used grades including types 405 (Fe-12Cr), 430 (Fe-17Cr), 434 (Fe-17Cr-1Mo), 446 (Fe-26Cr), and 406 (Fe-13Cr-3Al). Stabilized and free machining varieties exist in some cases, and it should be noted that new Fe-Cr-Mo ferritic alloys are being currently developed and introduced commercially.

Conditions Leading to Attack

From a practical point of view the intergranular corrosion behavior of these alloys is vastly different from the austenitic grades. Thermal treatments which produce susceptibility in the ferritic alloys are those used to minimize it in the austenitics and vice versa as shown in Table 1.

To be more specific, a sensitization treatment for a commercial purity ferritic stainless steel will involve heating above about 1700 F (927 C). Material heated above this temperature can show susceptibility to intergranular attack independent of whether it is water quenched or air cooled from that temperature. Slow cooling (for example, cooling in the

TABLE 1—*Contrasting effect of heat treatment on intergranular corrosion of ferritic (Type 430) and austenitic (Type 304) stainless steel.*

	Average Corrosion Rate, mpy,[a] in Boiling 65 percent HNO_3	
	Heat Treatment Prior to Test	
Alloy	1300 F, 1 h, WQ[b]	2000 F, 1 h, WQ[b]
Type 430	40[c]	400
Type 304	400	10[c]

[a] mpy = mils per year.
[b] WQ = water quenched.
[c] No significant intergranular attack.

furnace) from the sensitization temperature or annealing sensitized material at about 1450 F (788 C) for several minutes or hours will usually eliminate the problem. Note that for very low-carbon, low-nitrogen ferritic stainless steels water quenching can be beneficial as discussed later.

In the welded condition a ferritic stainless steel becomes sensitized in the weld metal itself and in the high-temperature heat affected zone, that is, greater than about 1700 F (927 C). Figure 13 shows welded Type 430 strip after an accelerated intergranular corrosion test and should be compared with the Type 302 in Fig. 5. Of course, some austenitic alloys (that is, the stabilized grades) can also show susceptibility to attack in high-temperature heat affected regions as discussed previously.

Still considering the practical aspects it should be noted that a sensitized ferritic stainless steel is susceptible to intergranular attack in a wider variety of media than a sensitized austenitic alloy. Environments generally considered quite mild (for example water) can attack sensitized Type 430 stainless [*84–86*]. Also, remedial measures used for austenitic alloys,

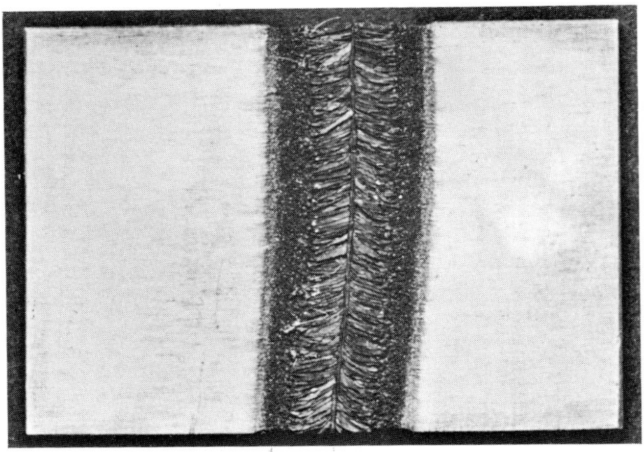

FIG. 13—*Intergranular attack in weld and high-temperature heat affected zones of ferritic stainless steel* (×2).

such as lowering the carbon content to 0.03 percent maximum, do not eliminate the problem and neither does the addition of stabilizing elements although they do minimize it. These factors increase the practical significance of intergranular corrosion susceptibility in this class of materials.

Ferritic stainless steels have received much less study and attention than the austenitics because of their smaller usage. There have been several studies in the past 20 years, however, and a marked increase in interest has developed in the past 3 years. This is a result of several factors. Periodic shortages and price increases in nickel, together with the fact that stress corrosion cracking is related closely to nickel content, have led to increased interest in alloys low in nickel. Also, continued advances in melting techniques have led to the possibility of economically making ferritic alloys sufficiently low in carbon and nitrogen to offset some of their poor as-welded mechanical and corrosion properties.

Mechanisms of Intergranular Attack in Ferritic Stainless Steels

Several mechanisms have been proposed. The apparent difference between the phenomena in ferritic compared to austenitic steels has led some investigators to believe the mechanisms of attack in these systems must be different, whereas others have been able to reconcile these differences into one common theory for all stainless steels.

By far the most work has been done on the Type 430 (Fe-17Cr) analysis. In this alloy, austenite can form at temperatures above 1700 F (927 C) in amounts dependent upon the precise analysis within the Type 430 range. When ferrite and austenite are present, chemical segregation of certain elements occurs due to their different solubilities in these two phases. For example, there is transfer of some carbon from the ferrite to the austenite, whereas the ferrite is enriched with chromium.

The correspondence in Type 430 between the temperature when austenite forms and that which must be exceeded to induce intergranular corrosion has naturally led to theories which relate intergranular corrosion susceptibility to austenite or the subsequent martensite that forms during cooling. Hochmann [*87*] proposed that the lower chromium content of the austenite compared to the adjacent ferrite cause it to be preferentially attacked. Kiefer [*88*] claimed that the second phase is attacked because it is martensite. It does seem reasonable that martensite (lower in chromium and higher in carbon) than the adjacent ferrite should be preferentially attacked in a test medium such as boiling 65 percent HNO_3, and this is indeed found to be the case [*89*]. It is very unlikely, however, that this is a necessary condition for intergranular attack because corrosion occurs in high-chromium alloys in which neither austenite nor martensite are detected. Also, even in alloys that contain martensite there may not be a continuous grain boundary network that would be necessary for both of the previously mentioned theories.

This lack of a continuous austenite network even when austenite is present can be actually used to explain an observation of Bäumel [*89*] and Baerlecken et al [*90*] that ferritic stainless steels containing austenite (martensite) can be more resistant to intergranular attack than completely ferritic material. The findings of Bäumel and Baerlecken et al are confirmed by the author's experience, an example of which is shown in Fig. 14. The composition of the two Type 430 alloys was varied to yield austenite in one weld but not in the other.

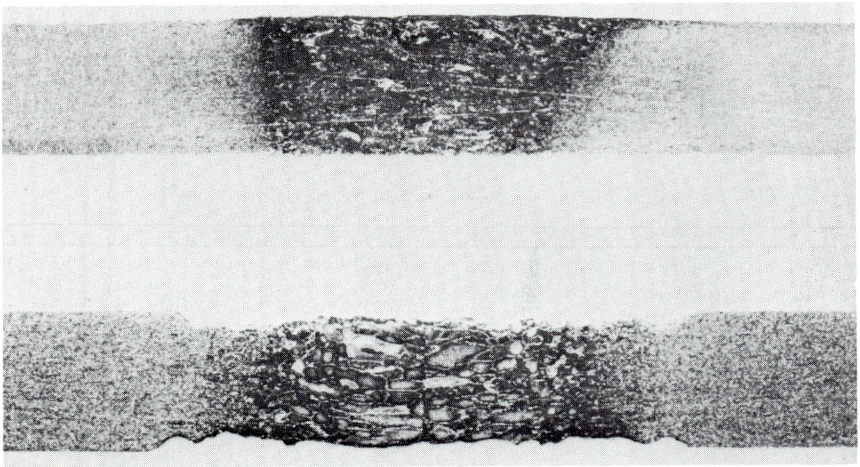

(*top*) Weld metal contains ferrite plus martensite.
(*bottom*) Weld metal essentially ferrite.

FIG. 14—*Polished and etched sections through welded Type 430 stainless steel after boiling 65 percent HNO_3 test* ($\times 10$). *Note less severe attack in two phase weld metal.*

Bäumel logically points up that the austenite formed at high temperatures acts as a sink for carbon. This decreases the supersaturation of carbon in ferrite so it is less prone to precipitate out at ferrite grain boundaries. Of course, during cooling, some precipitation occurs in the austenite (later martensite) and particularly at the ferrite-austenite interfaces. This precipitation does not result in intergranular corrosion provided the austenite does not form a continuous grain boundary network.

The formation of grain boundary carbides and nitrides has been used also to explain intergranular attack in ferritic stainless steels. Several roles have been ascribed to these precipitates. Houdremont and Tofaute [*91*] proposed that iron rich grain boundary carbides formed at about 1200 F (649 C) during cooling from temperatures above 1700 F (927 C). These iron rich carbides had low-corrosion resistance in many enviroments so leading to intergranular attack. Holding the steel at about 1400 F (760 C) caused chromium to diffuse into the carbides, increasing their corrosion resistance and reducing susceptibility to intergranular attack. The lack of evidence for these iron carbides, while chromium rich carbides and nitrides have been detected in susceptible material, is a serious drawback to this theory.

The role of grain boundary carbides and nitrides is explained differently by Lula et al [*92*] who attribute intergranular corrosion susceptibility to strains produced in the matrix adjacent to these precipitates. This strained material is postulated to be more anodic than the remaining base metal and grain boundary precipitates. The beneficial effects of annealing at about 1400 F (760 C) are explained on the basis of stress relief adjacent to the precipitates. The arguments against this theory are the same as discussed previously for austenitic steels.

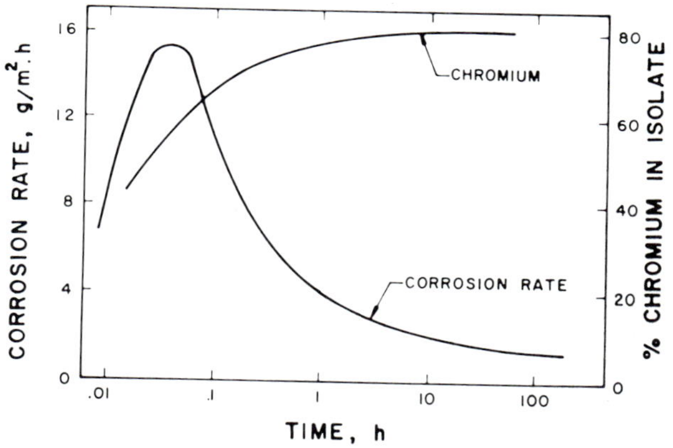

FIG. 15—*Variation of Type 430 stainless steel corrosion rate (in boiling 65 percent HNO_3) and isolated precipitate with time of sensitization at 600 C* [89].

Bäumel [*89*] explains the intergranular corrosion susceptibility of the ferritic stainless steels on the basis of chromium depletion adjacent to chromium rich grain boundary precipitates ($M_{23}C_6$) as is commonly proposed for the austenitic alloys. His main argument is based on the correspondence between intergranular attack and the chromium content of extracted precipitates as shown in Fig. 15. The maximum corrosion rate coincides with the greatest rate of increase in the chromium content of the isolate. This does support the depletion theory, although of course all the extracted precipitates were not necessarily from grain boundaries.

Baerlecken et al [*90*] also attribute the problem to chromium depletion adjacent to chromium carbides. More recently Demo [*93,94*], Bond and Lizlovs [*95,96*], and Hodges [*97*] have supported the chromium depletion theory and noted that nitrides (for example, Cr_2N) probably play a role also. All of the supporters of the chromium depletion theory concur that precipitation occurs in the temperature range of about 800 to 1700 F (427 to 927 C). Heating above this range dissolves carbides and nitrides (in the matrix as well as the grain boundaries) and supersaturates the ferrite. It seems likely also that some grain boundary segregation of carbon and nitrogen will occur at these higher temperatures. On cooling, the solubility of carbon and nitrogen is greatly lowered, and carbides and nitrides precipitate very rapidly. At the lower end of the precipitation range, say 800 to 1300 F (427 to 704 C), chromium depletion occurs unless the material is held for extended periods of time (hours) because chromium diffusion rates are low at these temperatures. At the higher end of the range chromium diffuses more rapidly and decreases the degree of chromium depletion and intergranular corrosion susceptibility. Some precipitate agglomeration would be expected also at these higher temperatures, and this in turn would decrease the possibility of continuous grain boundary attack. This behavior is shown in Fig. 16.

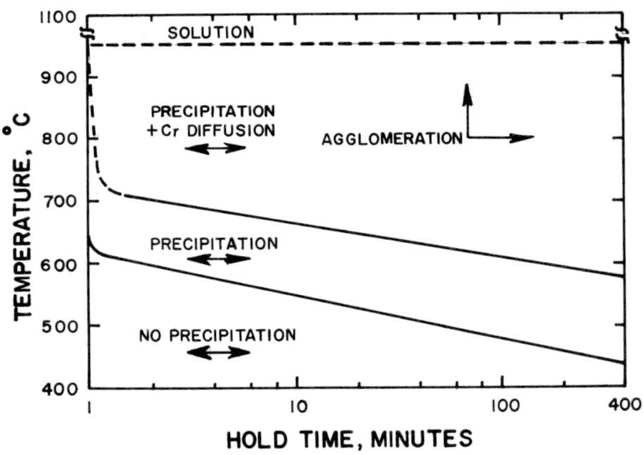

FIG. 16—*TTS curves for Fe-26Cr alloy (C + N = 180 ppm) treated 1100 C, 30 min, water quenched, ferric sulfate-sulfuric acid tests* [94].

By using the chromium depletion theory as the most feasible mechanism it can be seen that the ferritic stainless steels are behaving similarly to the austenitics. The main difference is that because of the very low solubility of carbon and nitrogen in ferrite (compared to austenite) the precipitation reactions occur much faster. This is demonstrated in Fig. 17.

The presence of molybdenum in the iron-chromium ferritic stainless steels appears to complicate the problem. Bond and Lizlovs [95] noted that the presence of 1 or 2 percent molybdenum in 18 percent chromium alloys resulted in intergranular attack in strongly oxidizing solutions in material furnace cooled from 1500 F (816 C). This problem did not show up in less oxidizing media (for example, copper-acidified copper sulfate test) and was attributed to the same metallurgical changes that cause embrittlement in these alloys after prolonged exposure at 700 to

1000 F ("885 F embrittlement"). Hodges [*98*], working with high-purity alloys, attributes the role of molybdenum to that of slowing down the diffusion of nitrogen so that nitride precipitation (and attendant chromium depletion) occur at longer times.

Avoiding Intergranular Corrosion in Ferritic Stainless Steels

The most satisfactory way of avoiding intergranular attack is to use these materials in the annealed condition, say 1450 F (788 C).

The addition of stabilizing elements (for example, titanium, columbium) does minimize although not completely eliminate the problem [*91,92,95, 105*]. Stabilizing elements increase the sensitizing temperature and can

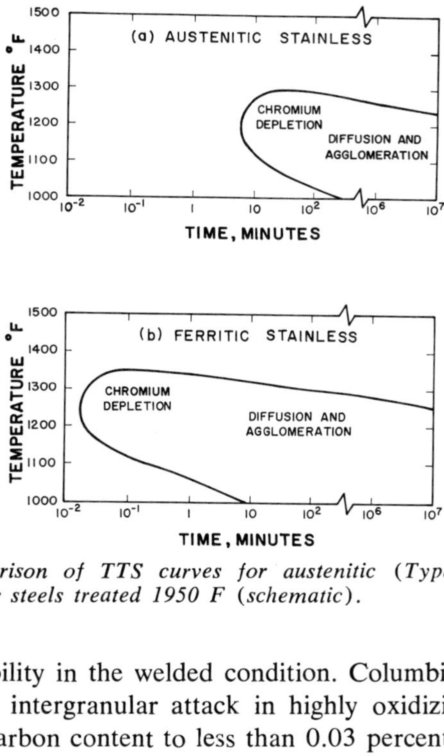

FIG. 17—*Comparison of TTS curves for austenitic (Type 304) and ferritic (Type 430) stainless steels treated 1950 F (schematic).*

decrease susceptibility in the welded condition. Columbium is more effective in preventing intergranular attack in highly oxidizing media [*95*].

Lowering the carbon content to less than 0.03 percent does not prevent attack as it does in the austenitics. Much higher purity (for example, <0.005C, <0.015N) does improve toughness [*87,99–102*] and reduce susceptibility to intergranular attack [*96–98,103,104*], but certain thermal treatments can make even these alloys very susceptible to attack. For example, Hodges [*97*] reports a corrosion rate of 439 μm/year for a 17Cr-0.002C-0.006N alloy water quenched from 1850 F (1010 C) and tested in boiling 65 percent HNO_3. The same alloy showed a rate of 5010 μm/year when air cooled.

Duplex Stainless Steels

Austenitic-ferritic stainless steels containing about 5 percent or more delta ferrite can have improved intergranular corrosion resistance compared to austenitic material with the same carbon content [*106–109*]. Colombier and Hochmann [*106*] discuss some of the probable reasons for the beneficial role of ferrite. They are as follows:

1. The total area of grain boundaries and phase interfaces is increased in the presence of ferrite, thereby reducing the quantity of carbide precipitates per unit area.
2. The ferrite is significantly higher in chromium than is the austenite. Therefore, it can lose chromium to a grain boundary precipitate and not become sufficiently depleted to be preferentially attacked.
3. Chromium diffuses more readily in ferrite than austenite so there is less chromium depletion.
4. The islands of ferrite present minimize the formation of continuous grain boundary precipitates in the austenite grain boundaries.

Item 4 is probably the predominant factor.

On the other hand, delta ferrite can be detrimental to corrosion resistance in certain media, particularly if a continuous network is present [*110–115*]. The problem is most prevalent in welds of the molybdenum containing alloys, Types 316 and 316L. Preferential attack may occur in the ferrite itself or at the ferrite-austenite interface and in the adjacent austenite, depending upon the alloy and the corrodent. The media causing attack are normally those with insufficient oxidizing power to induce passivity. Problems have occurred in urea service, acetic acid mixtures, and certain inorganic acid mixtures. Susceptibility to attack is readily revealed in boiling 50 percent by volume hydrochloric acid (HCl) [*115*]. Remedial measures include annealing after welding (some critical level of cold work between welding and annealing may be required) or specifying low-ferrite weld metal. The latter has the disadvantage that ferrite is beneficial in minimizing weld cracking, and it should be noted also that it is the distribution of the ferrite that is important from the corrosion standpoint and not just the total amount present.

Ferrite containing weldments of Types 316 and 316L also may be preferentially attacked by very strongly oxidizing media for example boiling 65 percent HNO_3. In this case the problem is probably related to transformation products of the ferrite (for example, χ- and σ-phase) [*110,114*].

Just as the presence of ferrite can be sometimes beneficial to the intergranular corrosion resistance of austenitic alloys the presence of austenite improves the ferritic grades. This was discussed earlier for Type 430 which can form austenite at elevated temperatures, that is, greater than about 1700 F (927 C). In that case the austenite transformed to martensite on cooling, but it should be noted that austenite is also beneficial in the higher chromium ferritic-austenitic alloys such as Type 329 (27Cr-5Ni-1.5Mo) in which the austenite does not transform to martensite [*106, 116–122*]. Figure 18 compares the as-welded corrosion resistance of low-and high-austenite material, and Fig. 19 shows the microstructures in the high-temperature heat affected zones.

Considerable work has been done in Europe and the USSR on Fe-21Cr-5Ni alloys, with or without molydenum and titanium [*116–119*], and 18Cr-8Mn-2Ni alloys [*120–122*]. It appears that these types of alloy (and based on the author's experience, Type 329 also) behave more like ferritic stainless steels than austenitics as far as intergranular corrosion

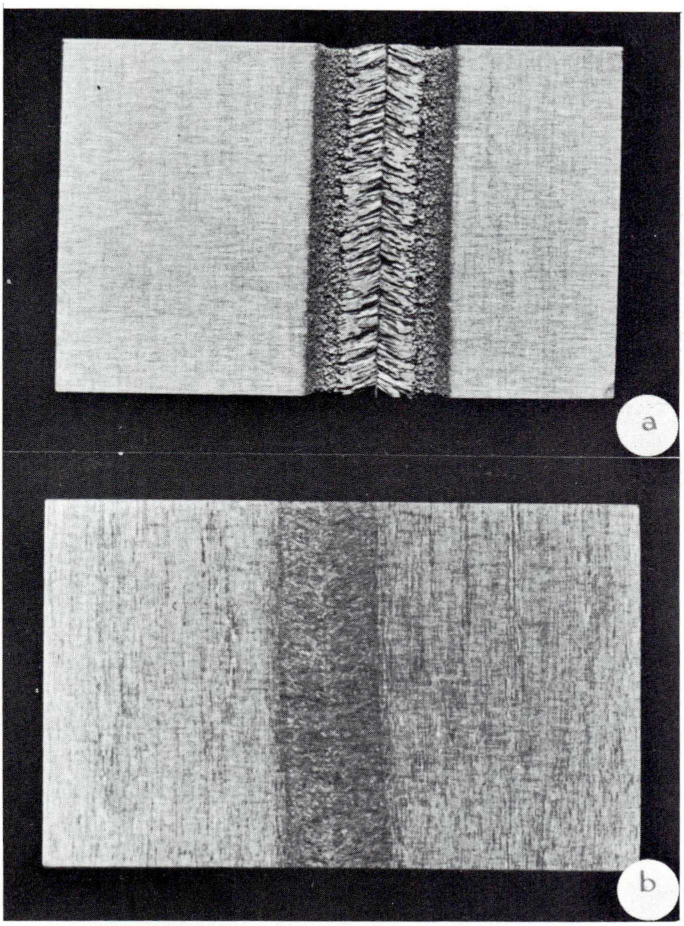

(*a*) Austenite content in base metal about 10 percent.
(*b*) Austenite content in base metal about 30 percent.

FIG. 18—*As-welded Type 329 stainless steel after 28-h exposure to 10 percent* HNO_3 *+ 3 percent HF at 70 C* (×2).

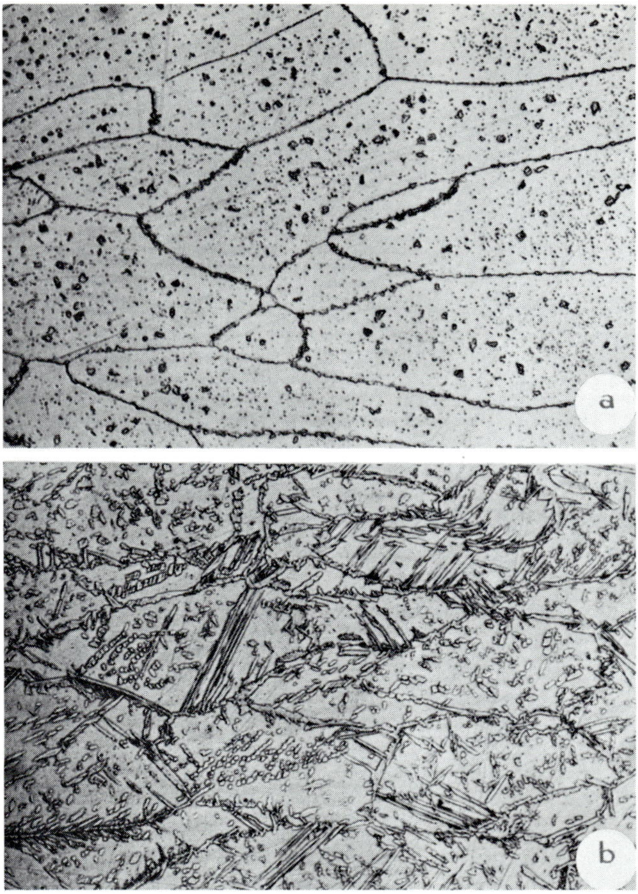

FIG. 19—*Weld metal microstructure for specimens shown in Fig. 18* ($\times$*100*). *Note larger amount of austenite in weld of the more resistant composition.*

is concerned. High-temperature treatments or welding can bring about susceptibility. Heat treatments which provide resistance to attack involve annealing at about 1750 F (954 C). Susceptibility can be induced by holding annealed material in the range of about 800 to 1550 F (427 to 843 C). Levin and Kochergina [*116*] note that 2 percent molybdenum suppresses the development of susceptibility when holding at 840 to 975 F (449 to 524 C). This is in agreement with the findings of Bond and Lizlovs [*95*] and Hodges [*98*] for purely ferritic alloys.

Stabilizing elements (especially titanium) are sometimes added to duplex stainless to further improve intergranular corrosion resistance. Some [*118,121*] claim this to be beneficial and others [*116,117*] detrimental. Since most published work on duplex steels is lacking in measurements of the amount and morphology (both of which are dependent upon prior mechanical and thermal treatment) of the two phases present it is very difficult to assess the role of elements such as carbon and titanium which, in addition to influencing carbide precipitation, also control the ferrite-austenite balance.

In summary, two-phase alloys have good intergranular corrosion resistance provided the amount and distribution of the second phase are such as to prevent continuous networks of grain boundary carbide.

High-Nickel Alloys

There are many commercially available, austenitic stainless alloys containing more than 15 percent chromium with nickel ranging from 9 percent (the common austenitic stainless steel level) to about 80 percent. Grades with 15 percent or more molybdenum will be discussed separately. First, Fe-Ni-Cr alloys (for example, Fe-20Cr-33Ni, Ni-15Cr-7Fe) will be considered, although it should be noted that many of them also contain a few percent of molybdenum and copper for added corrosion resistance. Major reasons for the use of these high-nickel alloys in corrosion service are their improved performance in reducing media and resistance to transgranular stress corrosion cracking [*123*]. The intergranular corrosion resistance of these high-nickel alloys has not been studied in such detail as the austentitic stainless steels, but the data available show clear similarities and differences in the two systems.

Both the stainless steels and nickel base alloys appear to be similar in that sensitization treatments which cause attack usually involve grain boundary carbide or intermetallic compound precipitation or both in the range 800 to 1600 F (427 to 871 C) [*124–134*]. Welding, stress relieving, and the slow cooling of thick sections are, therefore, potential sources of trouble. There is very little published information as to what service media will attack sensitized material (susceptible as determined by accelerated laboratory tests). Indications are that the high-nickel alloys are similar in this regard to the austenitic stainless steels. Their higher alloy content provides improved resistance in many media, but on the other hand they are often used in more severe corrodents in the first place.

Although the types of thermal treatment which cause attack are similar for high- and low-nickel alloys, nickel does have a significant effect on intergranular corrosion. Unfortunately, intergranular corrosion resistance in the sensitized condition usually decreases as the nickel content is raised [*3,124,125,127*]. One reason for this is that the solubility of carbon in austenite decreases as the nickel content increases. For relatively low-nickel levels Čihal [*3*] has determined the carbon content required to make Fe-Cr-Ni alloys susceptible as a function of nickel content (See Fig. 7). The influence of nickel in the range 13 to 46 percent is shown in Fig. 20. It should be noted that the evaluation test used for the data in Fig. 20 was not a severe one.

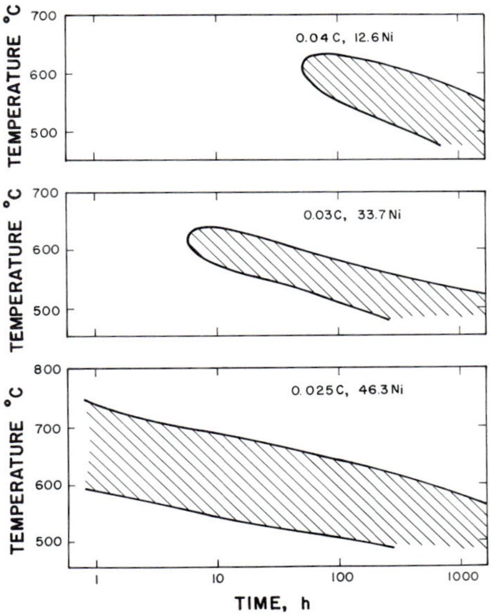

FIG. 20—*Influence of nickel content on TTS curves for Fe-20Cr alloys in $CuSO_4$-H_2SO_4 acid test* [125].

Practical implications of this decreased carbon solubility as nickel is increased are:

1. Maintaining carbon less than 0.03 percent does not minimize the problem to the extent it does in the 18Cr-9Ni alloys [*3,124,127*].

2. Much larger amounts of stabilizing elements (columbium, titanium) are needed to effectively minimize susceptibility [*124,125,128*] due to chromium carbides, and at nickel levels of about 40 percent or more it is reported that effective stabilization is not achievable with titanium alone [*129*].

The high-nickel alloys appear to be more sensitive to processing variables (prior mechanical working, annealing, etc.) than the common austenitic stainless steels [*126,130*]. Raymond [*129*] has emphasized the importance of grain boundary precipitate morphology in these alloys. Many of the high-nickel grades contain 2 to 3 percent or more molybdenum, and this further complicates matters (as in the stainless steels—compare Types 316 and 304) by introducing more complex carbides and intermetallic compounds such as σ-phase [*126,130–132*]. That molybdenum can be as important as carbon in these high alloys is demonstrated in Fig. 21. In spite of this, some of the alloys with nickel content just outside the stainless steel limits can be made to have resistance similar or approaching that of the common austenitic stainless steels by control of processing and analysis [*130*].

In the annealed and water-quenched condition high-nickel alloys are generally resistant to intergranular attack as are austenitic stainless steels.

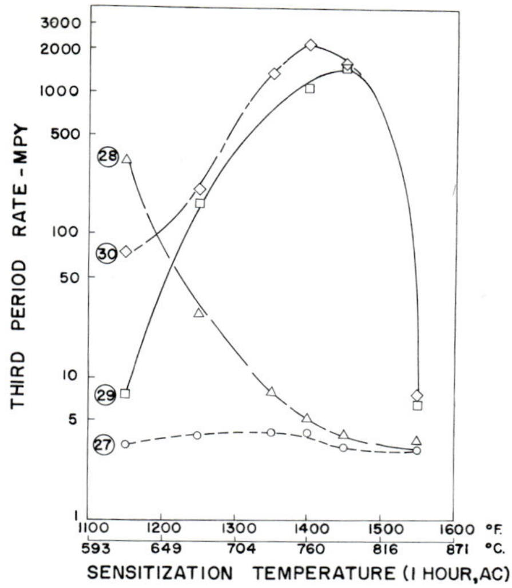

Heat Number	% C	% Mo
27	0.005	0.01
28	0.061	0.01
29	0.005	2.31
30	0.062	2.37

FIG. 21—*Effect of carbon, molybdenum, and sensitizing temperature on intergranular corrosion of Fe-20Cr-33Ni-3Cu alloys. Third 48-h period rate in boiling 65 percent* HNO_3 [130].

Like the latter they can be susceptible to attack in very strongly oxidizing media (for example, nitric acid plus sodium dichromate). This susceptibility is reported to increase with nickel content as shown in Fig. 22. Intergranular corrosion susceptibility is probably related to the incidence of intergranular stress-corrosion cracking (or perhaps more correctly—stress assisted intergranular corrosion) of Ni-15Cr-7Fe alloys in high-purity elevated temperature water [*135–138*].

In spite of their susceptibility to intergranular attack and reported cases of intergranular stress-corrosion cracking, high-nickel alloys find extensive use in both oxidizing and reducing media and in situations (chlorides, hydroxides) which cause transgranular cracking in the lower nickel stainless steels.

Nickel-Chromium-Molybdenum Alloys

Nickel-chromium-molybdenum (Ni-Cr-Mo) alloys are used because of their resistance to chloride pitting, stress corrosion cracking, and a wide variety of acid media which are corrosive to stainless steels. The most commonly used alloy contains about 56 percent nickel, 16 percent molybdenum, and 16 percent chromium, although higher chromium versions find some use in Europe and the USSR.

One of the major drawbacks of these materials has been their sus-

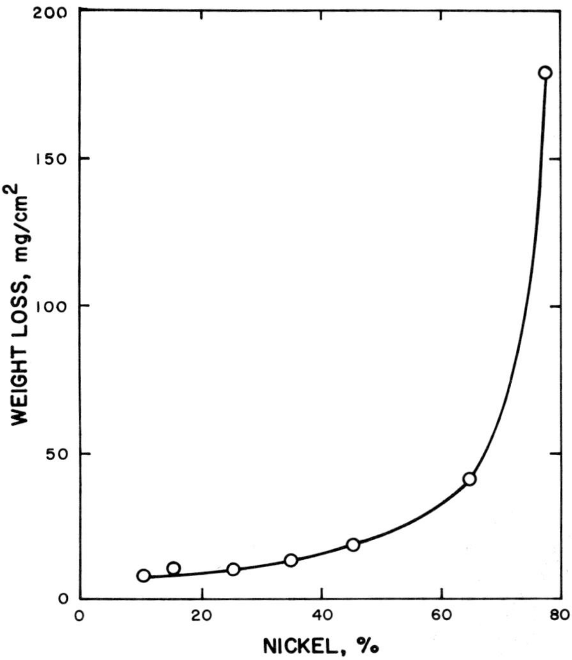

FIG. 22—*Influence of nickel content on intergranular corrosion of Ni-17Cr-Fe alloys. Tested 70 h in 5* N *HNO_3 + 1 g/liter Cr^{+6} at 108 C* [135].

ceptibility to intergranular corrosion if inadequately heat treated. The temperatures producing susceptibility are in the range of about 1100 to 2100 F (593 to 1149 C). Grain boundary precipitation occurs in short times, that is, in seconds or a few minutes at most. This means that welds are susceptible, and there is a potential problem that annealed material may be attacked if not cooled rapidly. Also, it is necessary to anneal above a certain temperature, for example, 2225 F (1219 C) while staying below temperatures of about 2300 F that cause incipient fusion. Intergranular corrosion has been detected in a variety of acid media, and the sensitizing temperature producing maximum susceptibility depends upon the corrodent [*139*] as shown in Fig. 23.

The mechanism of attack is usually attributed to the precipitation of grain boundary carbides and intermetallic compounds. These alloy systems have been studied in some detail by Streicher [*139*], Prötzl [*140*], Class et al [*141,142*], Samans et al [*143*], Babakov et al [*144*], and more recently by Leonard [*145*] and Hodge [*146*]. There is some disagreement as to the identity of the precipitated phases responsible for susceptibility.

It is generally agreed that annealed material resistant to attack contains randomly distributed M_6C carbides. Intermetallic compounds such as

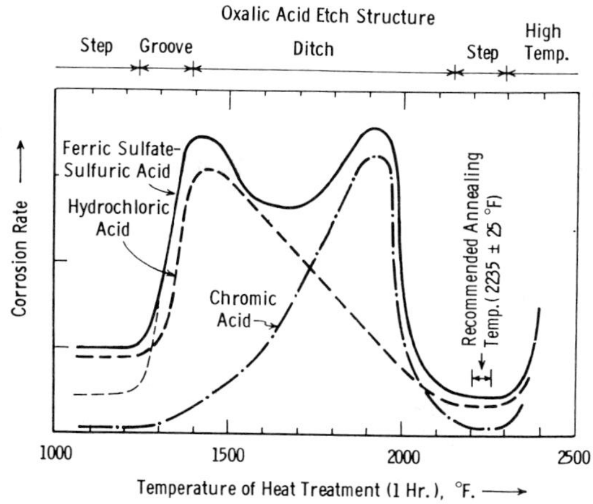

FIG. 23—*Schematic summary of relationship of heat treatment, etch structures and corrosion of wrought Ni-16Cr-16Mo alloys* [139].

P-phase [*145,147*], "7–6" compound [*143,145,148*], μ-phase [*146,148*], and σ-phase [*140–142,144,147*] have all been detected, and transformations from one form to another have been reported. These compounds contain some or all of the following elements—nickel, iron, cobalt, tungsten, molybdemum, chromium. It seems reasonable to conclude that they induce intergranular attack due to depletion of the adjacent matrix in elements such as chromium and molybdenum, although Hodge [*146*] has suggested that attack in oxidizing media is due to attack of the precipitate itself because of its high-molybdenum content. Chromium rich $M_{23}C_6$ carbides have been detected [*140–142,144,147*], and chromium depletion adjacent to this or some other chromium rich carbide may be a contributing factor to susceptibility. Mo_2C has been found [*140,144*], and molybdenum depletion adjacent to it is a possibility also.

Improved intergranular corrosion resistance can be obtained by limiting the silicon content in the alloy [*141,142,144*]. Also a low-carbon, low-silicon alloy is available, and although it does not eliminate susceptibility it does alleviate the problem considerably [*145*]. This reported improvement derived from low silicon and carbon lends support to a role for carbides as well as intermetallics. Streicher's observation (Fig. 23) that the sensitization temperature causing optimum attack in a reducing acid (hydrochloric) differed significantly from that for an oxidizing acid (chromic) is also consistent with there being more than one precipitate involved. Gräfen [*142*] saw somewhat similar behavior but suggests that only one phase (sigma) is involved and that its composition varies with sensitization temperature. The author's experience has been that carbon plays a significant role, and σ-phase is not the only problem.

The prime methods for avoiding intergranular attack in these alloy systems are to use material in the annealed plus water-quenched condition or to use a high-purity alloy. Thermal treatments to stabilize the deleterious intermetallic compounds have been proposed by Samans et al [*143*], and Streicher [*139*] notes that multiple short anneals can provide better resistance than a single longer time treatment.

Nickel-Molybdenum Alloys

These alloys do not contain chromium and thus differ from all the other materials discussed in this review. They are included because they are used in severe corrodents and do complement the previously discussed high-nickel grades. The most well-known alloy in this category contains about 65 percent nickel, 28 percent molybdenum, and 5 percent iron. It has excellent resistance to reducing media and in particular HCl.

The Ni-28Mo alloy can be susceptible to intergranular corrosion. For example, as-welded material is susceptible to intergranular attack in boiling 20 percent HCl and in some sulfuric acid (H_2SO_4) (for example, boiling 60 percent). The attack may occur in the weld metal itself, in the base metal almost immediately adjacent to the weld, and in a lower temperature zone which was heated in the region of 1450 F (788 C).

The intergranular corrosion resistance of this alloy system has been studied in some detail by Flint [*149*], Gräfen [*142*], and Svistunova and co-workers [*150–154*]. In the annealed condition, 2100 F (1149 C), the alloy contains randomly distributed M_6C type carbides of approximate composition Ni_3Mo_3C. As the temperature is increased toward the melting point there is a tendency for the molybdenum content of the M_6C to increase, for example, to Ni_2Mo_4C [*142,149*] and for Mo_2C to form [*142,149,152,153*]. Gräfen [*142*] also noted the presence of an iron-molybdenum type phase which contained more than 60 percent molybdenum at 2300 F (1260 C).

Intergranular corrosion in and immediately adjacent to weld metal is attributed generally to preferential attack of molybdenum depleted zones adjacent to high molybdenum M_6C and Mo_2C with the latter probably playing the major role.

Susceptibility in material exposed to temperatures of about 1450 F (788 C) is again attributed to molybdenum depletion, in this case adjacent to the intermetallic compound Ni_4Mo [*142,147,150,152,154*] predicted from the nickel-molybdenum phase diagram.

Attempts to totally eliminate intergranular corrosion susceptibility in weldments have been unsuccessful. Lowering carbon ($<$0.005 percent) has been reported to remove the high-temperature problem [*149, 153*] which is in line with the above mechanism relating the attack to carbides. Lowering the iron content (from 5 percent to less than 2 percent) [*149,153*] or adding a few percent tungsten [*152,154*] are also reported to be beneficial. Vanadium additions (1 to 2 percent) have been studied in some detail, with Flint [*149*] and Svistunova et al [*150*] claiming the addition to be beneficial. Svistunova et al [*153*] later show it can be detrimental which is consistent with the author's experience. Intergranular corrosion susceptibility produced after exposure to lower temperatures such as 1450 F (788 C) is not avoided by decreasing the carbon content as expected from the above mechanism relating susceptibility to Ni_4Mo. Vanadium has been reported to both retard [*150*] and accelerate [*153*] attack and tungsten to retard and narrow the temperature range producing it [*154*].

Commercial alloys completely immune to attack are not available. Some of the above alloys variations may minimize the problem, but the only completely satisfactory remedy for service in media capable of producing attack is to use annealed material.

Intergranular Corrosion Tests

Corrosion tests can be classified as: (1) service, (2) simulated service, or (3) accelerated. Most intergranular corrosion tests are a particular type of accelerated test, namely, quality control. In the latter, the test media may bear little relationship to the intended service environment. However, it is capable of detecting, in several hours or days, a metallurgical condition suspected or known to result in intergranular attack in service.

The development of intergranular corrosion tests has centered primarily on the austenitic stainless steels, but many of the methods developed can be applied to other stainless and high-nickel alloys. There are currently five established test methods for detecting susceptibility to intergranular attack in stainless steels and one etching technique which can be used as a screening test. All of these are ASTM standards and are listed in Table 2.

Although it is convenient and complimentary to use the inventor's name to describe a test, the practice is of dubious merit. For example, its pre-

TABLE 2—*Standard intergranular corrosion tests for stainless steels.*

ASTM Standard	Test Media	Common Test Names
A 393–63	boiling 5.7% $CuSO_4$ + 15.7% H_2SO_4	Strauss, Hatfield, Krupp
A 262–70, Practice A	10% $H_2C_2O_4$, electrolytic at ambient temperature	Streicher
A 262–70, Practice B	boiling 50% H_2SO_4 + $Fe_2(SO_4)_3$	Streicher
A 262–70, Practice C	boiling 65% HNO_3	Huey
A 262–70, Practice D	10% HNO_3 + 3% HF at 70 C	Warren
A 262–70, Practice E	same as A 393 but with specimen in contact with copper metal	Accelerated Strauss

cise origin may be in doubt and the test may be modified as the years go by, for example, the copper sulfate-sulfuric acid ($CuSO_4$-H_2SO_4) solution. Also, an inventor might have more than one test, for example, in addition to the ferric sulfate-sulfuric acid ($Fe_2(SO_4)_3$-H_2SO_4) test there is a pitting test named after Streicher, and the oxalic acid ($H_2C_2O_4$) etch test is also labelled by some as the Streicher test. For these reasons it is generally preferable to identify by the test media.

Copper Sulfate-Sulfuric Acid

Warren [*4*] has reviewed the history of this solution in some detail. The ability of $CuSO_4$-H_2SO_4 mixtures to detect susceptibility to intergranular corrosion in sensitized stainless steels was noted by Hatfield in 1926. The solution was used by Hatfield [*155*] as a test and also modified and promoted by Strauss [*11*] of the Krupp organization. Several further modifications were made to yield the current ASTM Standard Recommended Practice for Conducting Acidified Copper Sulfate Test for Intergranular Attack in Austenitic Stainless Steel (A 393–63).

The test specimen is immersed in the solution for 72 h. There is very little grain dropping so a weight loss evaluation is not pertinent. Intergranular attack can be evaluated by the ring test (when a severely attacked specimen is dropped on a hard surface it will not give a clear ring), electrical resistance measurements, or more commonly by bending the specimen to open up any attacked grain boundaries and observing at low magnification (5 to 15 magnification). Metallographic examination of sections from the tested specimen may be used to further evaluate the degree of attack. It should be noted that in European countries and on some material specifications in the United States the solution composition and severity of the bend to be given after the test often differ from ASTM A 393.

The fact that the method does not give a number (unless resistance measurements used) is a disadvantage. Also, it is not very severe and times longer than 72 h are often required to approach the degree of attack obtained with some of the other evaluation solutions. Plans are underway to drop this test from the ASTM standards because of its lack of severity.

Recently the test has been improved by placing the specimen in contact with copper in accordance with ASTM Standard Recommended Practice for Detecting Susceptibility to Intergranular Attack in Stainless Steel (A 262–70, Practice E). The copper increases the severity of the solution probably because it maintains the specimen potential in a region more aggressive to chromium depleted grain boundaries. It is only necessary to expose the specimen for 24 h, and in some European countries a similar test is run for 15 or 16 h.

The addition of copper to the solution was originated by Rocha in the mid1950s [*156*], and the work leading to the ASTM standard is described by Scharfstein and Eisenbrown [*157*]. There has been some disagreement as to whether contact between the specimen and the copper is necessary. Contact with copper (as compared to just the presence of solid copper in the solution) moves the specimen potential very slightly in the active direction and apparently increases the rate of attack [*25,157*]. This has been disputed by Schwenk [*158–160*] who claims contact is not necessary and suggests that the reported differences [*157*] are due to the use of cold finger condensers compared to his bulb condenser which in turn influences the oxygen content of the solution [*160*]. Since the ASTM standard specifies contact with copper and a bulb condenser it would seem to minimize these possible variations and represent the most severe test. A $CuSO_4$ test in 50 percent H_2SO_4 and with metallic copper in the solution but not in contact with the specimen also has been reported [*39,103*] as a more severe test.

Whether copper is added to the solution or not the $CuSO_4$-H_2SO_4 solution detects intergranular corrosion susceptibility that is primarily due to chromium carbides rather than intermetallics such as σ-phase [*4,25,32,161–163*].

Boiling 65 Percent Nitric Acid

In this test (ASTM Standard A 262, Practice C) the specimen is exposed for 240 h (usually five 48-h periods using fresh solution for each). Weight losses are used to calculate corrosion rates for each period. In susceptible material the rates will usually increase with increasing test periods because whole grains are lifted from the surface.

The solution and testing method were first reported by Huey [*164*], although he was not evaluating intergranular corrosion. It was later realized that the solution could be utilized in quality control work [*165*]. The influence of test variables has been studied in detail by Truman [*63*] and Streicher [*25*].

There are three potential disadvantages to the HNO_3 test:

1. It takes a long time and is expensive.

2. The degree of intergranular attack is accelerated by hexavalent chromium which builds up during the 48-h test periods. This can result in differing corrosion rates for different solution volume to specimen surface area ratios.

3. The solution is very sensitive to surface preparation [*25*], end grain [*25*] and structural changes other than chromium carbide, for example, it attacks σ-phase [*4,25,32,39,161–163*] and TiC [*41,42*].

Over the years these disadvantages have been countered with the argument that because the method is quantitative and severe it is the best one to use for quality control. For many years, it and the $CuSO_4$-H_2SO_4 test were the only ones available, and these arguments were somewhat valid. Now however, there are quantitative alternatives (for example, $Fe_2(SO_4)_3$-H_2SO_4 test), and also the $CuSO_4$-H_2SO_4 test has been made more severe with the addition of copper.

The expense of the test and the effects of Cr^{+6} can be minimized by using a multi sample tester (MST). These have been discussed in detail by DeLong [*166*], Truman [*63*], and Alger et al [*167*]. The acid is distilled continuously to remove Cr^{+6} as shown in Fig. 24, and as many as 60 specimens may be tested simultaneously. Because Cr^{+6} is removed continually in an MST the corrosion rates are lower than those obtained in a regular flask test. When the rates are low (that is, little susceptibility to intergranular corrosion) the difference is not usually significant [*168*]. However, when there is considerable susceptibility to intergranular attack a flask rate may be as much as 50 percent higher than that in an MST. The problem is complicated further by the fact that if one has a lot of highly susceptible material in an MST then the effects of Cr^{+6} buildup may not be completely eliminated.

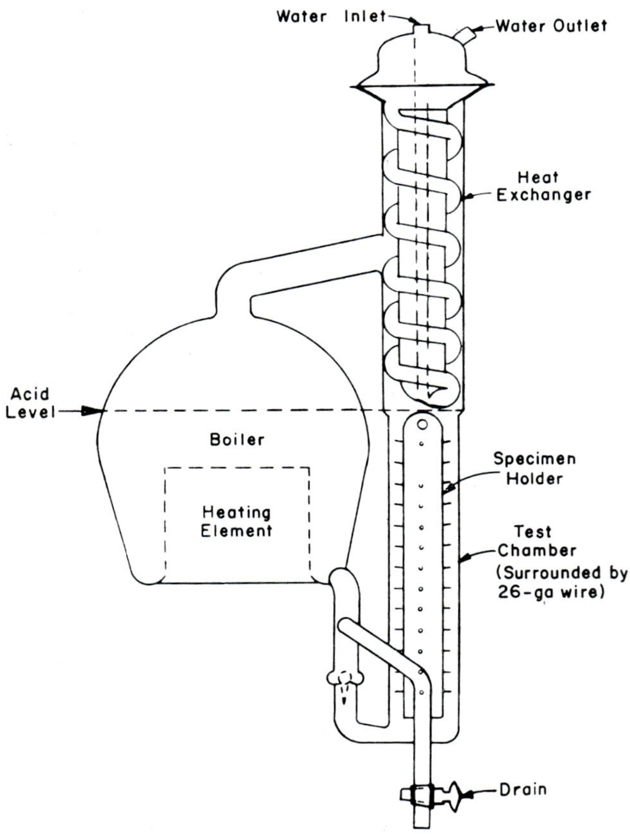

FIG. 24—*Multi sample tester for boiling 65 percent HNO_3 test* [167].

MST's are satisfactory for quality control work provided the material is good quality and shows a rate significantly below the acceptance limit (which is usually based on flask data). Material showing rates approaching the acceptance limit may have to be retested in a flask. MST's may be used also in research and are ideal for running a comparison series where all the specimens can be tested simultaneously. However, if one wishes to precisely compare the rates with past or future studies there can be complications for the reasons discussed above.

The 65 percent HNO_3 test is still popular in the United States but less so in Europe. With all its problems it would seem that except for evaluation prior to HNO_3 service its use will decrease in favor of the $Fe_2(SO_4)_3$-H_2SO_4 test and the $CuSO_4$-H_2SO_4 test with copper contact.

Ferric Sulfate-Sulfuric Acid

$Fe_2(SO_4)_3$ like $CuSO_4$, inhibits the general corrosion of stainless steels in H_2SO_4 [*169,170*]. Streicher [*25,170,171*], found that by adjusting the $Fe_2(SO_4)_3$-H_2SO_4 composition it was possible to inhibit the acid sufficiently to avoid a high rate of corrosion on the grain faces but attack grain boundary regions whose corrosion resistance had been lowered by chromium carbide precipitation. Because the solution produces grain dropping in susceptible austenitic stainless steels in 120 h or less, it is amenable to weight loss measurements and quantitative evaluation (ASTM Standard A 262, Practice B).

There are several practical aspects of the test worthy of note. First it is necessary to use a bulb condenser (not a finger condenser) and ensure a perfect flask/condenser seal. Otherwise, some evaporation may occur which increases the rate of general corrosion of the specimen. Second it is important to remove all traces of scale from the test specimen otherwise it may go active and yield a high rate of general corrosion throughout the test. Along the same lines it is important not to put the specimen into the acid before all the inhibiting ferric sulfate has dissolved. Another disadvantage is the necessity to boil HCl in the flask and condenser after the test to remove an iron oxide residue. An advantage over 65 percent HNO_3 is that corrosion products produced during the test do not influence the result.

The $Fe_2(SO_4)_3$-H_2SO_4 test is sensitive to grain boundary chromium depletion. It primarily detects problems associated with chromium carbides and, to a lesser extent, σ-phase in Type 321 (titanium stabilized), but not in the molybdenum bearing Type 316 steels [*39*]. It does not produce severe end grain attack of the type sometimes obtained in HNO_3.

The method is particularly suitable for evaluating high-chromium stainless steels (for example, 26 percent chromium ferritic [*123*]) and appears to be more sensitive than the HNO_3 test for detecting chromium depletion in these alloys. The test was applied by Streicher [*139*] to

Ni-Cr-Mo alloys and more recently has been applied to other high-nickel alloys [*126,130,145*]. Its application to all of these high-nickel alloys is covered in a new standard, ASTM Test for Detecting Susceptibility to Intergranular Attack in Wrought Nickel-Rich, Chromium-Bearing Alloys (G 28–72).

Nitric Acid-Hydrofluoric Acid

HNO_3-HF solutions have been used for many years to evaluate intergranular corrosion resistance in austenitic stainless steels. Warren [*172*] reviewed the history of this solution, and also performed an extensive evaluation of it. He was interested primarily in developing a quantitative test, for AISI Types 316 and 316L, which would detect intergranular corrosion susceptibility due to chromium carbide but not that due to σ-phase.

Warren's work, which included an evaluation of many production heats, resulted in the ASTM Standard A 262, Practice D, which consists of two 2-h exposure periods. Corrosion rates are calculated from weight loss data, and it is customary to compare these rates to those for material known to be resistant.

Disadvantages of the method are: (a) there is a high rate of general attack which varies greatly with alloy content; (b) maintaining a temperature at 70 C can be less convenient than using a boiling solution; and (c) hydrofluoric acid requires special handling, for example, no glass equipment. Even though the method is rapid it is not used widely for these reasons and because tests such as the $Fe_2(SO_4)_3$-H_2SO_4 and the copper accelerated, $CuSO_4$-H_2SO_4 tests are available.

Oxalic Acid Etch

$H_2C_2O_4$ is commonly used in metallography as an electrolytic etch for stainless steels. Streicher developed this etching technique into a procedure (ASTM Standard A 262, Practice A) for screening out material resistant to intergranular corrosion due to chromium carbides [*173*]. The test requires no costly specimen machining and can be run in less than 30 min.

The specimen is polished to a 000 finish and then etched for 1½ min at 1 A/in.^{2} in room-temperature 10 percent $H_2C_2O_4$. Based on the degree of attack at the grain boundaries the material is either deemed completely resistant to attack or must be subsequently tested with one of the previously described methods. Note that the etch test does not reject material—it only passes good quality material. The test is applicable to AISI Types 304, 304L, and probably other nonstabilized austenitics for all tests and for Types 316, 317, and their L grades for all tests except 65 percent HNO_3.

The oxalic acid test is very useful from both a time and economics standpoint. Unfortunately the percentage screened is quite low for one of the most common grades for which it is used (Type 304L). Streicher [*173*] reports it as 56.4 percent for the 65 percent HNO_3 test, that is, the oxalic acid etch test avoided the HNO_3 test only 56.4 percent of the time on "good" material. He has proposed a method of increasing this percentage, namely, to sensitize the $H_2C_2O_4$ acid etch specimen 20 min at 1250 F (677 C) rather than 60 min [*171*]. This obviously does increase the number of specimens passing the $H_2C_2O_4$ test, and the time more closely simulates that in most welding applications.

A screening test of this general type is needed for quality control of the increasingly used ferritic stainless steels. They can be sensitized during production by annealing at too high a temperature, and more importantly they are then susceptible to attack in relatively mild media.

Electrochemical Methods

Intergranular corrosion in stainless steels is predominantly under anodic control, and grain boundary attack is largely due to less anodic polarization at the grain boundaries [*25*]. It is not surprising, therefore, that materials very susceptible to intergranular attack can show anodic polarization behavior different to that of resistant material. Several examples of this have been reported [*16,95,174–180*], and typical behavior for austenitic and ferritic stainless steels is shown in Figs. 25 and 26, respectively.

Because the boundary area causing the increased current density in

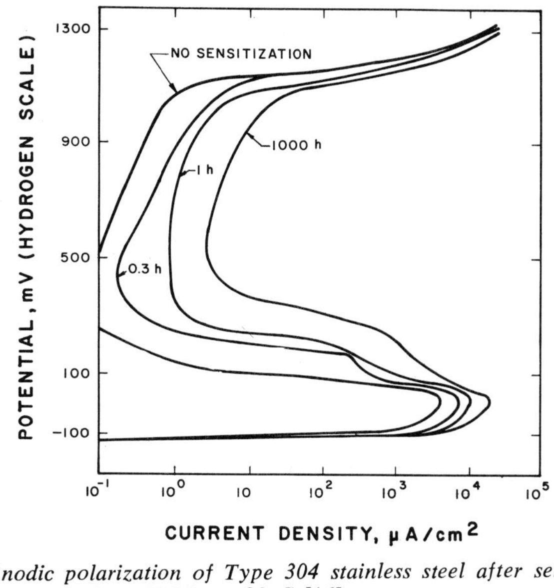

FIG. 25—*Anodic polarization of Type 304 stainless steel after sensitization from 0 to 1000 h at 650 C. 2* N *H_2SO_4 at 90 C* [16].

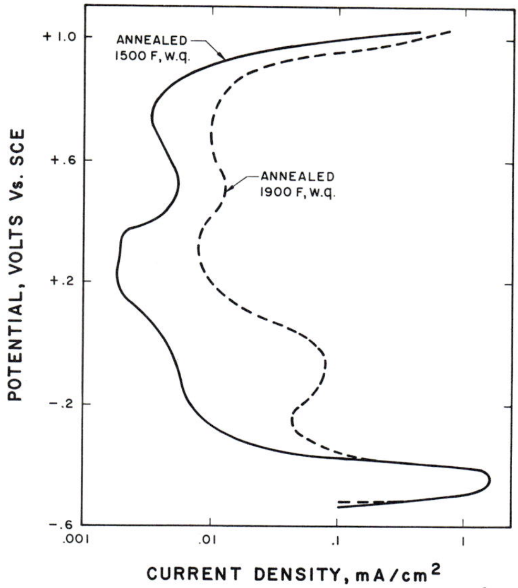

FIG. 26—*Anodic polarization of ferritic 18Cr-2Mo stainless steel in 5 percent H_2SO_4 at 24 C* [95].

Figs. 25 and 26 is very small, it is predicted that the chromium depleted areas adjacent to the grain boundaries actually have the anodic polarization behavior shown in Fig. 27. This is consistent with the anodic polarization curves in Fig. 28 which are typical for Fe-Cr alloys with chromium varied from 0 to 18 percent [*181,182*].

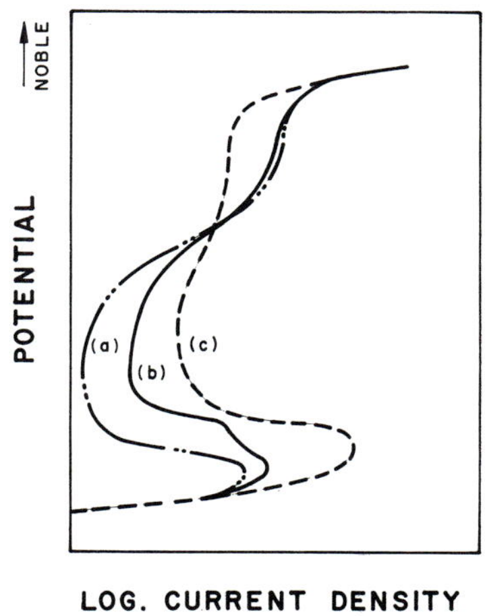

(*a*) Solution annealed.
(*b*) Sensitized.
(*c*) Chromium depleted zone.

FIG. 27—*Anodic polarization of Type 304 stainless steel in deaerated H_2SO_4* (*schematic*) [182].

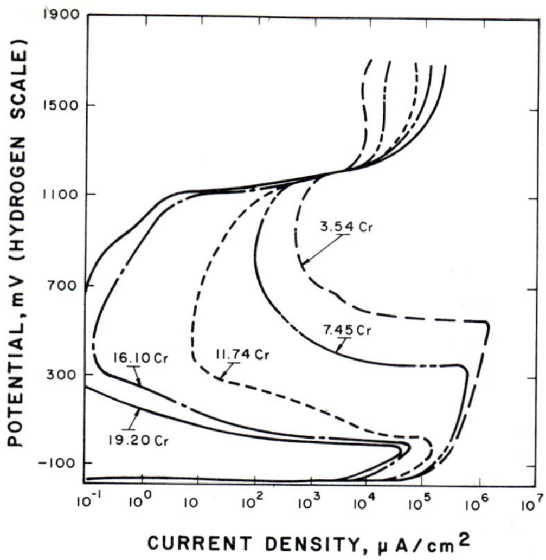

FIG. 28—*Effect of alloy chromium content on anodic polarization in 2* N *H_2SO_4 at 90 C. Nickel content in range 8.32 to 9.85 percent* [182].

Although anodic polarization curves, such as shown in Figs. 25 and 26, can distinguish between susceptible and resistant material they are probably not a very reliable technique unless the material is in an extremely susceptible condition. One reason for this is that the grain boundary areas constitute a very small part of the total surface whose behavior is reflected in the polarization curve. Also they do not distinguish between grain and grain boundary precipitation and do not take account of precipitate morphology. Another aspect is that they do not allow for the time factor often necessary for significant intergranular attack in service.

Anodic polarization curves can be useful in indicating which potential regions are most likely to cause attack. Then by performing long-term potentiostatic tests in these regions one can obtain potential—intergranular susceptibility relationships for the test media and time involved. This technique has been used by several [*16,175,176,177,179,183*] with detailed studies by Osozawa et al [*16*] and France and Greene [*183*]. Discussing the latter paper, Streicher [*184,185*] emphasizes that the potentiostatic tests are only good for the time they were run and the specific media (and temperature) involved. To demonstrate his argument he shows that intergranular attack can occur in extended periods of time under conditions claimed not to produce attack by France and Greene. Possible reasons for these differences have been suggested [*183*], but, taking care to avoid errors in experimental technique that could influence these results, the writer's experience confirms that of Streicher.

Another argument made by Streicher concerning the limitations of potentiostatic intergranular corrosion tests is that they can be achieved less expensively in other ways. For example, the standard corrosion tests previously discussed cover a wide range of potential (see Fig. 29), and each one is in effect a "chemical potentiostat." Therefore, one can select a standard test to represent potential ranges of interest.

Potentiostatic etch tests (several minutes in duration) have been claimed to be suitable for evaluating susceptibility to intergranular attack. The application of this technique has been reviewed by Linder [*187*]. A polished specimen is held (usually in H_2SO_4) at a potential just

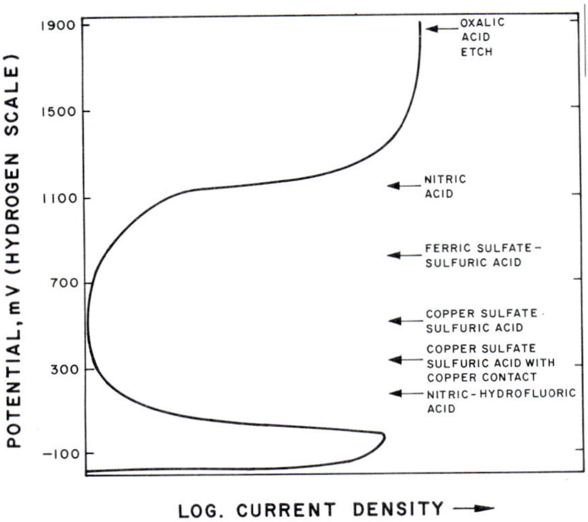

FIG. 29—*Potential regions for Type 304 stainless steel in standard test solutions (Table 2) compared to anodic polarization curve in H_2SO_4 (schematic).*

more noble than the critical passivation potential. It is recommended that such methods be carefully related to other quality control tests and service experience before putting too much reliance on them.

In short it would seem that electrochemical tests (other than simple etch tests for screening purposes) have little to offer over available standard accelerated test media particularly when one considers the cost of the equipment (potentiostat, etc.) needed for the former.

Other Evaluation Techniques

In most of the previously discussed quality control tests the method of evaluation was a weight loss which in turn was converted to an average corrosion rate. Ideally this rate is then compared to that for well annealed material (immune to intergranular attack). This procedure is generally satisfactory provided: (1) there is significant attack, (2) the corrodent is one which dislodges grains in susceptible material, and (3) the grain size is not unusually large.

The other most commonly used nondestructive evaluation technique is measurement of electrical resistance [*78,188*]. The method has the advantage of providing accurate "depth of attack"—"exposure time" information but does require specially shaped specimens (for example, wire or thin strip) and sensitive measuring apparatus. Another nondestructive technique for detecting the existence of intergranular corrosion is the use of dye penetrants. This is obviously attractive for in-service inspection or routine inspection during the overhaul of components prior to reassembly and return to service.

Commonly used destructive methods include metallography (usually to measure the depth of attack) and the measurement of a loss in mechanical strength as determined by bend or tension tests.

Etch tests, internal friction [*189*], magnetic measurements [*190*], Mössbauer spectroscopy [*191*], and thermoelectric methods [*192*] have all been used to investigate susceptibility to intergranular corrosion (that is, without doing a corrosion test). The latter four have not been investigated very thoroughly and at this time appear to have little to offer over the simpler, commonly used etch techniques.

Sensitization to Simulate Welding

Intergranular corrosion susceptibility is very often a result of welding. There are many problems involved in the corrosion evaluation of a weldment, and the conventional weight loss techniques are of limited value. A corroded weldment may show one or more of the forms of attack shown in Fig. 30. Suggested techniques for evaluating these types

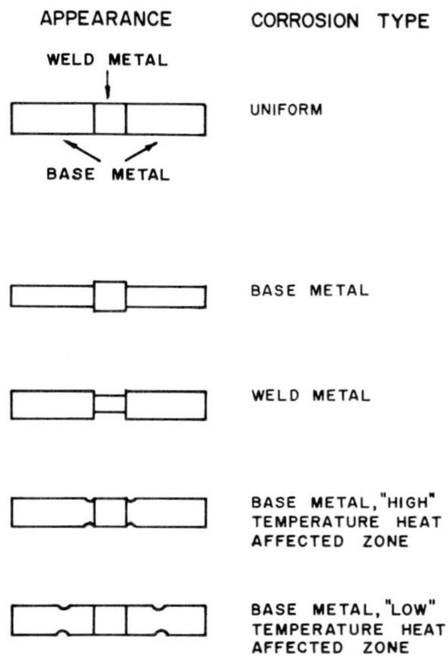

FIG. 30—*Types of weldment corrosion* [193].

of preferential attack are currently being studied by the Corrosion Subcommittee of the High Alloys Committee in the Welding Research Council [*193*]. It should be noted also that Pinnow and Moskowitz [*194*] have reviewed the corrosion of stainless steel weldments and collected a comprehensive bibliography on the subject.

Here we shall discuss the use of artificial heat treatments used to simulate welding problems. The one most commonly used for austenitic stainless is 1250 F (677 C), 1 h. For most welding applications this represents a much longer time at temperature than occurs during welding and is generally used with the justification that it is more severe than the latter. This is generally true as shown in Table 3 which summarizes data from a Welding Research Council test program [*6*].

TABLE 3—*Comparative severity of welding and heat treatment* [6].

AISI Type	Number of Racks Showing Intergranular Attack[a]	
	Welded	Heat Treated[b]
316L	0	11
316	1	15
317	6	22
318	1	21
302	15	19
304	1	10
304L	0	5
321	6	12
347	1	13

[a] Racks (24 specimens) exposed to a number of actual service environments for times ranging from 30 to 1600 days.
[b] Specimens heat treated for 1 to 4 h at 1100 to 1250 F.

It should be noted, however, that a 1250 F (677 C) treatment on annealed material may bear little relationship to welding if some metallurgical changes occur in the temperature range between the annealing temperature and the melting point. For example a stabilizing carbide may be solutioned during welding thus making the weld and high-temperature heat affected zones much more susceptible to sensitization than annealed material. This can be simulated by giving a double "sensitization," say 2300 F (1260 C) + 1250 F (677 C). One has to be careful not to change the bulk analysis (for example, by loss of carbon, nitrogen, or chromium) during the high-temperature treatment.

To simulate the weldment corrosion of ferritic stainless steels, sensitization consists of heating to above about 1700 F (927 C), that is, the sensitization is above the annealing temperature which is typically 1450 F (788 C). The rate of cooling from the sensitization temperature can be very important in these alloys because the actual damaging precipitation occurs (very rapidly) at lower temperatures such as 1100 F (593 C).

The simulation of welding in some of the nickel base alloys is even more difficult than for the stabilized austenitic stainless grades. This is because there may be a combination of deleterious precipitates—some formed by exposure at intermediate temperatures such as 1200 to 1600 F (649 to 871 C) and others formed (or dependent upon exposure at) temperatures approaching the solidus temperature.

Sensitization treatments are used obviously with credibility to study stress relieving, but to simulate welding they leave much to be desired. Unless a correlation is known between weld performance and a specific sensitization treatment it is usually preferable to test weldments. Of course in many cases this also has limitations because it may be difficult to simulate bulk welding techniques (for example, multipass) on specimens suitable for corrosion testing.

Relationship Between Intergranular Corrosion and Other Forms of Localized Corrosion

Stress-corrosion cracking is sometimes intergranular. In such cases it is possible that the only role of the stress is to open up the crack allowing ready access of corrodent through corrosion product which might otherwise stifle the corrosion reaction. In such a case the phenomenon might more correctly be called "stress accelerated intergranular corrosion." It is probable that many, but not necessarily all, cases of intergranular stress corrosion cracking fit into this category. Specific examples for some of the alloy systems discussed in this review are:

1. Polythionic acid cracking in stainless steels [*195*].
2. Cracking of Ni-Cr alloys in high-temperature water [*135–138*].
3. Cracking of annealed stainless steels in high-temperature water [*196–198*].
4. Cracking of sensitized stainless steels in marine environments.
5. Cracking of sensitized stainless steels in caustic environments [*199, 200*].

In all of these examples, factors which minimize intergranular corrosion susceptibility also improve stress corrosion cracking resistance but do not necessarily eliminate cracking.

Pitting can be connected with intergranular corrosion in at least two ways. First an intergranularly corroded surface can act as a crevice to develop the environmental conditions for pitting as shown for an austenitic stainless steel in Fig. 31. In this example some intergranular corrosion occurred during the pickling (HNO_3-HF) of a carburized surface prior to seawater service. Conversely the environmental conditions within a pit may cause intergranular attack which propagate the pit even though the bulk solution does not attack the grain boundaries. Also, Streicher [*73*] has noted that even in the absence of intergranular

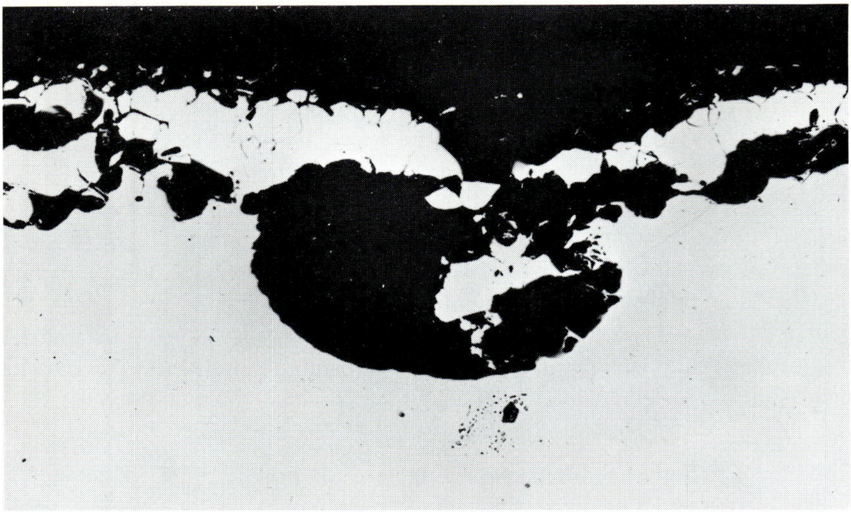

FIG. 31—*Corrosion pitting initiated by intergranular corrosion* (×*100*).

attack, pits in austenitic stainless steels tend to initiate at grain boundaries.

References

[*1*] McLean, D., *Grain Boundaries in Metals,* Oxford University Press, London, 1957.

[*2*] Ebling, H. F. and Scheil, M. A. in *Advances in the Technology of Stainless Steels and Related Alloys, ASTM STP 369,* American Society for Testing and Materials, 1963, pp. 275–282.

[*3*] Čihal, V., *Protection of Metals,* PTNMA, Vol. 4, No. 6, Nov./Dec. 1968, pp. 563–577.

[*4*] Warren, Donald, *Corrosion,* SBIIA, Vol. 15, April 1959, pp. 213t–220t.

[*5*] Auld, J. R. in *Advances in the Technology of Stainless Steels and Related Alloys, ASTM STP 369,* American Society for Testing and Materials, 1963, pp. 183–199.

[*6*] "Intergranular Corrosion of Chromium-Nickel Stainless Steels—Final Report," *Bulletin,* Welding Research Council, WRCBB, No. 138, New York, 1969.

[*7*] Bain, E. C., Aborn, R. H., and Rutherford, J. B., *Transactions,* American Society for Steel Treating, TASTA, Vol. 21, 1933, pp. 481–509.

[*8*] Mahla, E. M. and Nielsen, N. A., *Transactions,* American Society for Metals, TASEA, Vol. 43, 1951, pp. 290–322.

[*9*] Kinzel, A. B., *Transactions,* Metallurgical Society of American Institute of Mining, Metallurgical, and Petroleum Engineers, TMSAA, Vol. 194, 1952, pp. 469–488.

[*10*] Philibert, J., Henry, G., Roberts, M., and Plateau, J., *Revue de Métallurgie,* REMEA, Vol. 58, 1961, p. 557.

[*11*] Strauss, B., Schottky, H., and Hinnuber, J., *Zeitschrift Anorganische Chemie,* ZACMA, Vol. 188, 1930, p. 309–324.

[*12*] Hatwell, H., "Study of the Precipitation of Chromium Carbide Responsible for Intercrystalline Corrosion in Austenitic Stainless Steels," *3rd Annual Colloquium on Metallurgy, Corrosion of Steel,* British Iron and Steel Industry Translation No. 2771, June 1962.

[13] Stickler, R. and Vinckier, A., *Transactions,* American Society for Metals, TASEA, Vol. 54, 1961, pp. 362–380.
[14] Stickler, R. and Vinckier, A., *Corrosion Science,* CRRSA, Vol. 3, 1963, pp. 1–8.
[15] Stefanides, N. V., *Transactions,* American Society for Steel Treating, TASTA, Vol. 19, 1931, pp. 742–746.
[16] Osozawa, K., Bohnenkamp, K., and Engell, H. J., *Corrosion Science,* CRRSA, Vol. 6, 1966, pp. 421–433.
[17] Bäumel, A., Bühler, H. E., Schüller, H. J., Schwaab, P., Schwenk, W., Ternes, H., and Zitter, H., *Corrosion Science,* CRRSA, Vol. 4, 1964, pp. 89–103.
[18] Wilson, F. G., *British Corrosion Journal,* BCRJA, Vol. 6, May 1971, pp. 100–108.
[19] Philibert, J., Crussard, C., Wache, X., and Gerber, M., *Comptes Rendus,* Vol. 251:2, 1960, pp. 1289–1291.
[20] Alm, S. and Kiessling, R., *Journal,* Institute of Metals, JIMEA, Vol. 91, 1963, p. 190.
[21] Gellings, P. J. and deJongh, M. A., *Corrosion Science,* CRRSA, Vol. 7, 1967, pp. 413–421.
[22] Stawström, C. and Hillert, M., *Iron and Steel Institute,* Jan. 1969, pp. 77–85.
[23] Tedmon, C. S., Jr., Vermilyea, D. A., and Rosolowski, J. H., *Journal,* Electrochemical Society, JESOA, Vol. 118, No. 2, Feb. 1971, pp. 192–202.
[24] Čihal, V. and Kašová, Irena, *Corrosion Science,* CRRSA, Vol. 10, 1970, pp. 875–881.
[25] Streicher, M. A., *Electrochemical Society,* Vol. 106, 1959, pp. 161–180.
[26] Dillon, C. P., *Corrosion,* SBIIA, Vol. 16, Sept. 1960, pp. 433t–440t.
[27] Savkina, L. Ya. and Fel'dgandler, E. G., *Metal Science and Heat Treatment,* MHTRA, Nos. 11–12, Nov./Dec. 1968, pp. 857–860.
[28] Scharfstein, L. R. in *Effects of Residual Elements on Properties of Austenitic Stainless Steels, ASTM STP 418,* American Society for Testing and Materials, 1967, pp. 90–104.
[29] Lennartz, G., *Mikrochimica Acta,* MIACA, Vol. 3, 1965, pp. 405–428, British Iron and Steel Industry Translation No. 4639, Iron and Steel Institute, London, 1965.
[30] Binder, W. O., Brown, C. M., and Franks, F., *Transactions,* American Society for Metals, TASEA, Vol. 41, 1949, pp. 1301–1346.
[31] Binder, W. O. and Brown, C. M. in *Evaluation Tests for Stainless Steels, ASTM STP 93,* American Society for Testing and Materials, 1950, pp. 146–171.
[32] Warren, D., *Corrosion,* SBIIA, Vol. 15, No. 5, May 1959, pp. 221t–232t.
[33] Castro, R., Hochmann, J., and Granjon, H., *Revue de Métallurgie,* REMEA, Sept. 1965, pp. 847–850.
[34] Bourrat, J. and Hochmann, J., *Bulletin du Cerde d'Etudes des Metaux,* Vol. 9, Dec. 1964, pp. 285–303, British Iron and Steel Industry Translation 4129, Iron and Steel Institute, London, 1965.
[35] Franks, R., *Corrosion Handbook,* Uhlig, ed., Wiley, New York, 1948, p. 160.
[36] Brown, M. H., DeLong, W. B., and Myers, W. R. in *Evaluation Tests for Stainless Steels, ASTM STP 93,* American Society for Testing and Materials, 1950, pp. 103–119.
[37] Rosenberg, S. J. and Darr, J. H., *Transactions,* American Society for Metals, TASEA, Vol. 41, 1949, pp. 1261–1288.
[38] Phillips, F. J., *Transactions,* American Society for Metals, TASEA, Vol. 39, 1947, pp. 891–906.
[39] Streicher, M. A., *Corrosion,* SBIIA, Vol. 20, No. 2, Feb. 1964, pp. 57t–72t.
[40] Levin, I. A. (ed.) *Intercrystalline Corrosion and Corrosion of Metals Under Stress,* Consultants Bureau, New York, 1962.

[41] Čihal, V., Kašová, I., and Kubelka, J., *Metaux Corrosion-Industries,* MTUXA, No. 529, Sept. 1969, p. 281.
[42] Schwaab, P., Schwenk, W., and Ternes, H., *Werkstoffe und Korrosion,* WKKOA, Vol. 16, 1965, p. 844.
[43] Levitin, V. V. and Syreishchikova, *The Physics of Metals and Metallography,* PHMMA, Vol. 7, No. 2, 1959, pp. 152–154.
[44] Farrell, J. W. and Rosenthal, P., *Metal Progress,* MEPOA, Feb. 1960, pp. 101–103.
[45] Voeltzel, J. and Henry, G., *Revue de Métallurgie,* Part I *Mémoires,* RVMMA, Vol. 61, Feb. 1964, pp. 123–128, British Iron and Steel Industry English Translation 4493, Iron and Steel Institute, 1965.
[46] Rundell, G. R., "Austenitic Stainless Steel," U. S. Patent 3,306,736, Feb. 1967.
[47] Costello, T. M., Pinnow, K. E., and Moskowitz, A., *Materials Protection,* MAPRA, Vol. 8, Nov. 1969, pp. 15–18.
[48] Bungardt, K. and Lennartz, G., *Archiv für das Eisenhüttenwesen,* AREIA, Vol. 34, July 1963, pp. 531–546, British Iron and Steel Industry English Translation 3439, Iron and Steel Institute, London, 1963.
[49] Plumensi, J. P. and Kohn, A., "Distribution of Boron in 18–10 Stainless Steel as Detected by Autoradiography," *Comptes Rendus,* Vol. 258, 1964, pp. 5651–5653.
[50] Franks, R., Binder, W. O., and Thompson, J., *Transactions,* American Society for Metals, TASEA, Vol. 47, 1955, pp. 231–259.
[51] Renshaw, W. G., and Lula, R. A., *Transactions,* American Society for Testing and Materials, Vol. 56, 1956, pp. 866–889.
[52] Warren, Donald, *Corrosion,* SBIIA, Vol. 16, March 1960, pp. 119t–130t.
[53] Gooch, T. G., *Metal Construction and British Welding Journal,* MCBWA, Vol. 1, No. 12, Dec. 1969, pp. 20–25.
[54] Jesper, H. and Wessling, W., *Werkstoffe und Korrosion,* WKKOA, 16 June 1965, pp. 453–465, British Iron and Steel Industry Translations 4433, Iron and Steel Institute, 1965.
[55] Hamstead, A. C. and Van Delinder, L. S., *Corrosion,* SBIIA, Vol. 15, March 1959, pp. 147t–157t.
[56] Medvedev, Yu S. and Tomilina, T. D., *Protection of Metals,* PTNMA, Vol. 5, No. 4 July-Aug. 1969, pp. 348–352.
[57] Uhlig, H. H., *Transactions,* Electrochemical Society, TESOA, Vol. 87, 1945, pp. 141–154.
[58] Tedmon, C. S., Jr., Vermilyea, D. A., and Broecker, D. E., *Corrosion,* SBIIA, Vol. 27, No. 3, March 1971, pp. 104–106.
[59] Newell, H. D., *Transactions,* American Society for Steel Treating, Vol. 19, 1932, pp. 673–732.
[60] Čihal, V., *Hutnicke Listy,* HUTLA, Vol. 12, No. 1, 1966, pp. 73–82.
[61] Čihal, V., *Protection of Materials,* PTNMA, Vol. 2, No. 2, March-April 1966, pp. 127–133.
[62] Levin, I. A. and Kaloshina, Z. M., *Protection of Materials,* PTNMA, Vol. 4, No. 3, May-June 1968, pp. 235–237.
[63] Truman, J. E., *Journal of Applied Chemistry,* JACHA, Vol. 4, May 1954, pp. 273–283.
[64] Coriou, H., Hure, J., and Plante, G., *Electrochemica Acta,* Vol. 5, 1961, pp. 105–111.
[65] Chaudron, G., EURAEC-976 Quarterly Report No. 6, Oct./Dec. 1963.
[66] Coriou, H., Desestret, A., Grall, L., and Hochmann, J., *Comptes Rendus,* Vol. 254, 1962, p. 4467.
[67] Coriou, H., Desestret, A., Grall, L., and Hochmann, J., *Revue de Métallurgie,* REMEA, Vol. 61, 1964, p. 177.
[68] Coriou, H., Grall, L., Mahieu, C., and Gay, D., *Comptes Rendus des Seances de l'Academie des Sciences,* CRSAA, Paris, Vol. 258, March 1964, pp 2566–2569.

[69] Coriou, H., Desestret, A., Grall, L., and Hochmann, J., *Corrosion et Anticorrosion,* CRACA, Vol. 14, No. 4, 1966, pp. 163–169.
[70] Camus, A., Desestret, A., Froment, M., and Guiraldenq, P., *Comptes Rendus des Seances de l'Academie des Sciences,* CRSAA, Paris, Vol. 265, Jan. 1967, pp. 425–428.
[71] Desestret, A., Epelboin, I., Froment, M., and Guiraldenq, P., *Corrosion Science,* CRRSA, Vol. 8, 1968, pp. 225–234.
[72] Desestret, A., Froment, M., and Guiraldenq, P. in *Extended Abstracts,* 4th International Congress on Metallic Corrosion, 7–14 Sept. 1969, Amsterdam, pp. 122–124. Proceedings to be published by National Association of Corrosion Engineers, Houston, Tex.
[73] Streicher, M. A., *Journal,* Electrochemical Society, JESOA, Vol. 103, July 1956, pp. 375–390.
[74] Aust, K. T., Armijo, J. S., and Westbrook, J. H., *Transactions,* American Society for Metals, TASEA, Vol. 59, 1966, pp. 544–556.
[75] Aust, K. T., Armijo, J. S., Koch, E. F., and Westbrook, J. H., *Transactions,* American Society for Metals, TASEA, Vol. 60, 1967, pp. 360–372.
[76] Aust, K. T., Armijo, J. S., Koch, E. F., and Westbrook, J. H., *Transactions,* American Society for Metals, TASEA, Vol. 61, 1968, pp. 270–277.
[77] Aust, K. T., *Transactions,* Metallurgical Society of American Institute of Mining, Metallurgical, and Petroleum Engineers, TMSAA, Vol. 245, 1969, pp. 2117–2126.
[78] Armijo, J. S., *Corrosion,* SBIIA, Vol. 21, July 1965, pp. 235–244.
[79] Armijo, J. S., *Corrosion,* SBIIA, Vol. 24, Jan. 1968, pp. 24–30.
[80] Armijo, J. S. and Wilde, B. E., *Corrosion Science,* CRSSA, Vol. 8, 1968, pp. 649–664.
[81] Vermilyea, D. A. and Tedmon, C. S., *Transactions,* MTGTBF, Vol. 1, 1970, p. 1076.
[82] Rhodin, T. N., *Corrosion,* SBIIA, Vol. 12, 1956, p. 123t.
[83] Mahla, E. M. and Nielsen, N. A., *Transactions,* Electrochemical Society, JESOA, Vol. 93, No. 1, 1948, pp. 1–17.
[84] Espy, R. H., "How Composition and Welding Conditions Affect Corrosion Resistance of Weldments in Type 430 Stainless Steel," preprint of paper presented at NACE Meeting Cleveland, March 1968, National Association of Corrosion Engineers, Houston, Tex.
[85] Lincoln, R. A. and Pruger, T. A., *Iron Age,* IRAGA, Vol. 171, 26 Feb. 1953, pp. 127–131.
[86] Herbsleb, G. and Schwenk, W., *Werkstoffe und Korrosion,* WKKOA, Vol. 19, No. 2, 1968, pp. 103–113.
[87] Hochmann, J., *Revue de Métallurgie,* RVMMA, Vol. 48, 1951, pp. 734–758.
[88] Kiefer, G. C., *Engineering Station News,* Vol. 22, June 1950, p. 21, Ohio State University, Columbus, Ohio.
[89] Bäumel, A., *Archiv für das Eisenhüttenwesen,* AREIA, Vol. 34, Feb. 1963, pp. 135–146, British Iron and Steel Industry Translation 3287, Iron and Steel Institute, London, 1963.
[90] Baerlecken, E., Fisher, W. A., and Lorenz, K., *Stähl und Eisen,* STEIA, Vol. 81, No. 12, 1961, pp. 768–778.
[91] Houdremont, E. and Tofaute, W., *Stahl und Eisen,* STEIA, Vol. 72, No. 10, 1952, pp. 539–545.
[92] Lula, R. A., Lena, A. J., Kiefer, G. C., *Transactions,* American Society for Metals, TASEA, Vol. 46, 1954, pp. 197–230.
[93] Demo, J. J., "Effect of High Temperature Exposure on the Corrosion Resistance and Ductility of AISI 446 Stainless Steel," NACE Preprint No. 23, 1968 Conference, Cleveland, National Association of Corrosion Engineers, Houston, Tex.
[94] Demo, J. J., *Corrosion,* SBIIA, Vol. 27, Dec. 1971, pp. 531–544.

[*95*] Bond, A. P. and Lizlovs, E. A., *Journal,* Electrochemical Society, JESOA, Vol. 116, Sept. 1969, pp. 1305–1311.
[*96*] Bond, A. P., *Transactions,* Metallurgical Society of American Institute of Mining, Metallurgical, and Petroleum Engineers, TMSAA, Vol. 245, Oct. 1969, pp. 2127–2134.
[*97*] Hodges, R. J., *Corrosion,* SBIIA, Vol. 27, March 1971, pp. 119–127.
[*98*] Hodges, R. J., *Corrosion,* SBIIA, Vol. 27, April 1971, pp. 164–167.
[*99*] Binder, W. O. and Spendelow, H. R., Jr., *Transactions,* American Society for Metals, TASEA, Vol. 43, 1951, pp. 759–772.
[*100*] Lowe, J. H. C., d'A.Hunt, C., and Harker, H. H., *Transactions,* International Vacuum Metallurgy Conference, 1969.
[*101*] Oppenheim, R. and Laddach, H., *DEW Technishe Berichte,* Vol. 11, 1971, pp. 71–77.
[*102*] Wright, R. N., *Welding Journal,* WEJUA, Vol. 50, 19 Oct. 1971, pp. 434s–440s.
[*103*] Streicher, M. A., "The Effect of Heat Treatment on the Corrosion of 16% & 25% Cr Stainless Steels," NACE Preprint No. 25, 1968 Conference, Cleveland, National Association of Corrosion Engineers, Houston, Tex.
[*104*] Parker, E. A., *Advances in the Technology of Stainless Steels and Related Alloys, ASTM STP 369,* American Society for Testing and Materials, 1965, pp. 348–355.
[*105*] Perkins, R. A., discussion to Ref *92*, pp. 225–229.
[*106*] Colombier, L. and Hochmann, J., *Stainless and Heat Resisting Steels,* St. Martins Press, New York, 1968.
[*107*] Payson, P., *Transactions,* American Institute of Mining and Metallurgical Engineers, TAMDA Vol. 100, 1932, pp. 306–328.
[*108*] Masi, A. and Mancini, F., *Steel International,* STINB, March/April 1969, pp. 67–69.
[*109*] Kadlecek, P. E., *Materials Protection and Performance,* Vol. 10, June 1971, pp. 25–30.
[*110*] Bloom, F. K. and Carruthers, M. E., *Evaluation Tests for Stainless Steels, ASTM STP 93,* American Society for Testing and Materials, 1950, pp. 87–100.
[*111*] Carruthers, M. E., *Welding Journal, WEJUA,* Vol. 38, 1959, pp. 259s–267s.
[*112*] Espy, R. H., *Metal Progress,* MEPOA, Vol. 86, Sept. 1964, pp. 109–115.
[*113*] Stenas, G., *Proceedings,* 5th Scandinavian Corrosion Congress, Copenhagen, 1968, Vol. 1, pp. 13–2 to 13–22, Danish Corrosion Centre, Copenhagen, Denmark.
[*114*] Daeman, R. A. and Dept, F., *Welding Journal,* WEJUA, Vol. 49, 1970, pp. 33s–40s.
[*115*] Gurry, R. W., Jacob, E. M., and Allen, S. H., *Industrial and Engineering Chemistry Product Research and Development,* IEEDB, Vol. 10, No. 1, 1971, pp 112–119.
[*116*] Levin, I. A. and Kochergina, D. G., *Protection of Metals,* PTNMA, Vol. 1, May/June, 1965, pp 225–231.
[*117*] Vyklicky, M. and Voboril, J., *Kovove Materialy,* KOMAA, Vol. 3, No. 3, 1965, pp. 232–245.
[*118*] Babakov, A. A., Fel'dgandler, E. G., Kareva, E. N., and Savkina, L. Y., *Protection of Metals,* PTNMA, Vol. 3, March-April 1967, pp. 117–122.
[*119*] Kakhovskii, N. I., Langer, N. A., Yushchenko, K. A., and Chalyuk, G. I., *Avtomaticheskaia Svavka,* AVSVA, No. 12, 1964, pp. 30–37.
[*120*] Zhadan, T. A., *Protection of Metals,* PTNMA, Vol. 4, Sept./Oct. 1968, pp. 500–503.
[*121*] Zhadan, T. A. and Gulyaev, A. P., *Sbornik Trudov. Tsentral'nyi Nauchoissledovatel'skii. Institut Chernoi Metallurgii,* STIIA, Moscow, 1968, No. 65, pp. 34–37.

[*122*] Zhadan, T. A., *Protection of Metals,* PTNMA, Vol. 6, Nov.-Dec. 1970, pp. 631–632.
[*123*] Copson, H. R., *Physical Metallurgy of Stress Corrosion Fracture,* Rhodin, T. N., ed., Metallurgical Society Conference, Vol. 4, Interscience, New York 1959, pp. 247–267.
[*124*] Posysaeva, I. I., Babakov, A. A., and Petrovskaya, V. A., *Metal Science and Heat Treatment,* MHTRA, No. 7–8, July-Aug. 1969, pp. 556–559.
[*125*] Babakov, A. A., Zakharov, Y. V., Levin, F. L., and Golovin, A. I., *Protection of Metals,* PTNMA, Vol. 5, No. 3, May/June 1969, pp. 292–294.
[*126*] Brown, M. H., *Corrosion,* SBIIA, Vol. 25, Oct. 1969, pp. 438–443.
[*127*] Gulyaev, A. P. and Tokareva, T. B., *Metal Science and Heat Treatment,* MHTRA, Nos. 1 and 2, Jan./Feb. 1971, pp. 117–123.
[*128*] Levin, F. L., Babakov, A. A., Abramov, A. A., and Zakharov, Y. V., *Metal Science and Heat Treatment,* MHTRA, No. 5, May 1970, pp. 373–377.
[*129*] Raymond, E. L., *Corrosion,* SBIIA, Vol. 24, June 1968, pp. 180–188.
[*130*] Henthorne, M. and DeBold, T. A., *Corrosion,* SBIIA, Vol. 27, June 1971, pp. 255–262.
[*131*] Copson, H. R., Hopkinson, B. E., and Lang, F. S., *Proceedings,* American Society for Testing and Materials, Vol. 61, 1961, pp. 879–889.
[*132*] Scharfstein, L. R. and Maniar, G. N., *British Corrosion Journal,* BCRJA, Vol. 1, July 1965, pp. 36–41.
[*133*] Tedmon, C. S., Jr., and Vermilyea, D. A., *Corrosion,* SBIIA, Vol. 27, Sept. 1971, pp. 376–381.
[*134*] Harris, J. A. and Scarberry, R. C., *Journal of Metals,* KNZKA, Sept. 1971, pp. 45–49.
[*135*] Coriou, H., Grall, L., Mahieu, C., and Pelas, M., *Corrosion,* SBIIA, Vol. 22, Oct. 1966, pp. 280–287.
[*136*] Copson, H. R. and Economy, G., *Corrosion,* SBIIA, Vol. 24, March 1968, pp. 55–65.
[*137*] Economy, G., "Alloys Resistant to Stress-Corrosion Cracking in Leaded High Purity Water," U.S. Patent 3,573,901, 6 April 1971.
[*138*] Copson, H. R. and van Rooyen, D., "Nickel-Chromium Alloys Resistant to Stress-Corrosion Cracking," U.S. Patent 3,574,604, 13 April 1971.
[*139*] Streicher, M. A., *Corrosion,* SBIIA, Vol. 19, Aug. 1963, pp. 272t–284t.
[*140*] Prötzl, M., *Archiv für das Eisenhüttenwesen,* AREIA, Dec. 1958, pp. 745–756.
[*141*] Class, I., Gräfen, H., and Scheil, E., *Zeitschrift für Metallkunde,* ZEMTA, Vol. 53, May 1962, pp. 283–293.
[*142*] Gräfen, H. in *Advances in the Technology of Stainless Steels and Related Alloys, ASTM STP 369,* American Society for Testing and Materials, 1963, pp. 223–234.
[*143*] Samans, C. H., Meyer, A. R., and Tisinai, G. F., *Corrosion,* SBIIA, Vol. 22, Dec. 1966, pp. 336–345.
[*144*] Babakov, A. A., Svistunova, T. V., and Chermenskaya, N. F., *Metal Science and Heat Treatment,* MHTRA, Nos. 11–12, Nov./Dec. 1965, pp. 781–785.
[*145*] Leonard, R. B., *Corrosion,* SBIIA, Vol. 25, May 1969, pp. 222–228.
[*146*] Hodge, F. G., Preprint, Paper No. 82 Annual Meeting NACE, Chicago, March 1971, National Association of Corrosion Engineers, Houston, Tex.
[*147*] Schüller, H. J., Schwaab, P., and Albrecht, K., *Zeitschrift für Metallkunde,* ZEMTA, Vol. 57, April 1966, pp. 312–317.
[*148*] Svistunova, T. V., Chermenskaya, N. F., Smirnova, A. V., Yakovleva, E. F., Stegnukhina, L. V., and Zharkova, D. N., *Metal Science and Heat Treatment,* MHTRA, Nos. 7–8, July/Aug. 1969, pp. 613–616.
[*149*] Flint, G. N., *Journal,* Institute of Metals, JIMEA, Vol. 87, 1958–59, pp. 303–310.
[*150*] Svistunova, T. V., Doronin, V. M., Kruzhkov, V. I., Topilin, V. V., Dzugutov, M. Ya., Vinogradov, Yu. V., Chermenskaya, N. F., and Kordonov, B. A., *Stal,* STALB, Aug. 1966, pp. 659–662.

[*151*] Babakov, A. A., Svistunova, T. V., and Chermenskaya, N. F., *Proceedings,* 3rd International Congress on Metallic Corrosion, Moscow 1966, available from National Association of Corrosion Engineers, Houston, Tex., Vol. 2, pp. 380–389.

[*152*] Pavlov, S. S. and Svistunova, T. V., *Metal Science and Heat Treatment,* MHTRA, No. 11, Nov. 1968, pp. 861–865.

[*153*] Svistunova, T. V., Runova, Z. K., and Sakharnov, A. A., *Metal Science and Heat Treatment,* MHTRA, No. 5, May 1970, pp. 359–363.

[*154*] Pavlov, S. S. and Svistunova, T. V., *Metal Science and Heat Treatment,* MHTRA, No. 10, Oct. 1970, pp. 830–832.

[*155*] Hatfield, W. H., discussion of paper by Pfeil, L. B. and Jones, D. G., *Journal,* Iron and Steel Institute, JISIA, Vol. 127, 1933, pp. 380–383.

[*156*] Rocha, H. J., discussion of paper by Brauns, E. and Pier, G., *Stähl und Eisen,* STEIA, Vol. 75, 1955, p. 586.

[*157*] Scharfstein, L. R. and Eisenbrown, C. M. in *Advances in the Technology of Stainless Steels and Related Alloys, ASTM STP 369,* American Society for Testing and Materials, 1965, pp. 235–239.

[*158*] Schwenk, W. in *Advances in Technology of Stainless Steels and Related Alloys, ASTM STP 369,* American Society for Testing and Materials, 1965, p. 239.

[*159*] Brauns, E. and Schwenk, W., *Stähl und Eisen,* STEIA, Vol. 84, No. 13, June 1964, pp. 787–789.

[*160*] Herbsleb, G. and Schwenk, W., *Corrosion Science,* SCRRA, Vol. 7, 1967, pp. 501–511.

[*161*] Bäumel, A., *Stähl und Eisen,* STEIA, Vol. 84, June 1964, pp. 798–807.

[*162*] Ternes, H. and Schwenk, W., *Archiv für das Eisenhüttenwesen,* AREIA, Vol. 36, Feb. 1965, pp. 99–108, British Iron and Steel Industry Translation 4456, Iron and Steel Institute, London.

[*163*] Schwaab, P., Schwenk, W., and Ternes, H., *Werkstoffe und Korrosion,* WSKRA, Vol. 16, Oct. 1965, pp. 844–852.

[*164*] Huey, W. R., *Transactions,* American Society for Steel Treating, Vol. 18, 1930, pp. 1126–1143.

[*165*] Brown, M. H., *Bulletin,* American Society for Testing and Materials, ASTBA, No. 169, Oct. 1950, pp. 200–209.

[*166*] DeLong, W. B. in *Evaluation Tests for Stainless Steels, ASTM STP 93,* American Society for Testing and Materials, 1949, pp. 211–214.

[*167*] Alger, J. V., Roberts, E. C., Lent, R. P., Anderton, G. W., *Bulletin,* American Society for Testing and Materials, ASTBA, No. 214, May 1956, pp. 57–60.

[*168*] Tisinai, G. F. and Samans, C. H., *Bulletin,* American Society for Testing and Materials, ASTBA, May 1959, pp. 64.

[*169*] Hatfield, W. H., *Transactions,* Electrochemical Society, TESOA, Vol. 46, 1924, pp. 297–312.

[*170*] Streicher, M. A., *Corrosion,* SBIIA, Vol. 14, Feb. 1958, pp. 59t–70t.

[*171*] Streicher, M. A., *Bulletin,* American Society for Testing and Materials, ASTBA, No. 229, April 1958, pp. 189–198.

[*172*] Warren, Donald, *Bulletin,* American Society for Testing and Materials, ASTBA, No. 230, May 1958, pp. 199–210.

[*173*] Streicher, M. A., *Bulletin,* American Society for Testing and Materials, ASTBA, No. 188, Feb. 1953, pp. 35–38.

[*174*] Brauns, E. and Pier, G., *Stähl und Eisen,* STEIA, Vol. 75, No. 9, 1955, pp. 579–586, Henry Brutcher English Translation 3662, Altadena, Calif.

[*175*] Schüller, H. J., Schwaab, P., and Schwenk, W., *Archiv für das Eisenhüttenwesen,* AREIA, Vol. 33, Dec. 1962, pp. 853–862, British Iron and Steel Industry Translation 3177, Iron and Steel Institute, London, 1963.

[*176*] Schwenk, W., Schüller, H. J., and Schwaab, P., *Werkstoffe und Korrosion,* WSKRA, Vol. 15, Aug. 1964, p. 621–631.

[*177*] Herbsleb, G., *Werkstoffe und Korrosion,* WSKRA, Vol. 19, March 1968, pp.

204–209, British Iron and Steel Industry Translation 6640, Iron and Steel Institute, London, 1969.
[*178*] Herbsleb, G., *Archiv für das Eisenhüttenwesen,* AREIA, Vol. 41, Sept. 1970, pp. 875–881.
[*179*] Gooch, T. G., *British Welding Journal,* BRWJA, Nov. 1968, pp. 10–22.
[*180*] Gooch, T. G., Honeycombe, J., and Walker, P., *British Corrosion Journal,* BCRSA, Vol. 6, July 1971, pp. 148–154.
[*181*] Kossyi, G. G. and Kuzub, V. S., *Zavodskaya Laboratoriya,* ZVDLA, (English Translation available, Industrial Laboratory), Vol. 31, May 1965, pp. 582–584.
[*182*] Osozawa, K. and Engell, H. J., *Corrosion Science,* CRRSA, Vol. 6, 1966, pp. 389–393.
[*183*] France, W. D., Jr., and Greene, N. D., *Corrosion Science,* CRRSA, Vol. 8, 1968, pp. 9–18.
[*184*] Streicher, M. A., *Corrosion Science,* CRRSA, Vol. 9, 1969, pp. 53–56.
[*185*] Streicher, M. A., *Corrosion Science,* CRRSA, Vol. 11, 1971, pp. 275–276.
[*186*] France, W. D., Jr., and Greene, N. D., *Corrosion Science,* CRRSA, Vol. 10, 1970, pp. 379–382.
[*187*] Linder, B. in *Proceedings,* 5th Scandinavian Corrosion Congress, 1969, pp. 6.1–6.27.
[*188*] Rutherford, J. J. B. and Aborn, R. H., *Transactions,* Iron and Steel Division, American Institute of Mining & Metallurgical Engineers, TAMDA, Vol. 100, 1932, pp. 293–301.
[*189*] Prikhod'ko, V. N. and Savitskaya, A. N., "Susceptibility of 18–8 Type Steel to Intergranular Corrosion Determined by Method of Internal Friction," *Izvestiia Vuz-Chem. Met.,* Nov. 1961, pp. 185–192, Henry Brutcher English Translation 5473, Altadena, Calif., 1962.
[*190*] Belenkova, M. M. and Mikheev, M. N., *Defektoskopiya,* (in Russian) DEFKA, No. 5, 1967, pp. 65–75.
[*191*] Albritton, O. W., *Corrosion,* SBIIA, Vol. 24, Dec. 1968, pp. 389–392.
[*192*] Shadrunova, A. P., "Use of Thermoelectric Method for Investigating the Processes Taking Place in Austenitic Steels," *Izvestiia Vuz-Chem. Met.,* Oct. 1962, pp. 107–110, Henry Brutcher English Translation 5779, Altadena, Calif.
[*193*] Henthorne, M., "Corrosion Testing of Weldments," to be published.
[*194*] Pinnow, K. E. and Moskowitz, A., *Welding Journal,* WEJUA, June 1970, pp. 278s–284s.
[*195*] Samans, C. H., *Corrosion,* SBIIA, Vol. 20, 1964, pp. 256t–262t.
[*196*] Armijo, J. S., Low, J. R., Jr. and Wolff, U. E., *Nuclear Applications,* NUAPA, Vol. 1, 1965, p. 462.
[*197*] Pashos, T. J., Duncan, R. N., Williamson, H. E., Arlt, W. H., Baroch, C. J., and Hoffman, J. P., *Transactions,* American Nuclear Society, TANSA, Vol. 7, 1964, p. 416.
[*198*] Pickett, A. E., "Investigation of Intergranular Cracking of Nonsensitized Austenitic Stainless Steels in Circulating Autoclaves," EURAEC-GEAP-5495, U. S. Atomic Energy Commission, Contract AT(04-3)-189, June 1967.
[*199*] Wanklyn, J. N. and Jones, D., *Journal of Nuclear Materials,* JNUMA, Vol. 1, 1959, p. 154.
[*200*] Snowden, P. P., *Journal,* Iron and Steel Institute, JISIA, Vol. 194, 1960, pp. 181–189.

B. W. Lifka[1] *and D. O. Sprowls*[1]

Significance of Intergranular Corrosion in High-Strength Aluminum Alloy Products

REFERENCE: Lifka, B. W. and Sprowls, D. O., **"Significance of Intergranular Corrosion in High-Strength Aluminum Alloy Products,"** *Localized Corrosion—Cause of Metal Failure, ASTM STP 516,* American Society for Testing and Materials, 1972, pp. 120–144.

ABSTRACT: Most heat treatable aluminum alloys are, to some degree, susceptible to intergranular corrosion, depending upon the metallurgical treatments that have been given to the metal. Susceptibility to intergranular attack results from a localized decomposition of solid solution at the grain boundaries and the formation of identifiable precipitates in the grain boundaries. Hence, any alloying addition, thermal treatment or mechanical treatment that influences precipitation of alloying constituents can influence the susceptibility to intergranular attack. The exact nature of the localized corrosion path and the specific roles of grain boundary precipitates and the parent solid solutions in the grain margins depend upon alloy composition. The actual significance of the intergranular corrosion tendency with regard to the serviceability of an aluminum alloy product varies with the alloy and the temper.

Products of alloys such as 2024-T3, 7075-T6, and 7178-T6 become susceptible to intergranular attack usually as a result of heat treating procedures that are less than the optimum, and in such a condition they may be susceptible to stress-corrosion cracking or exfoliation corrosion. Such products can be artificially aged to produce special tempers, 2024-T8, 7075-T76, and 7178-T76 that provide a high resistance to stress-corrosion cracking and to exfoliation. Although products in these tempers may show some susceptibility to intergranular attack in an accelerated corrosion test, such as specified in MIL-H-6088E, and in some natural environments, this intergranular attack does not indicate a poor service performance.

Hence, it should be understood that predictions of the serviceability of a high-strength aluminum alloy product can be made on the basis of an intergranular corrosion test only in certain instances, chiefly for thin sections of aluminum-copper alloys in the naturally aged temper (T3, T4) and aluminum-zinc-magnesium-copper alloys in the maximum strength temper (T6). The most reliable predictions can be obtained from appropriate stress-corrosion cracking and exfoliation tests; for some alloys other criteria, such as electrical conductivity and potential measurements, are feasible.

KEY WORDS: corrosion, intergranular corrosion, grain boundaries, exfoliation corrosion, temper (metallurgy), grain structure, aluminum alloys, electrochemistry, stress corrosion cracking, pitting

[1] Senior research engineer and section head, respectively, Chemical Metallurgy Division, Aluminum Company of America, New Kensington, Pa. 15068.

Intergranular corrosion is attack on a metal that proceeds along grain boundaries at a considerably faster rate than through adjoining grain bodies. Most of the high-strength, heat-treatable aluminum alloys are to some extent, depending upon the metallurgical treatments employed, susceptible to intergranular corrosion. The mechanism of the corrosion is electrochemical and can be related to precipitation in the vicinity of the grain boundaries. The specific electrochemical roles of the grain boundary precipitates and the adjacent solid solutions vary from one alloy system to another [*1–3*].[2]

This paper reviews the causes of susceptibility to intergranular corrosion in the 2XXX and the 7XXX aluminum alloys, and places in perspective the susceptibility of the various tempers with regard to product serviceability.

Service experience has shown that the principal corrosion problems with high-strength aluminum alloys in aircraft and areospace vehicles have resulted from stress-corrosion cracking (SCC) and exfoliation corrosion [*4–5*]. Although these are generally intergranular phenomena, it has been shown [*6–8*] that the serviceability of aluminum alloy products cannot be related in a general way to the susceptibility to intergranular corrosion or to any other single characteristic such as the microstructure, tensile strength, or toughness. Corrosion test data from laboratory and outdoor exposure tests are used to illustrate the fallacies in trying to judge corrosion performance of high-strength aluminum alloys by an accelerated intergranular corrosion test.

Historical Background

The embrittling effect of extensive intergranular attack in thin (0.012-in.) duralumin (3.5Cu-0.5Mg-0.3Mn) sheet used as covering material in one of the early "all metal" airplanes was reported in 1922 by Rawdon et al [*9*]. Subsequent studies at the National Bureau of Standards [*10*] indicated that "the practical means for preventing this deterioration of duralumin consists, first, in increasing the resistance of the material toward intercrystalline corrosion by the proper heat treatment of the duralumin, and second, in protecting the surface against corrosive attack by a suitable coating." Materials problems caused by exfoliation corrosion, SCC, or corrosion fatigue of the early alloys and tempers were identified with intergranular corrosion, and the blame often laid to faulty heat treatment, particularly with the quality of quench following solution heat treatment. Thus, problems resulting from intergranular corrosion of duralumin came to be associated with "improper heat treatment," and in 1944 an accelerated corrosion test for susceptibility to intergranular corrosion was incorporated into a U. S. Govern-

[2] The italic numbers in brackets refer to the list of references appended to this paper.

ment specification [*11*] for the heat treatment of aluminum alloys. Hence, it is not surprising that the occurrence of intergranular attack in an aluminum alloy product today is likely to be a matter of concern among aerospace engineers.

The Bureau of Standards, around 1925, demonstrated that sprayed coatings of commercially pure aluminum provided excellent corrosion protection to duralumin sheet. Such treatments, however, were not considered practical for large sheets. Subsequently E. H. Dix, Jr., at Alcoa Research Laboratories established methods to metallurgically clad commercial aluminum to both sides of a 2017-T4 (then known as 17S-T) sheet to obtain outstanding corrosion protection [*12*]. In 1928 Dix published an article [*13*] describing the improved resistance to corrosion of Alclad 2017-T4. The great value and versatility of this concept has been demonstrated by the widespread use of alclad sheet. In many of today's aircraft designs, however, the advantages of alclad products cannot be realized because parts are machined frequently from thick plate and extruded sections.

While a properly controlled solution heat treatment in some instances may have been an appropriate preventative for the ills of intergranular corrosion of the thin sheet used in the early aircraft designs, there is no way to avoid susceptibility to intergranular corrosion by regulating the heat treatment and quenching of the relatively thick plate, extrusions, and forgings used in modern aircraft. Therefore, beginning in the 1950s and continuing to date, special tempers [*14–16*] were introduced to assure greater resistance to exfoliation and SCC. With many of the new exfoliation and stress-corrosion resistant alloys and tempers, the significance of a degree of susceptibility to intergranular corrosion is considerably different from that for the early duralumin and superduralumin alloys.

Al-Cu (2XXX) Type Alloys

Electrochemical Relationships in the Microstructure

Copper is an important alloying addition to give strength to aluminum, and because of its appreciable solubility in aluminum the aluminum-copper (Al-Cu) alloys respond to solution heat treatment. Brown et al [*17*] showed that the retention of 5 percent copper in solid solution also changes the electrode potential of aluminum about 200 mV in the cathodic direction. Thus, the subsequent precipitation of copper rich constituents from the supersaturated solid solution alters the potential in the anodic direction, and pronounced electrochemical relationships can develop in the microstructure. Initial precipitation of discrete particles occurs in grain boundaries and copper depleted regions develop adjacent to the boundaries. The copper rich grain boundary precipitates are also relatively cathodic, and in a corrosive electrolyte such as sodium chloride solution electrochemical attack occurs in the copper depleted grain boundary regions. Other alloying elements such as magnesium, silicon, manganese, cadmium, zirconium, and vanadium are added to the 2XXX alloys to provide various mechanical properties. The susceptibility to intergranular corrosion of all of the alloys, however, is influenced primarily by copper concentration gradients in the Al-Cu solid solution in the grain boundary regions.

Brown and Mears [*18*] devised a method to demonstrate the electrochemical mechanism of intergranular corrosion by measuring the electrochemical potentials and electrical current flow between grain boundary zones and grain bodies of large grained materials. Using this procedure, the change in potential at the grain boundaries and grain centers of a high-purity Al-4Cu alloy with time of aging was measured [*19*]. The data in Fig. 1 indicate that precipitation occurred more rapidly at the grain boundaries than within the grains, with the boundaries becoming anodic to the grain centers. The maximum difference in potential between the grain boundaries and grain centers, as shown by the lower curves occurred after aging for about 4 to 8 h at 191 C (375 F). With more extended heating, the precipitation within the grain centers began to approach that at the grain boundaries until virtually complete precipitation occurred, both at the grain boundaries and within the grains. Thus, the difference in potential between the two was reduced almost to zero. It is shown also in Fig. 1 that the precipitation, both in the grains and

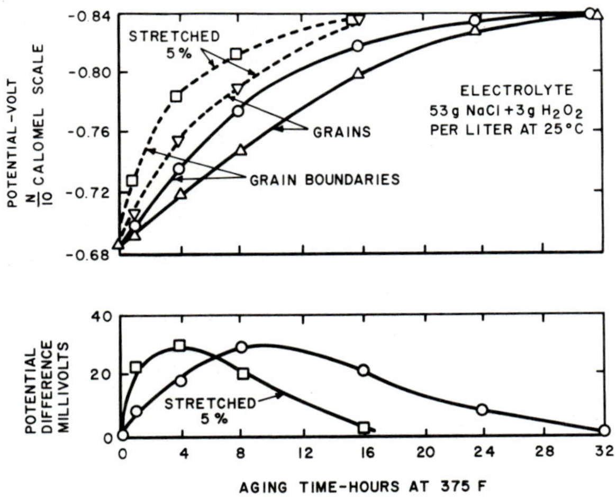

FIG. 1—*The potentials of the grains and grain boundaries of an aluminum alloy containing 4 percent copper which was heat treated at 930 F quenched in cold water and aged at 375 F* [19].

the grain boundaries, was accelerated by plastic deformation (5 percent stretch).

Hunter et al [*20*], using specimens thinned to 1000 Å for transmission electron microscopy, succeeded in relating the path of intergranular attack to microstructural features of aluminum-copper-magnesium (Al-Cu-Mg) alloy 2024. Figure 2 shows an electron transmission micrograph illustrating intergranular corrosion of the copper depleted grain margins. Susceptibility to pitting attack associated with a high degree of precipitation is shown in Fig. 3. This alignment of pits along grain boundaries would appear as a broad form of intergranular attack at lower (light microscope) magnifications.

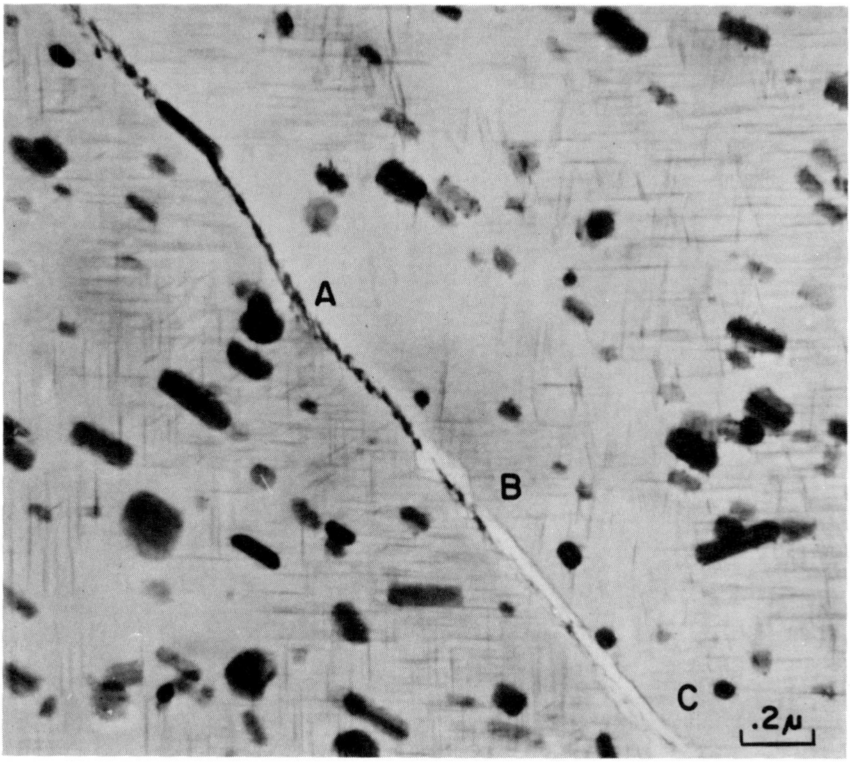

FIG. 2—*Electron transmission micrograph of a thin metal specimen of an Al-Cu-Mg alloy (2024) corroded in NaCl-H_2O_2 solution. The metal had been solution heat treated, quenched in cold water, and aged 4 h at 191 C (375 F). Intergranular attack (shown by the white path) between points* B *and* C *followed the Cu-depleted grain margins visible between* B *and* A. *Particles of the relatively cathodic grain boundary precipitate are visible extending from* B *(surrounded by corrosion) to the top of the print. A Widmanstätten pattern of* S′ *precipitate has formed in the grains. The large dark particles scattered through the grains are relatively insoluble constituents of Al-Mn and Al-Fe-Si compounds. In this condition 2024 alloy is highly susceptible to stress-corrosion cracking* [20].

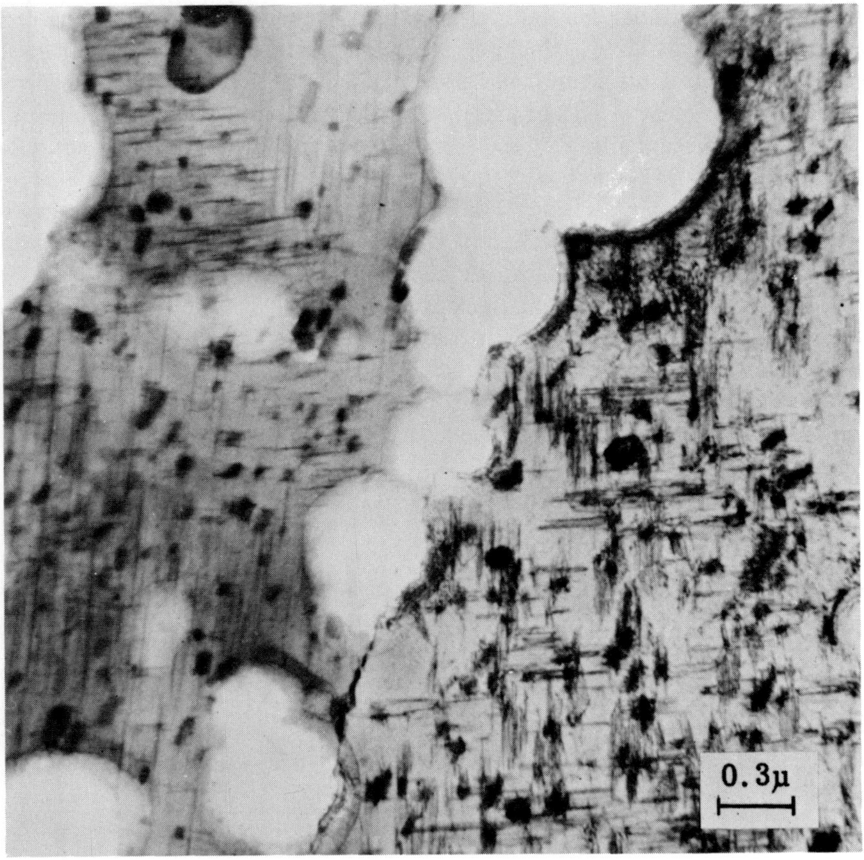

FIG. 3—*Similar to Fig. 2 except that the metal was aged 16 h at 191 C (375 F). Shows rounded pits in the grain body and along a grain boundary. Note also the advanced degree of precipitation in the grain bodies. In this condition 2024 alloy is highly resistant to stress-corrosion cracking* [20].

Metallurgical Treatments Affecting Susceptibility to Intergranular Corrosion

The principal factors affecting precipitation and susceptibility to intergranular corrosion of Al-Cu alloy products are total alloy content, rate of quench after solution heat treatment, plastic deformation after quenching, and exposure to elevated temperatures.

Quenching Rate—The effect of cooling rate during quenching upon the resistance to intergranular corrosion and to stress-corrosion cracking has been studied by various investigators [*21–23*]. With Al-Cu alloys in the naturally aged tempers (T3, T4) maximum resistance to corrosion is obtained by rapid quenching typified by prompt immersion in cold water. However, this procedure can cause excessive distortion of formed parts, and so at one time, air and fog quenches were employed to minimize distortion. Such slow quenching produced undesirable susceptibility to intergranular corrosion and was not permitted by Government Specifications [*11*]. However, in the case of artificially aged (T6, T8 tempers) Al-Cu alloys, rapid quenching is not essential for maximum resistance to corrosion.

The contrasting effects of cooling rate during quench on the susceptibility to intergranular attack and on the exfoliation and SCC[3] performance of 2024 alloy sheet in the naturally and artificially aged tempers are shown in Fig. 4. The relationship of SCC and exfoliation of the T42 temper sheet to susceptibility to intergranular attack in the accelerated test in sodium chloride-hydrogen peroxide (NaCl-H_2O_2) solution is very striking. Equally striking, however, is the absence of a similar relationship for the T62 temper. Although the T62 temper sheet went through

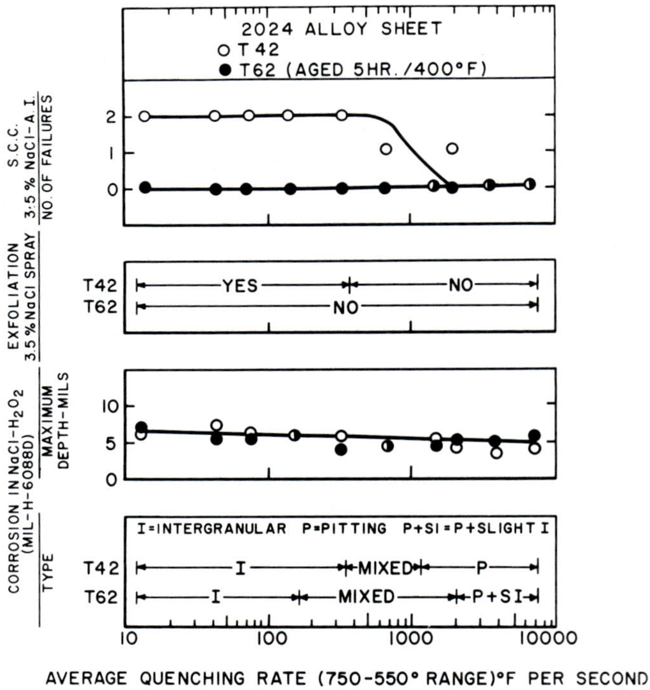

FIG. 4—*Comparative effects on the corrosion and stress corrosion resistance of 2024-T42 and T62 tempers of sheet as influenced by quenching rate.*

[3] The stress corrosion tests in this experiment were conducted on 0.064-in.-thick preformed specimens [*24*] stressed in frames as described by Sager et al [*25*] and exposed to 3.5 percent NaCl solution by alternate immersion (60-min cycle with 10-min immersion followed by 50-min drying in air). The stressed preform is a very severe test in which a high elastic stress is superimposed upon residual stresses introduced by forming, giving a total sustained stress that is greater than the proportional limit of the material.

a variation of type of attack with cooling rate similar to that of the T42 temper sheet, there was no SCC or exfoliation of any of the items in the T62 temper. Note in Fig. 5 the similarity in appearance of the intergranular attack developed by the $NaCl\text{-}H_2O_2$ test (MIL-H-6088D) in both the T42 and T62 temper sheet quenched with a cooling rate of 120 F/s. Compare this with the difference in type of attack of the two tempers when the sheet was highly stressed and exposed to 3.5 percent sodium chloride (NaCl) solution by alternate immersion. It is significant that even with a very high stress, but in a less aggressive environment, the rate of propagation of intergranular attack in the T62 temper sheet was slow compared to that in the T42 temper sheet.

It is impossible commercially to quench any but relatively thin sections fast enough to avoid susceptibility to intergranular corrosion. These data, therefore, indicate that aging the metal to a T62 (or T8) temper improves the resistance to exfoliation and SCC even though susceptibility to intergranular corrosion is not eliminated.

Artificial Aging—Although solution heat treated and quenched Al-Cu alloys are hardened at room temperature, this hardening phenomenon involves the formation of Guinier-Preston (GP) zones in the grain bodies, with no detectable precipitate formation in the grain boundaries. Artificial aging by heating the metal at a suitable temperature (usually in the 149 to 204 C (300 to 400 F) range) initiates precipitation in the grain boundaries and stimulates extensive precipitation in the grain bodies. As was discussed previously in connection with Fig. 1, if the metal is aged for a prolonged time there would be no susceptibility to grain boundary corrosion. However, it has been found unnecessary to extend the aging treatment to that point to achieve practical immunity to SCC and to exfoliation corrosion. In other words, artificially aged products of alloys such as 2011, 2021, 2024, and 2219 when aged by the usually recommended practices (Specification MIL-H-6088E, 5 Feb. 1971) will provide maximum resistance to corrosion and stress corrosion even though they may be susceptible to intergranular attack. In these instances the differences in potential between the grain boundary regions and the grain bodies are relatively small, and thus there is very little driving force for intergranular attack. Exceptions are 2014-T6 and 2024-T6; the aging practices in common use for these alloys were selected to produce maximum strength, and a considerable extension of the aging time and reduction in strength is required to achieve maximum corrosion resistance.

Artificial Aging to Avoid Corrosion Problems in Natural Environments

Two examples of the use of artificially aged tempers of Al-Cu-Mg alloy 2024 to avoid corrosion problems will be discussed next. Rolled plate of 2024-T351 generally is susceptible to intergranular attack in the standard type of attack test (MIL-H-6088E) and to exfoliation corrosion in out-

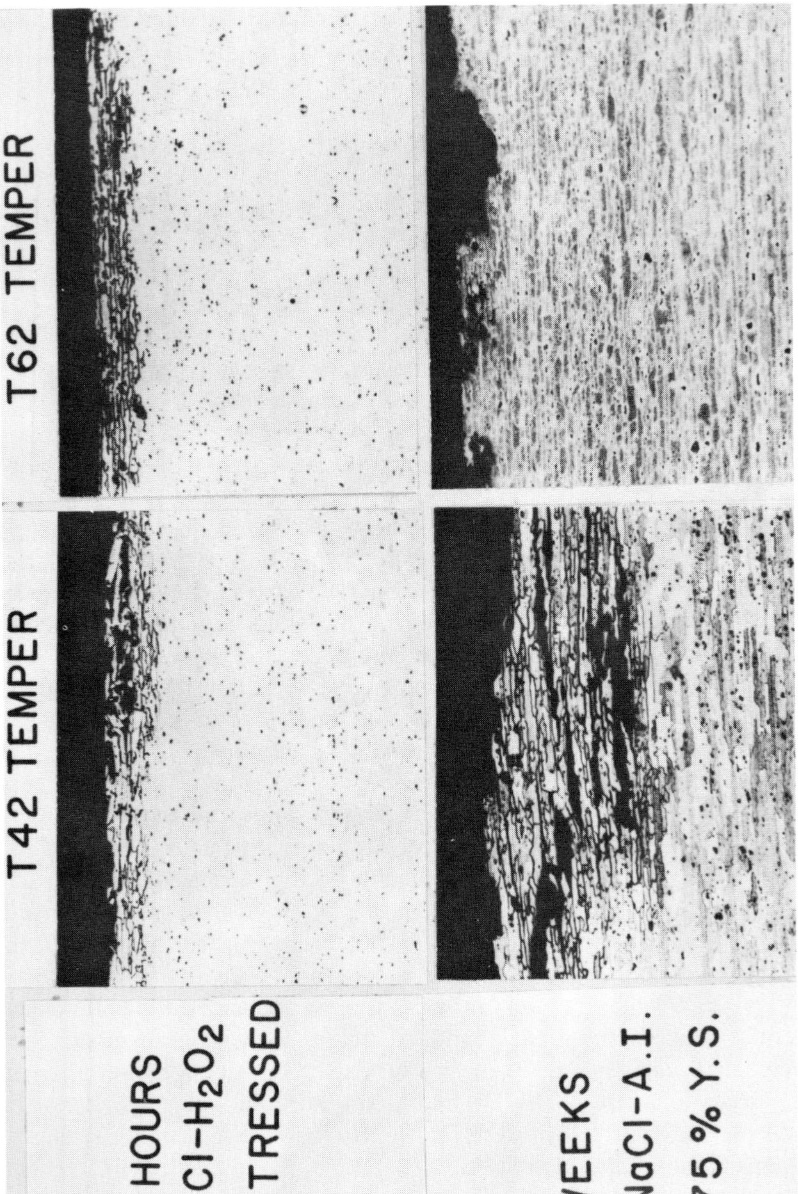

FIG. 5—*Metallographic cross sections of corroded specimens of 2024 alloy sheet heat treated and quenched with a cooling rate of 120 F/s* ($\times 100$).

door atmospheres, particularly in marine environments. When this plate is artificially aged to the T851 temper by the recommended treatment of 12 h at 191 C (375 F) it still is susceptible to intergranular attack (Fig. 6), but not to exfoliation corrosion. The photographs in Fig. 7 illustrate the high resistance of 2024-T851 alloy plate to exfoliation in marine environments after extended periods of exposure.

In the second example, 2024-T3 alloy drawn tube swaged for a short distance on the ends stress corrosion cracked during 5-months' exposure in a seacoast atmosphere, whereas another piece of the same swaged tube that had been aged to the T8 temper (12 h at 191 C (375 F)) showed no sign of SCC after eight years in the seacoast atmosphere (Fig. 8). Again it was noted that the NaCl-H_2O_2 corrosion test revealed a somewhat similar susceptibility to intergranular attack for both tempers, but the type of attack and the stress corrosion performance in the natural environment was very different (Fig. 9).

Significance of Susceptibility to Intergranular Attack

Although only a slight susceptibility to intergranular attack can result in marked susceptibility to exfoliation and to stress-corrosion cracking of the as-quenched and naturally aged tempers (T3, T4) of Al-Cu alloys, it was shown that the same is not true for the artificially aged tempers (T6, T8). It may be seen from Fig. 1 that if the aging treatment is not quite at the optimum period there will be a small difference in potential between grains and grain boundaries. Metal in this condition may be susceptible to a mixture of relatively wide intergranular and pitting attack but not necessarily to exfoliation or to SCC. The NaCl-H_2O_2 accelerated corrosion test tends to exaggerate the intergranular corrosion of artificially aged products, whereas in other environments the tendency for pitting is more pronounced. Thus, it may be concluded that the use of such accelerated intergranular corrosion tests has no significance as a criterion of the corrosion performance to be expected of artificially aged products of Al-Cu alloys.

Al-Zn-Mg-Cu (7XXX) Type Alloys

The second family of alloys comprising the high-strength, heat-treatable, aluminum alloys are the aluminum-zinc-magnesium-copper (Al-Zn-Mg-Cu) alloys (7XXX series) that provide the highest strength in aluminum. Alloy 7075 is a typical example and much of the following discussion is based on experience with this alloy.

Mechanism of Intergranular Corrosion

Some controversy exists over the actual mechanism by which intergranular corrosion occurs in the Al-Zn-Mg-Cu alloys [*26*]. Under some heat treat conditions, precipitation occurs as closely spaced particles

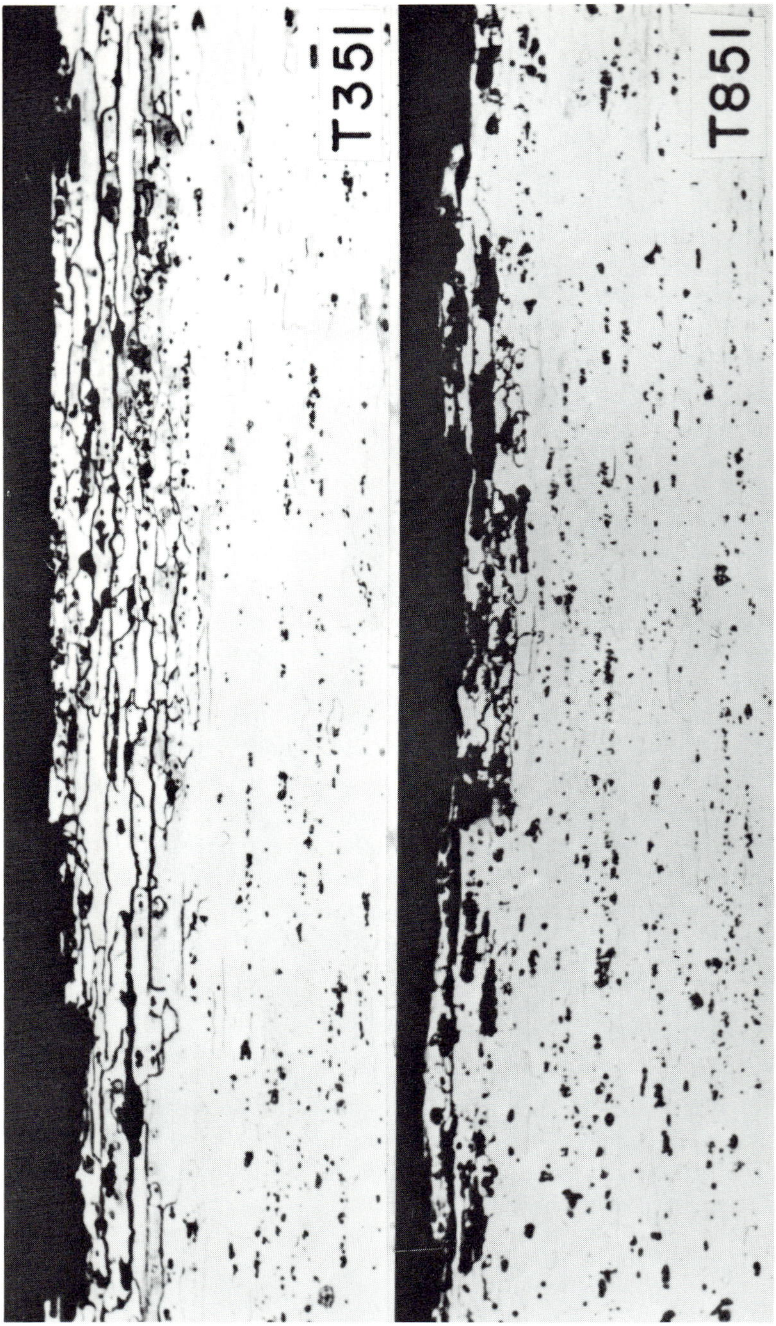

FIG. 6—*Intergranular attack developed by the standard $NaCl$-H_2O_2 accelerated corrosion test (MIL-H-6088E) in both the T351 and the T851 tempers of mill heat treated 2024 alloy plate 0.340 in. thick (as polished,* $\times$*100).*

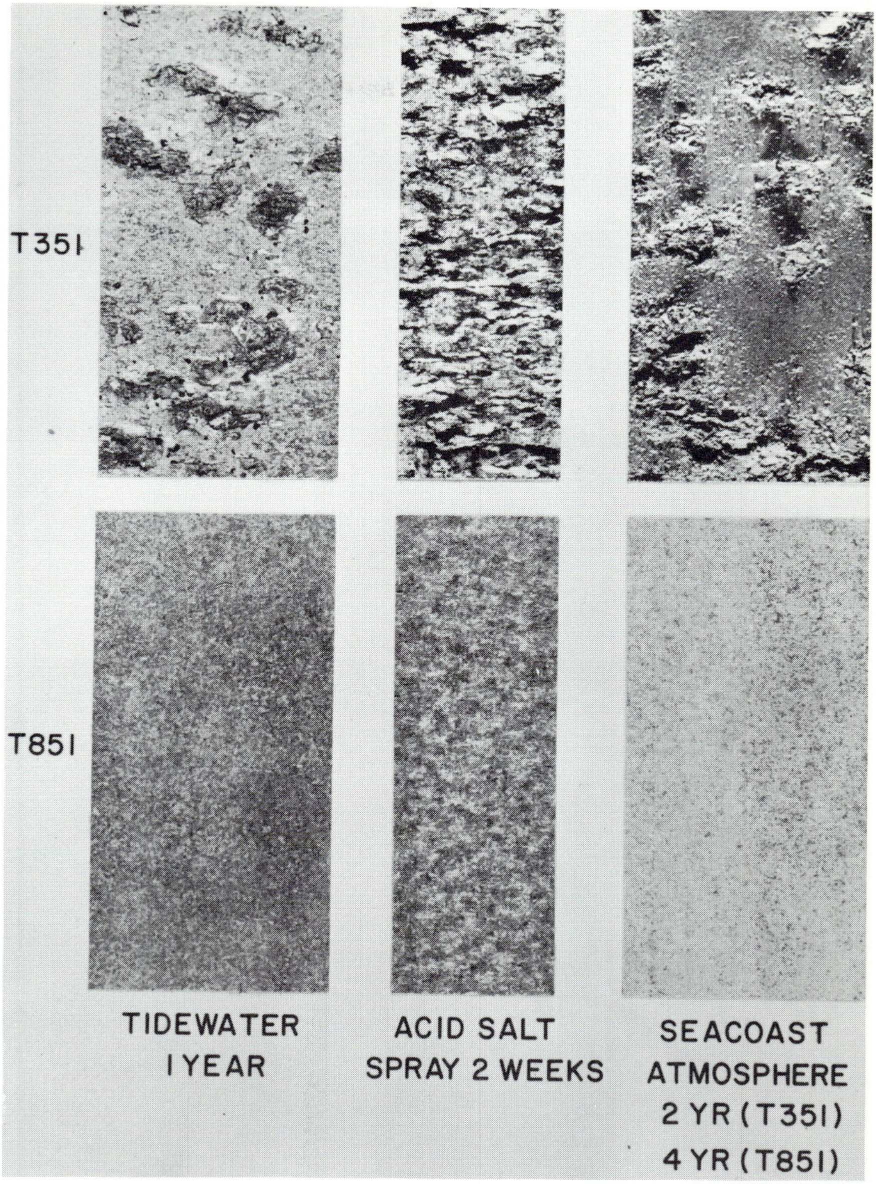

FIG. 7—*Midplane (T/2) surfaces of 1-in.-thick 2024 alloy plate exposed as indicated, showing the high resistance to exfoliation of the T851 temper compared to the T351 temper* (×*1*).

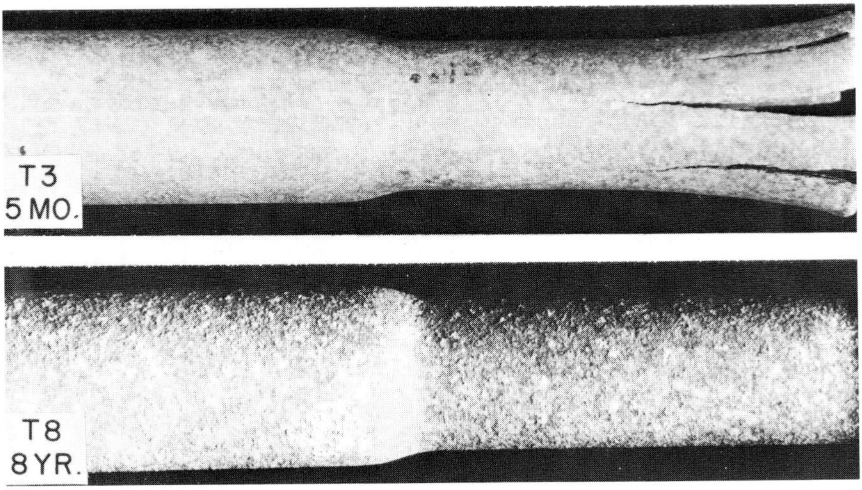

FIG. 8—*Swaged ends of 1-in.-outside-diameter drawn tubes exposed to the seacoast atmosphere at Point Judith, R. I. The 2024-T3 tube was susceptible to intergranular attack and stress corrosion cracked within five months because of the residual stress developed by the swaging. The 2024-T8 tube, also susceptible to intergranular attack, withstood 8-years' exposure with no visible or microscopic stress-corrosion cracking.* (*Corrosion products from weathering not removed*).

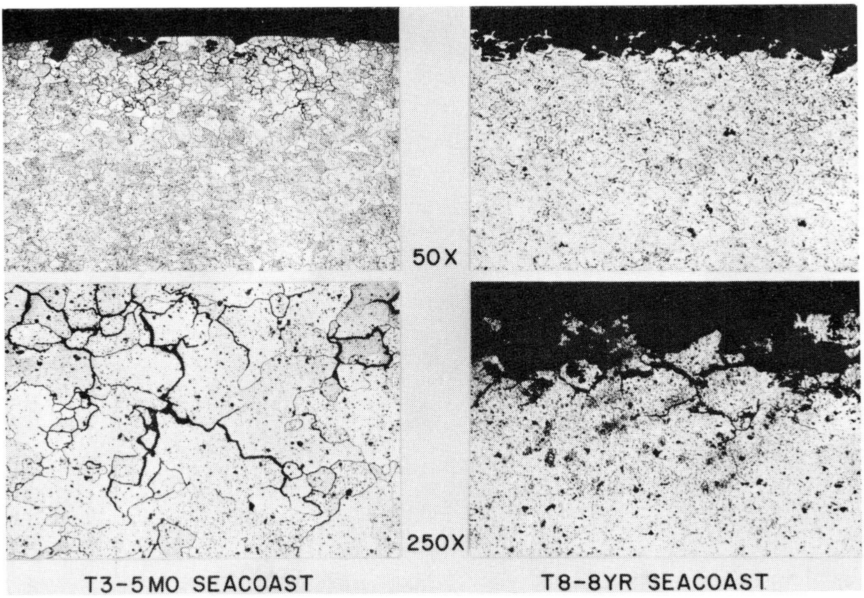

FIG. 9—*Metallographic cross sections* (*Keller's etch*) *of corroded 2024 alloy tubes shown in Fig. 8. Contrast the shallow mixture of intergranular and pitting corrosion present in the T8 temper tube after eight years with the sharp penetrating intergranular attack in the T3 temper tube after only five months* (*reduced 50 percent for reproduction*).

along grain boundaries. Although the grain boundary precipitates have been identified as *M* or *M′* phase ($MgZn_2$), they may contain substantial percentages of copper. Because the particles are too small to be analyzed, their electrochemical characteristics cannot be determined accurately. In a corrosive environment, the most anodic constituent corrodes preferentially. It was theorized, therefore, that intergranular attack in this family of alloys could result from selective attack of anodic constituents precipitated in the boundaries. This was the theory advanced by Golubev [*27*] and by Dix [*3*] and the generally accepted one until about 1961 [*28*].

As early as 1947, however, Fink and Willey [*29*] had suggested that the partial depletion of copper (along with zinc and magnesium) from the solid solution at the grain boundary region causes the grain margins to become anodic to both the grain bodies and precipitates in the boundaries, much the same as for the Al-Cu system. With the advent of transmission electron microscopy, Hunter et al [*30*] studied the corrosion in NaCl-H_2O_2 solution of thin foils of slowly quenched 7075-W sheet, They observed extremely selective corrosion along the grain margins, leaving particles of grain boundary precipitate unattacked, as shown in Fig. 10. This supports the mechanism proposed by Fink and Willey. It is possible, though, that either mechanism may be operative, depending upon the chemical composition and thermal history of the particular specimen involved.

As-Quenched (W) Temper

The Al-Zn-Mg-Cu alloys in the freshly heat treated (as-quenched) condition are unstable and spontaneously age at room temperature. This condition, designated the W temper, is rarely used commercially, except to facilitate certain forming operations, because of the unstable mechanical properties and because it is also highly susceptible to exfoliation and SCC. The factors governing resistance to corrosion are much the same as for the commercial T6 temper discussed next.

Artificially Aged Tempers

T6 Temper—The T6 temper designation is given to materials that have been artificially aged to their maximum mechanical strength. This aging is performed at relatively low temperatures of about 100 to 135 C (212 to 275 F). The principal hardening precipitation is a coherent clustering of solute atoms to form GP zones that have little or no effect on the resistance to intergranular corrosion in NaCl-H_2O_2 solution (MIL-H-6088E). With slow quenching, however, small amounts of a transition precipitate *M′* form in the grain boundaries, and the metal in either W or T6 temper becomes susceptible to intergranular corrosion. Consequently, the cooling rate during quenching from the solution heat-treat temperature is the prime factor as to whether such alloys corrode inter-

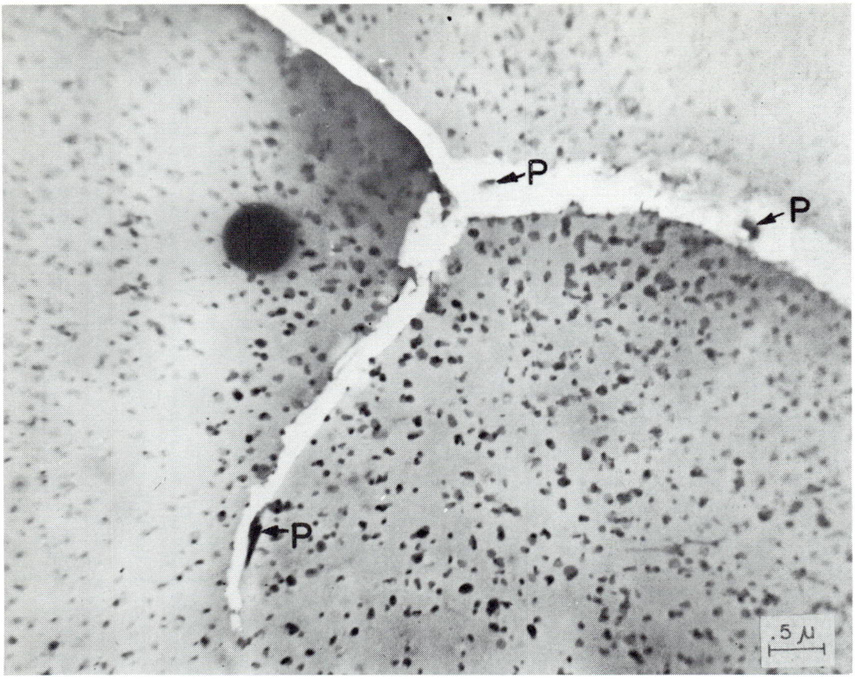

FIG. 10—*Electron transmission micrograph of a thin metal specimen of Al-Zn-Mg-Cu-Cr alloy (7075-W) corroded in NaCl-H_2O_2 solution. The bulk metal was solution heat treated and then quenched in boiling water to cause a susceptibility to intergranular corrosion. This micrograph shows selective corrosion that has proceeded along the edges of three adjacent grains. The upper left and the lower boundaries lie at about 90 deg to the surface and the corrosion has followed a comparatively narrow path. The upper right boundary, on the other hand, lies at a relatively low angle to the specimen surface and the corrosion path appears to be wide. Particles marked* P *are grain boundary precipitates left intact and uncorroded. The small dark particles scattered through the grains are the relatively insoluble chromium rich constituents. In this condition (7075-W) or when artificially aged to the T6 temper, 7075 alloy sheet is susceptible to stress-corrosion cracking* [30].

granularly. The general considerations for a fast cooling rate to prevent intergranular susceptibility are much the same as previously discussed for Al-Cu alloys, with three exceptions.

The first difference is that the minimum cooling rate to avoid susceptibility to intergranular corrosion is appreciably lower for Al-Zn-Mg-Cu alloys, of the order of 300 F/s as shown by Fig. 11 [*31*].

Figure 11 also illustrates the second difference, a reversion to pitting corrosion for Al-Zn-Mg-Cu alloys at very slow cooling rates. This probably results from gross overall precipitation during quenching, thereby minimizing regions of large differences in solution potential and achieving an improved resistance to SCC and exfoliation. Because of this phenomenon, plots of corrosion performance of Al-Zn-Mg-Cu alloys versus cooling rates generally show a region of lowest performance in contrast with Al-Cu alloys that tend to worsen continually with decreasing quench rates.

The third difference is that cold work introduced between quenching and artificial aging has negligible effect on both the strength and corrosion resistance of Al-Zn-Mg-Cu alloys in the T6 temper.

T7X Tempers.—The T7 temper designation is given to alloys that are aged beyond their maximum strength to provide a specified degree of resistance to exfoliation or SCC, or both. For example, material in the T76 temper is very resistant to exfoliation and provides improved stress corrosion resistance in the short-transverse direction over that in the T6 temper; while material in the T73 temper [*32*] is essentially immune to both exfoliation and SCC.

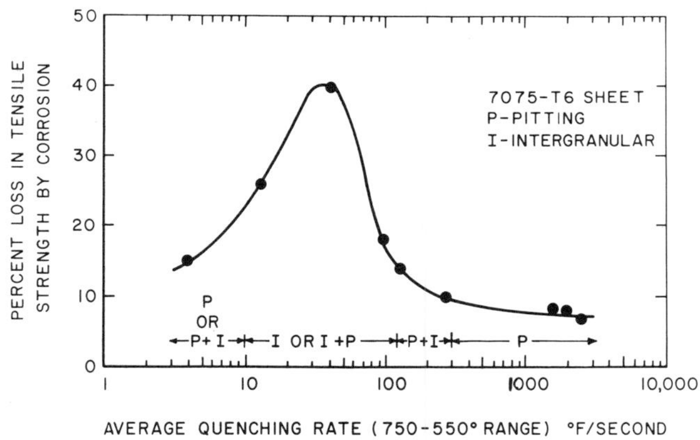

FIG. 11—*Effect of quenching rate on the inherent type of attack and the corrosion resistance of 7075-T6. Unstressed, transverse specimens from 0.064-in. sheet exposed 3 months to 3.5-percent NaCl alternate immersion.*

The T7 aging is a two step process. The first step involves heating at T6 aging temperatures, 100 to 135 C (212 to 275 F) and often the actual T6 age is used. The second step is performed commercially at temperatures between 160 and 185 C (320 and 365 F). A similar aging effect can be accomplished with a single high temperature furnace setting, provided slow (less than 1 F/min) heating rates are used to permit sufficient low-temperature aging during the heat-up cycle [*33*].

Aging at temperatures higher than those that produce maximum strength causes a growth in the zone size and the formation of the metastable *M′*-phase, Mg (Al, Cu, Zn)$_2$ (Figs. 12 and 13). The effect of this aging process is reflected in the electrochemical potential of the material. The potential of rapidly quenched 7075-W sheet aged at 121 C (250 F) shifts about 75 mV in the cathodic direction after 24 to 36 h of aging

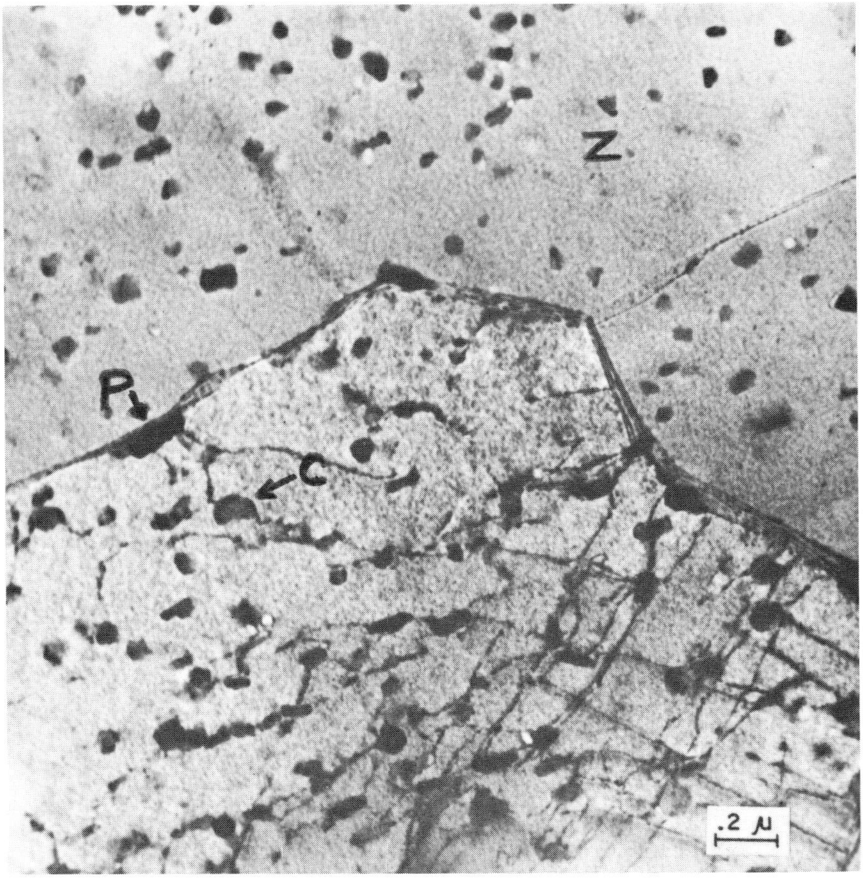

FIG. 12—*Thin foil of 7075-T6 plate. Minute zones are visible in the grains (black specks* Z) *and numerous precipitate particles* (P) *are present in the grain boundaries. Dislocations are pinned to chromium rich constituent* (C) *and to particles of grain boundary precipitate* (P). *Short-transverse specimens susceptible to stress-corrosion cracking at 25 percent yield strength* [30].

and remains unchanged thereafter (Fig. 14). Isothermal aging at temperatures of 135 and 149 C (275 and 300 F) produces essentially the same result but in a shorter time. However, at temperatures around 160 C (320 F) a second precipitation process is indicated, the potential first shifting in the cathodic direction and then in the anodic direction after about 8 h of aging. This change in the anodic direction is more pronounced on T73 material (Fig. 15) which is aged 24 h at 121 C (250 F) and then aged for 8 h at 177 C (350 F). The first change of potential with aging corresponds with precipitation of zinc from solid solution at the lower temperature. The second aging step reduces the copper as well as zinc in solid solution in the grain bodies with the result that the grains

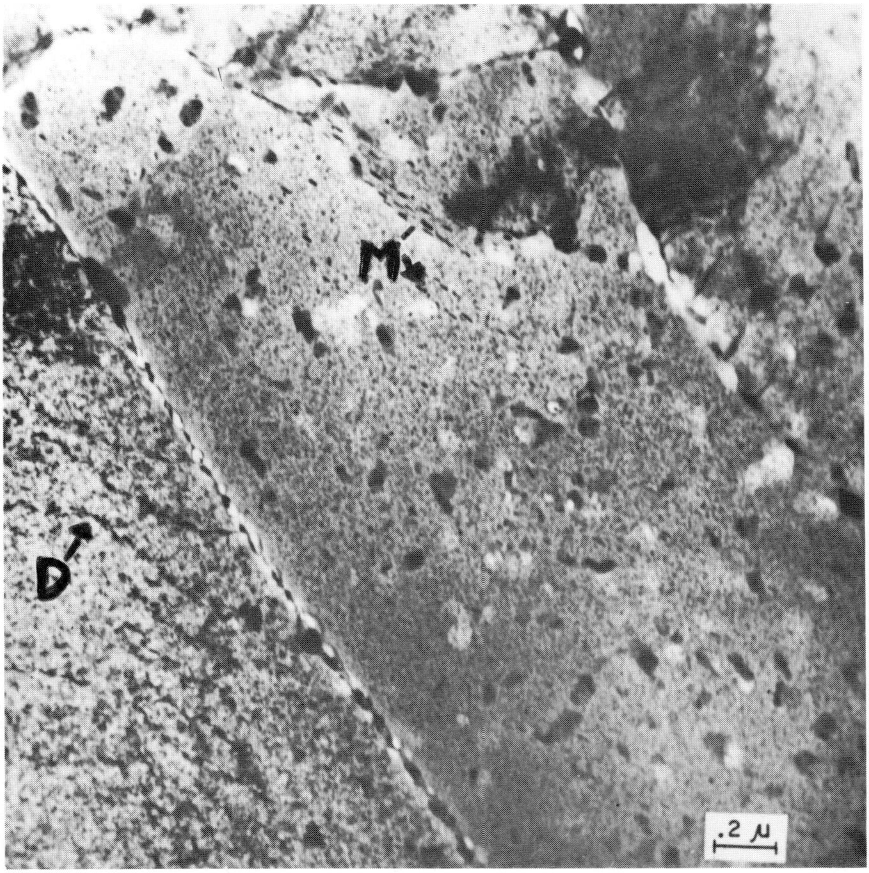

FIG. 13—*Thin foil of 7075-T73 plate. Compared to the 7075-T6 (Fig. 12) the zones are larger and the spacing between them greater. Tiny platelets of* M' *precipitate* (M') *also are present. Residual quenching dislocations* (D) *are still visible. The density of the grain boundary precipitate has increased, and there is a very narrow region devoid of zones immediately adjacent to the boundary precipitate. Short transverse specimens resistant to stress-corrosion cracking at 75 percent yield strength* [30].

and grain boundary regions attain similar potentials, thereby providing improved resistance to intergranular attack.

Corrosion tests of 7075-T6 sheet reheated at temperatures of 160 to 216 C (320 to 420 F) have shown [*34*] that for each temperature there is a period of heating that causes high susceptibility to intergranular attack that is eliminated by continued heating. The length of heating required to eliminate the intergranular susceptibility varied with the quenching rate following heat treatment and, to some extent, with the amount of cold work introduced before aging. As such, freedom from intergranular susceptibility in the T7 tempers is a function of the amount

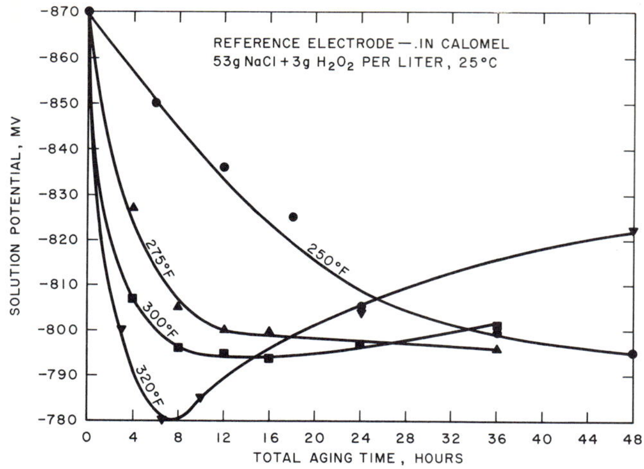

FIG. 14—*Extent of isothermal artificial aging as indicated by potential.*

of second step aging and adverse effects of slow quench on the T6 performance are overcome by this aging.

Significance of Susceptibility to Intergranular Attack

The mere presence of intergranular attack in 7XXX series alloys is not a reliable criterion as to whether exfoliation or SCC will occur in natural environments. Hence, determination of the inherent type of attack in an accelerated test is of limited usefulness and must be interpreted in the light of the metallurgical history of the metal.

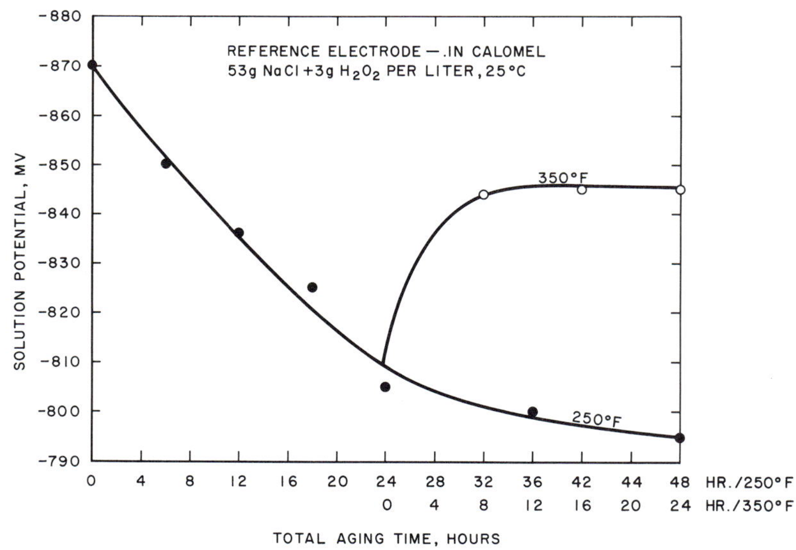

FIG. 15—*Extent of two step artificial aging as indicated by potential.*

T6 Temper—Generally, items that are susceptible to intergranular attack are more prone to exfoliate than corresponding items with susceptibility only to pitting corrosion. Such is not the case, though, for the resistance to SCC. For example, in tests similar to those illustrated in Fig. 11, but evaluating both exfoliation and stress corrosion resistance, susceptibility to exfoliation was encountered over the entire range of cooling rates resulting in intergranular susceptibility, whereas SCC was encountered only at those quench rates (40–50 F/s) for which reduction in tensile strength by corrosion was at a maximum. In another investigation, the resistance to exfoliation and SSC in seacoast atmosphere was evaluated for 13 production lots of 0.063- and 0.080-in. gage 7075-T6 sheet that had been tested previously for suceptibility to intergranular attack (MIL-H-6088E). The atmospheric specimens were located in close proximity to the type of attack specimens. The SCC test specimen was a transverse, plastically deformed sheet specimen of the type described previously in connection with Fig. 4. Although all 13 lots had shown some susceptibility to intergranular corrosion, only 8 lots developed exfoliation during the 4-year seacoast exposure and only 4 lots incurred SCC.

A more serious drawback to the use of the type of attack method to predict service performance is the fact that some materials, which are susceptible only to pitting corrosion when exposed in the absence of stress, can still be susceptible to SCC, particularly when stressed in the short-transverse or transverse direction. Table 1, excerpted from an investigation conducted under government sponsorship [*35*], shows that high resistance to SCC in the transverse direction of 2½-in.-diameter rod was provided only by 7075-T7351 and 2219-T87 even though all three 7XXX alloys were tested, and both of the 2XXX alloys had shown pitting corrosion in the type of attack test.

T76 Temper—The interpretation for the T76 temper is somewhat analogous to the artificially aged Al-Cu alloys. The initial aging at the elevated temperature induces susceptibility to intergranular attack which is eliminated by continued aging. However, commercial practices are

TABLE 1—*Lack of correlation between type of attack and SCC performance 2 1/2-in.-diameter cold finished rolled rod* [35].

Alloy	Type of Attack (MIL-H-6088D) Surface	Type of Attack (MIL-H-6088D) Interior	SCC Performance in Seacoast Atmosphere[a]
2014-T651	P	P	failed at 25% YS[b] (15 ksi)
2219-T87	P + SI	P	ok at 75% YS (39 ksi)
7079-T651	P	P	failed at 25% YS (17 ksi)
7178-T651	P	P	failed at 25% YS (18 ksi)
7075-T7351	P	P	ok at 75% YS (44 ksi)

[a] 1/8-in.-diameter transverse tension specimens exposed seven years at Point Comfort, Tex.

[b] YS = yield strength.

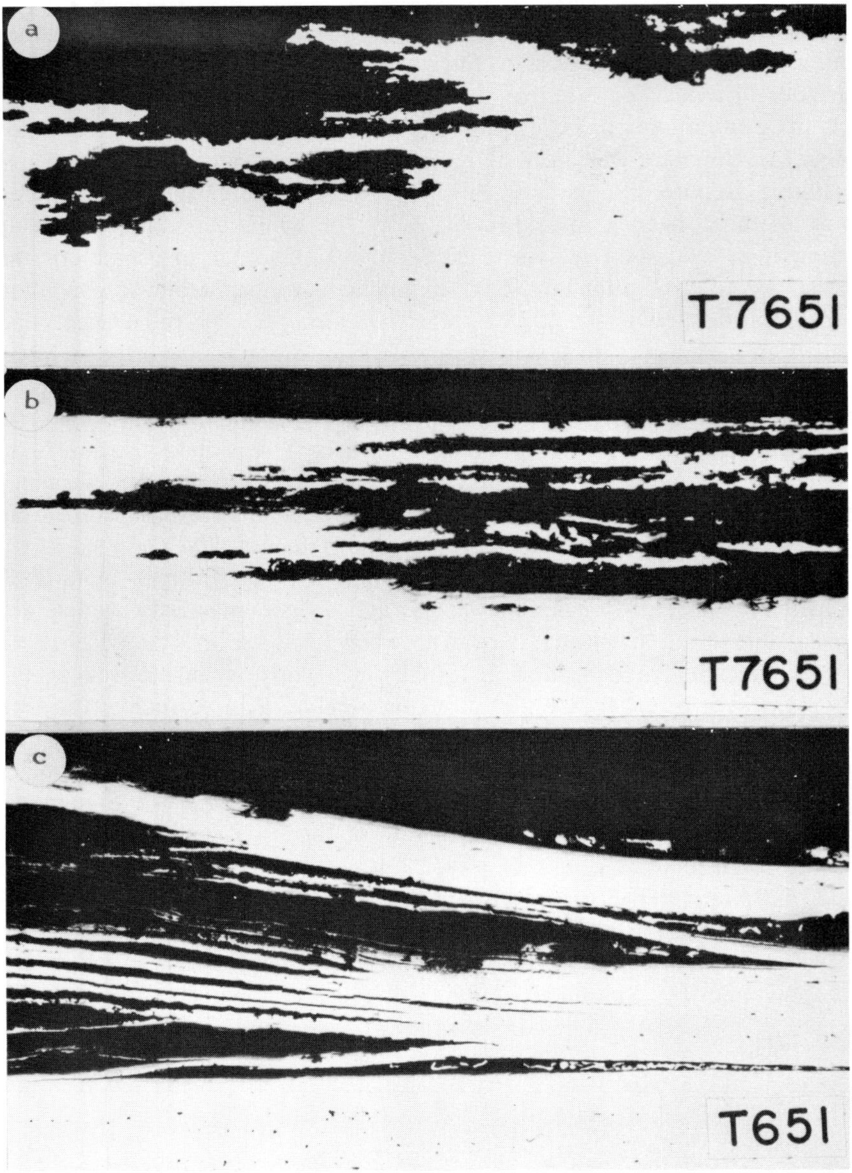

(*a*) T7651 specimen showing primarily pitting corrosion with a slight intergranular tendency.
(*b*) T7651 specimen showing broad, blunt intergranular corrosion but with general grain dissolution as well.
(*c*) T651 specimen showing sharp, penetrating intergranular attack with no appreciable corrosion of the grains.

FIG. 16—(*As Polished,* ×*79*). *Metallographic cross sections from 0.375-in.-thick 7178 plate after 1-week exposure to the accelerated exfoliation test, acidified* (*pH 3*) *5 percent NaCl spray at 49 C* (*120 F*).

aimed at maintaining the maximum strength that is consistent with a high resistance to exfoliation. The amount of aging necessary to produce the required resistance to exfoliation normally is not enough to eliminate completely intergranular susceptibility. Thus, both the T76 and T6 tempers may show intergranular attack (Fig. 16), even though they differ markedly in their resistance to exfoliation (Fig. 17). Intergranular attack in the T76 products may be either very slight and associated primarily with pitting corrosion (Fig. 16*a*), or else it will be relatively broad and blunt, resulting in general solution of the grains as well as attack along the boundaries (Fig. 16*b*). In contrast, intergranular attack in T6 material tends to occur solely at the boundary region and is much finer and penetrating (Fig. 16*c*). In some cases, however, there is little, if any, metallographic difference in the intergranular attack of the two tempers. It does not appear practical, therefore, to set a metallographic standard limiting the amount of intergranular attack that can be tolerated in the T76 temper. As such, determination of the inherent type of attack is of no real significance in predicting serviceability of T76 temper products.

T73 Temper—In the case of the T73 temper, the amount of second step aging should be sufficient to eliminate all vestiges of intergranular susceptibility so that only pitting corrosion occurs. Consequently, determination of the inherent type of attack can be of some assistance in de-

FIG. 17—*T/10 plane panels of 0.500-in.-thick 7075 plate after 1-week exposure to the acidified 5-percent NaCl spray test required by T76 specifications. The T7651 specimen typifies the high resistance to exfoliation afforded by this temper* ($\times 1/2$).

ducing whether or not a full T73 condition was achieved. If appreciable intergranular attack is present, it can be concluded that the material is not in a full T73 condition; but this does not necessarily imply that such material is susceptible to exfoliation or SCC. Also the converse is not true, in that the presence of only pitting corrosion does not guarantee a T73 condition.

Summary

The susceptibility of aerospace heat-treatable aluminum alloys to intergranular corrosion and how this relates to exfoliation and stress corrosion performance in natural environments has been discussed. The intent has been to show that susceptibility to intergranular attack in an accelerated corrosion test does not preclude, *per se,* reliable serviceability. As final support for this statement, Table 2 shows that the relative rating for resistance to exfoliation and SCC of the principal aerospace alloys and tempers does not follow a consistent trend with regard to whether or not intergranular attack can occur.

Historically, conventional type of attack tests of unstressed specimens have been used to judge the adequacy of heat treatment for thin sections of Al-Cu alloys in the naturally aged temper (T3, T4) and of Al-Zn-Mg-Cu alloys in the maximum strength temper (T6). For these situations such tests have value. The extension of this criterion to other tempers, however, is not justified. The limitation of this criterion was noted as long ago as the year 1949 by Golubev [*27*]. In the case of thick sections, maltreatment is not implied necessarily because even the best known heat treat methods cannot produce rapid enough quench-cooling rates to prevent intergranular susceptibility. With regard to the artificially aged Al-Cu alloys (T6, T8) and Al-Zn-Mg-Cu alloys in the T76 temper, the aging practice itself may result in some susceptibility to intergranular attack, but there is no relationship between this attack and resistance to exfoliation or stress corrosion cracking. The adequacy of such tempers has to be judged by actual exfoliation or stress corrosion tests or other criteria which can be calibrated against such tests [*36–39*].

References

[*1*] Dix, E. H., Jr., *Transactions,* American Institute of Mining and Metallurgical Engineers, TAMMA, Vol. 137, 1940, p. 11.

[*2*] Dix, E. H., Jr., and Brown, R. H., *Metals Handbook,* 1948 edition, American Society for Metals, Metals Park, Ohio, 1948, p. 228.

[*3*] Dix, E. H., Jr., *Transactions,* American Society for Metals, TASEA, Vol. 42, 1950, p. 1057.

[*4*] Malloy, A. M., *Naval Aviation News,* May 1965, p. 26.

[*5*] Leak, J. S., *Materials Protection and Performance,* Vol. 10, No. 1, 1971, p. 17.

[*6*] Sprowls, D. O. and Brown, R. H., *Metal Progress,* MEPOA, Vol. 81, No. 4, 1962, p. 79; Vol. 81, No. 5, 1962, p. 77.

[*7*] Lifka, B. W., Sprowls, D. O., and Kaufman, J. G., *Corrosion,* SBIIA, Vol. 23, No. 11, 1967, p. 335.

TABLE 2—*Comparisons illustrating poor relationship between corrosion performance and susceptibility to intergranular corrosion.*

Alloy	SCC Rating[a]	Susceptible to Exfoliation	Probable Susceptibility to Intergranular Corrosion (MIL-H-6088E)
6061-T6	A	no	yes
2024-T3,T4	B[b]	no[b]	no[b]
2219-T3	B[b]	no[b]	no[b]
7075-T6	B[b]	no[b]	no[b]
7075-T73	B	no	no
2011-T8	B	no	yes
2021-T8	B	no	yes
2024-T72	B	no	yes
2024-T8	B	no	yes
2219-T8	B	no	yes
7075-T76	C	no	yes
7175-T736	C	no	yes
7178-T76	C	no	yes
2024-T6	C	yes	yes
2011-T3	D	yes	yes
2024-T3,T4	D	yes	yes
2219-T3	D	yes	yes
7075-T6	D	yes	yes
7175-T66	D	yes	yes
7079-T6	D	yes	yes
7178-T6	D	yes	yes

[a] Ratings are based on service experience (excluding special chemicals or elevated temperatures) and on 4-year exposures to seacoast or industrial atmospheres or 84 days to 3.5 percent NaCl alternate immersion.

A—No known instance of SCC in service or in laboratory tests.

B—No known instance of SCC in service; SCC may occur in extreme laboratory tests of short transverse specimens.

C—SCC not anticipated in service or in laboratory tests at sustained tension below typical design stresses or residual stresses resulting from heat treatment, welding, or controlled assembly stresses.

D—Instances of SCC in service unlikely with metal stressed parallel to direction of grain flow, but occasionally experienced with sustained tension in the transverse or short-transverse directions.

[b] Rating only for thin sections which can be heat treated and quenched to achieve a high cooling rate.

[*8*] DiRusso, E., *Alluminio e Nuova Metallurgia,* ANOMA, Vol. 39, No. 3, 1970, p. 117.

[*9*] Rawdon, H. S., Krynitsky, A. I., and Berliner, J. F. T., *Chemical and Metallurgical Engineering,* CMENA, Vol. 26, No. 4, 1922, p. 154.

[*10*] Rawdon, H. S., *Industrial and Engineering Chemistry,* IECHA, Vol. 19, No. 5, 1927, p. 613.

[*11*] Army-Navy Aeronautical Specification AN-QQ-H-186a, 10 Feb. 1944, "Heat Treatment of Aluminum Alloys, Process For," (Superseded by MIL-H-6088E, 5 Feb. 1971.

[*12*] Dix, E. H., U. S. Patent 1,856,089, 28 June 1928.

[*13*] Dix, E. H., *Aviation,* AVTNA, Vol. 25, 1928, p. 26.

[*14*] Nock, J. A., Jr., Holt, M., and Sprowls, D. O., *Metal Progress,* MEPOA, Vol. 80, No. 3, Sept. 1961, p. 87.

[*15*] Baker W. A., Bryant, A. J., Durham, R. J., and Elkington, R. W., *Journal of the Royal Aeronautical Society,* JRAOA, Vol. 70, 1966, p. 757.

[*16*] Rotsell, W. C. and Long, J. R., *Metal Progress,* MEPOA, Vol. 91, No. 5, May 1967, p. 106.

[17] Brown, R. H., Fink, W. L., and Hunter, M. S., *Transactions,* American Institute of Mining and Metallurgical Engineers, TAMMA, Vol. 143, 1941, p. 115.
[18] Brown, R. H. and Mears, R. B., *Transactions,* Electrochemical Society, TESOA, Vol. 74, 1938, p. 495.
[19] Mears, R. B., Brown, R. H., and Dix, E. H. in *Stress-Corrosion Cracking of Metals,* published jointly by American Society for Testing and Materials and American Institute of Mining and Metallurgical Engineers, 1944, p. 329.
[20] Hunter, M. S., Frank, G. R., and Robinson, D. L. in *Proceedings,* Second International Congress on Metallic Corrosion, published by National Association of Corrosion Engineers, 1963, p. 102.
[21] Dix, E. H., Jr., in *Physical Metallurgy of Aluminum Alloys,* American Society for Metals, Cleveland, 1949, p. 214.
[22] Ketcham, S. J. and Haynie, F. H., *Corrosion,* SBIIA, Vol. 19, No. 7, 1963, p. 242t.
[23] Ketcham, S. J., *Corrosion Science,* CRRSA, Vol. 7, No. 6, 1967, p. 305.
[24] *Stress Corrosion Testing, ASTM STP 425,* American Society for Testing and Materials, 1967, p. 7.
[25] Sager, G. F., Brown, R. H., and Mears, R. B. in *Stress-Corrosion Cracking of Metals,* published jointly by American Society for Testing and Materials and American Institute of Mining and Metallurgical Engineers, 1944, p. 255.
[26] Sprowls, D. O. and Brown, R. H. in *Proceedings,* International Conference on Fundamental Aspects of Stress-Corrosion Cracking, published by National Association of Corrosion Engineers, 1969, p. 466.
[27] Golubev, A. I., *Journal of Physical Chemistry* (*Russian*), JPCUA, Vol. 23, No. 9, 1949, p. 1116.
[28] Dix, E. H., Brown, R. H., and Binger, W. W., *Metals Handbook,* 8th edition, Vol. I, American Society for Metals, Metals Park, Ohio, 1961, p. 918.
[29] Fink, W. L. and Willey, L. A., *Metals Technology,* METVA, Vol. 14, No. 8, 1947.
[30] Hunter, M. S. and Robinson, D. L. et al in *Proceedings,* International Conference on Fundamental Aspects of Stress-Corrosion Cracking, Ohio State University, published by National Association of Corrosion Engineers, 1968, p. 495.
[31] Lifka, B. W. and Sprowls, D. O. in *Aluminum,* Vol. I., K. R. Van Horn, ed., American Society for Metals, Metals Park, Ohio, 1967, p. 140.
[32] Sprowls, D. O. and Nock, J. A., Jr., U. S. Patent 3,198,676, 3 Aug. 1965 (Alcoa).
[33] Thompson, D. S., Singleton, O. R., McGowan, R. D., and Spangler, G. E., *Metal Progress,* MEPOA, Vol. 98, No. 3, 1970, p. 82.
[34] Adenis, D., Develay, R., and Guilhaudis, A., *Memoires Scientifique de la Revue de Metallurgie,* MRMTA, Vol. 64, No. 12, 1967, p. 1095.
[35] Lifka, B. W. and Vandenburgh, D. G., Final Report–Government Contract NAS 8–5340, Control No. 1–4–50–01167–01(1f), CPB-02–1215–64, 1966, Phase I, pp. 3–20.
[36] King, W., Lifka, B. W., and Willey, L. A., *Materials Evaluation,* MAEVA, Vol. 23, No. 2, 1965, p. 89.
[37] Horst, R. L., Hollingsworth, E. H., and King, W., *Corrosion,* SBIIA, Vol. 25, No. 5, 1969, p. 199.
[38] Horst, R. L., Jr., and Lifka, B. W., *Corrosion,* SBIIA, Vol. 26, No. 3, 1970, p. 111.
[39] Hagemaier, Donald, *Nondestructive Testing,* NODTA, Vol. 21, No. 5, 1963, p. 317.

Robert Baboian[1]

Investigations of Galvanically Induced Localized Corrosion

REFERENCE: Baboian, Robert, **"Investigations of Galvanically Induced Localized Corrosion,"** *Localized Corrosion—Cause of Metal Failure, ASTM STP 516,* American Society for Testing and Materials, 1972, pp. 145–163.

ABSTRACT: Modern electrochemical theories and techniques are used to explain galvanically induced localized corrosion of metals. Examples used are low-carbon steel, aluminum and aluminum alloys, and stainless steels. The relevance of the galvanic series of metals to predict this behavior and the value of critical pitting potentials and critical protection potentials are discussed.

KEY WORDS: corrosion, galvanic corrosion, pitting, concentration cell corrosion, corrosion prevention, polarization (charge separation), anodic polarization, potential theory, corrosion environments, oxidation reduction reactions, electrolytes

Galvanic effects occur when two or more metals are in electrical and electrolyte contact. To predict their behavior in galvanic couples, a galvanic series of metals can be constructed according to their potentials measured in a specific environment. Usually the corrosion rate of the more noble metal is decreased, and that of the more active metal is increased in a couple.

In addition to potential differences, polarization behavior of metals is also important in predicting galvanic effects. For example, metals which corrode uniformly when electrically isolated may undergo localized corrosion when galvanically coupled. Aluminum alloys and stainless steels which normally have a protective film over a wide range of potentials are susceptible to pitting and crevice corrosion in sodium chloride (NaCl) solutions. Galvanic couples of these alloys with other metals can either inhibit or induce localized corrosion.

This paper describes methods of predicting the occurrence of galvanically induced localized corrosion. Mixed potential theory and polarization techniques are used on several stainless steels and aluminum alloys. Couples of these alloys with other metals are evaluated, and supporting data obtained in long-term experiments are presented.

[1] Manager, Corrosion Laboratory, Metallurgical Materials Division, Texas Instruments Inc., Attleboro, Mass. 02703.

Theory

Electrode Kinetics

Materials which contain several different metals and alloys are susceptible to galvanic corrosion. The difference in electrochemical potential between two or more dissimilar metals in electrical and electrolyte contact causes electron flow between them. Attack of the more noble metal or metals is usually decreased, and corrosion of the more active metal is usually increased. A galvanic series of metals can be constructed according to their potential measured in a particular environment [*1*].[2] However, the information concerning a galvanic couple obtained from such a series is of limited value because the series is affected strongly by environmental factors such as type of electrolyte, temperature, agitation, etc. In addition, under the influence of galvanic coupling, appreciable polarization of the metals may occur, producing a protective film on the metal surface or causing breakdown of an already existing protective film. This effect is commonly observed with stainless steel. Thus, an overall characterization of each metal in the galvanic couple is necessary to evaluate the behavior of the metals in a particular corrosive environment.

Galvanic corrosion of metals can be treated by application of the mixed potential theory first described by Wagner and Traud [*2*]. The theory is based on two simple hypotheses: (1) any electrochemical reaction can be divided into two or more oxidation or reduction reactions, and (2) there can be no net accumulation of electrical charge during an electrochemical reaction.

Under the simplest circumstance, metallic corrosion would involve only two reactions—oxidation and reduction. The corrosion of iron in sulfuric acid (H_2SO_4) involves the anodic dissolution of iron and the evolution of hydrogen. This is demonstrated by the polarization curves for iron in 0.52 N H_2SO_4 in Fig. 1 [*3*]. The first hypothesis of the mixed potential theory is satisfied if one considers that each reaction has its own reversible potential and polarization parameters. The second hypothesis, that the total rate of oxidation equals the total rate of reduction, is only satisfied at the intersection E_{corr}, the corrosion or "mixed" potential. At this point the rate of iron dissolution is equal to the rate of hydrogen evolution. The potential is so displaced from the equilibrium potential that the reverse reactions occur at a negligible rate and do not influence the corrosion rate.

In Fig. 1, the data indicate that iron will corrode at a rate of about 0.5 mA/cm^2 and will exhibit a potential of about −0.52 V versus the saturated calomel electrode (SCE).

[2] The italic numbers in brackets refer to the list of references appended to this paper.

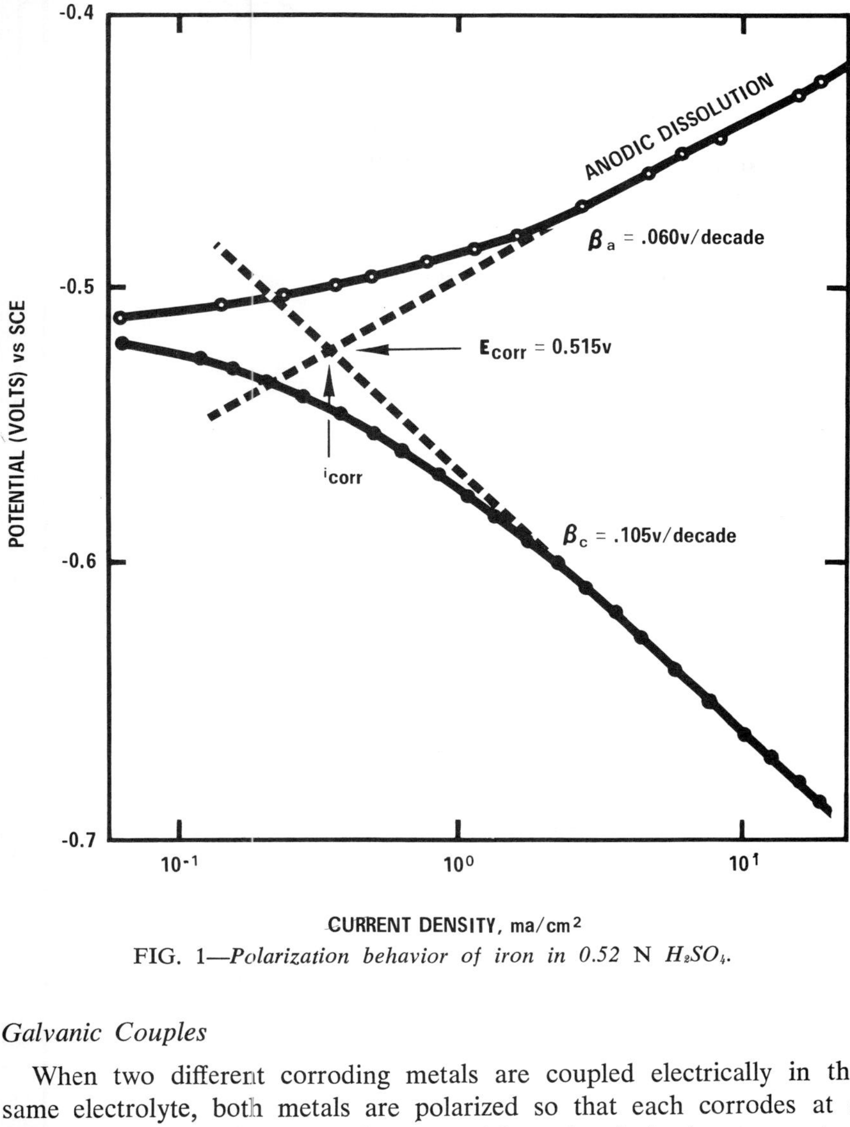

FIG. 1—*Polarization behavior of iron in 0.52* N H_2SO_4.

Galvanic Couples

When two different corroding metals are coupled electrically in the same electrolyte, both metals are polarized so that each corrodes at a new rate. In Fig. 2 the corrosion potentials and polarization parameters for uncoupled Metals A and B are shown. Metal A is more noble than Metal B in that the equilibrium potential is less negative. When the mixed potential theory is applied to the individual reactions (A/A^+,H_2/H^+, B/B^+,H_2/H^+) the uncoupled corrosion rates are $i_{corr,A}$ for Metal A and $i_{corr,B}$ for Metal B. When equal areas of Metals A and B are coupled, the resultant mixed potential of the system $E_{corr,AB}$ is at the intersection where the total oxidation rate equals the total reduction rate. The rate of oxidation of the individual coupled metals is such that Metal A corrodes

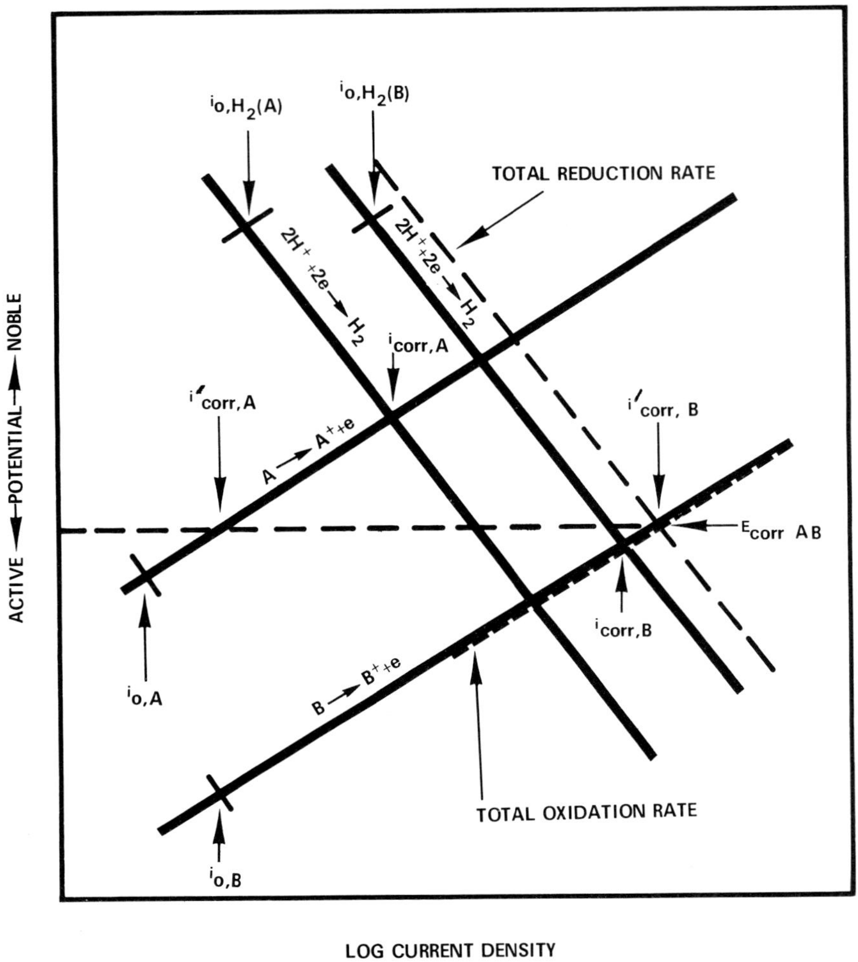

FIG. 2—*Mixed potential behavior of galvanically coupled metals.*

at a reduced rate $i'_{corr,A}$ and Metal B corrodes at an increased rate $i'_{corr,B}$.

The information required to predict the corrosion behavior of galvanically coupled Metals A and B is shown in Fig. 2. In addition to the anodic polarization curves for Metals A and B, it is necessary to measure the mixed potential of the galvanic couple ($E_{corr,AB}$) under actual environmental conditions because the nature of the cathodic reactions can have a marked influence on the mixed potential.

Experimental

Polarization Measurements

Polarization measurements were made using the cell similar to that described previously [*4*]. Purified argon was passed through a fritted glass bubbler to continuously deaerate the solution. The potentiodynamic method (continuous potential scan and current recording) described previously was used in this work [*4–6*]. The current from the potentiostat to the auxiliary electrode is measured as a potential across a precision resistor which is selected to provide the required logarithmic converter input voltage. The filtered output from the logarithmic converter is plotted on the X-axis, and the working electrode potential versus the saturated calomel electrode is plotted simultaneously on the Y-axis. In this investigation potentials were scanned at a rate of 0.6 V/h.

All measurements were made in 5 percent NaCl solution. Solutions were prepared by dissolving reagent grade NaCl in distilled water and neutralizing with hydrochloric acid or sodium hydroxide. Measurements were made in deaerated solution at a pH of 7 and 30 C. Disk-shaped specimens for polarization measurements were punched from commercial grade sheet material in the mill annealed condition. The specimen surface was polished with 600-grit silicon carbide paper, followed by ultrasonic cleaning in detergent solution, and finally rinsed with distilled water. The specimen was mounted in the type of holder described previously by Myers et al [*7*]. In use, the specimen holder exposed only Teflon, and 1 cm^2 of specimen to the electrolyte. This design avoids solution contamination and crevice effects, while allowing the use of flat metal specimens.

Mixed Potential Measurements

Mixed potentials for electrically isolated metals and galvanically coupled metals were measured using a six-channel Orion digital printer system. Potentials measured versus the SCE were printed hourly over the period of time required.

The 5 percent NaCl solution was prepared as described previously for the polarization work. Measurements were made in air saturated solutions using a constant temperature bath to maintain the solution temperature at 30 C $\pm$ 0.1 C.

Rectangular specimens of commercial grade metals in the mill-annealed condition were ground flat and polished through 600-grit silicon carbide paper. Specimens for galvanic coupling were:

1. Partially immersed in a parallel configuration with a 0.5-in. (1.27-cm) spacer and connected external to the solution interface so that the desired area ratio was obtained.
2. Pressure roll bonded to form bimetal or trimetal strip.
3. Drilled and connected with nylon nuts and bolts.

Methods 1 and 2 were used for mixed potential measurements only. Specimens were masked peripherally at the air-solution interface. Method 3 was used to observe the type of corrosion and weight loss of specimens. Measurements were made after ultrasonic cleaning in detergent and distilled water.

Results and Discussion

Frequently, metals which normally corrode uniformly will undergo severe localized corrosion when galvanically coupled to another metal. Examples include galvanic pitting or edge corrosion in metallic coatings. Although this effect normally can be predicted by referring to the galvanic series of metals, more specific information on this behavior can be obtained by referring to the polarization behavior of the metals involved.

Anodic polarization curves for DHP copper and 1006 low-carbon steel (LCS) in 5 percent NaCl solution are shown in Fig. 3. The corrosion potentials of the isolated metals, obtained under actual corroding conditions in aerated 5 percent NaCl solution, are used to determine the rates of corrosion $i_{corr,Cu}$ and $i_{corr,Fe}$ from the polarization curves in Fig. 3.

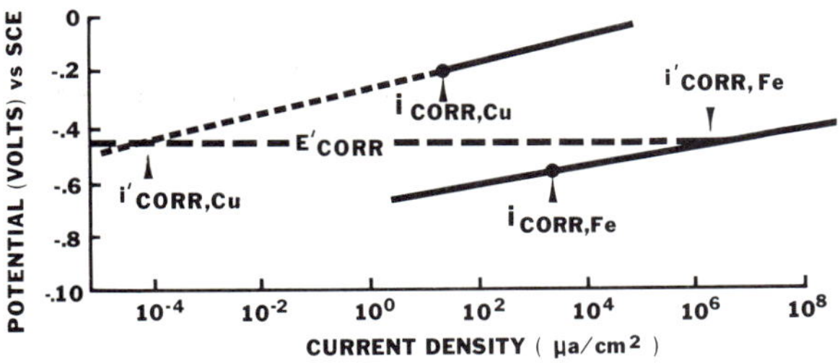

FIG. 3—*Potentiodynamic polarization curves for DHP copper and 1006 LCS in argon saturated 5 percent NaCl solution at 30 C (0.6 V/h scan rate).*

When copper and LCS are coupled galvanically under these conditions, the mixed potential is E'_{corr}, and, from the polarization curves in Fig. 3, the rate of corrosion of copper, $i'_{corr,Cu}$, is decreased and the rate of corrosion of iron, $i'_{corr,Fe}$, is increased.

Increasing the copper/LCS area ratio in the galvanic couple shifts the mixed potential (E'_{corr}) to less negative values so that galvanic corrosion of LCS is accelerated.

The corrosion behavior of copper coated 1006 LCS can be used to illustrate galvanically induced localized corrosion. Normally, LCS corrodes uniformly. However, at pores or damage sites in copper coatings on LCS and at coating edges, extreme localized corrosion of the LCS occurs. This behavior is illustrated by the results of total immersion experiments in 5 percent NaCl.

The photomicrograph of a cross section in Fig. 4*a* shows galvanic pitting of LCS at a damage site in the copper coating. Galvanically induced tunneling occurred at the coating edge, as shown in Fig. 4*b*.

Data obtained from copper coated 1006 LCS total immersion experiments are listed in Table 1. Notice the shift in mixed potential for isolated 1006 LCS and copper-coated 1006 LCS (Cu/LCS area ratio, 100/1) is 0.14 V. From the Tafel slope (0.06 V) for 1006 LCS in Fig. 3, the

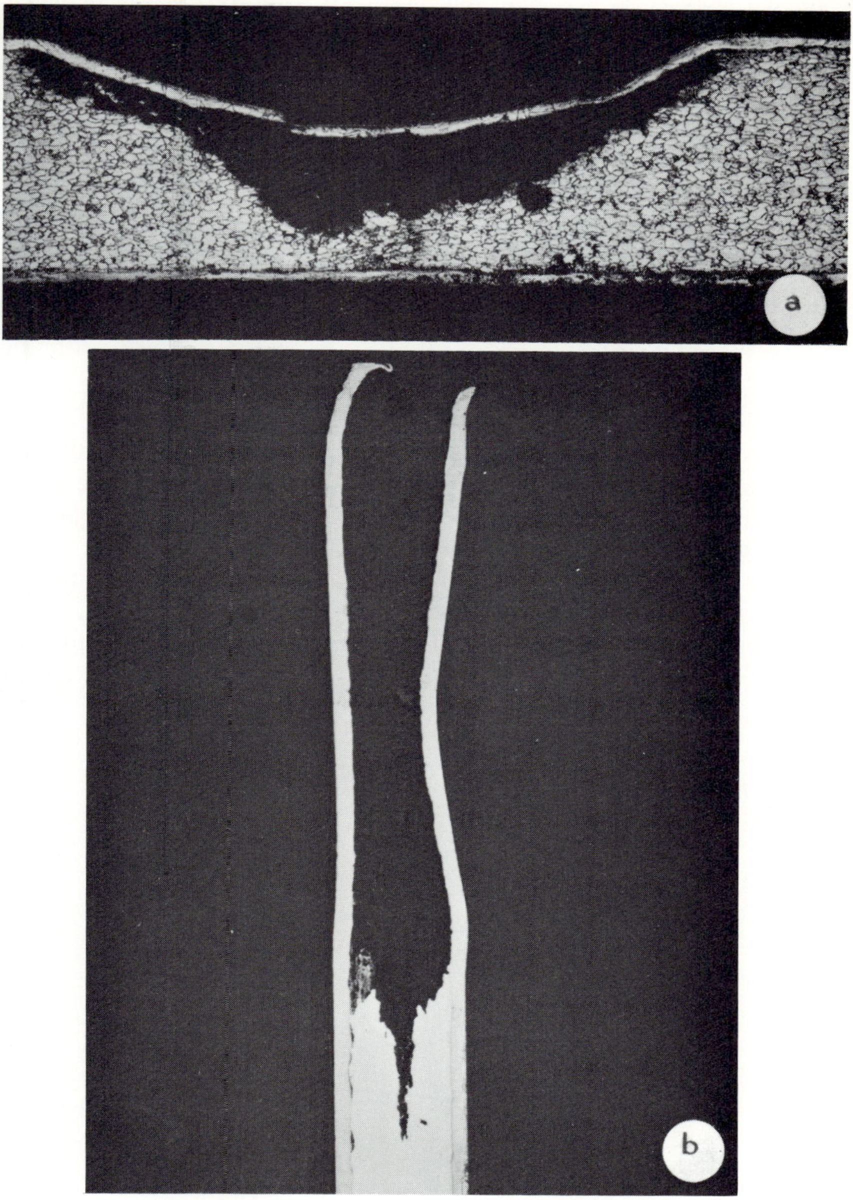

(*a*) Galvanic pitting at damage site in copper coating.
(*b*) Galvanic tunneling at exposed 1006 LCS edge.

FIG. 4—*Photomicrographs of cross sections* (×50) *of Cu/1006 LCS/Cu strip showing corrosion behavior after total immersion in aerated 5 percent NaCl solution at 30 C for 62 days.*

predicted increase in corrosion rate of LCS is over two orders of magnitude. This is in agreement with the measured increase listed in Table 1.

In contrast to LCS, aluminum and aluminum alloys are susceptible to localized corrosion in NaCl solutions [*8*]. The problem is most severe under galvanic conditions due to breakdown of the protective film on the aluminum surface. This effect is explained by referring to the polarization behavior of aluminum.

TABLE 1—*Galvanic corrosion (5 percent NaCl) of 1006 LCS.*

Metal System	Area Ratio	Mixed Potential, V versus SCE	Corrosion Rate, mpy[a]
1006 LCS	. . .	−0.61	5
Cu/1006 LCS	100/1	−0.47	570

[a] mpy = mils per year

A typical polarization curve for aluminum and aluminum alloys is shown in Fig. 5. The solid line represents anodic potentiodynamic polarization from active to more noble potentials. The dashed line represents reversal of polarization direction back to more active potentials.

Below E_c, the critical breakdown potential, the corrosion rate remains low and unchanged over a wide range of potentials. Above E_c, pitting and crevice corrosion occurs due to the breakdown of the protective film on the metal surface and is represented by the rapid increase in current

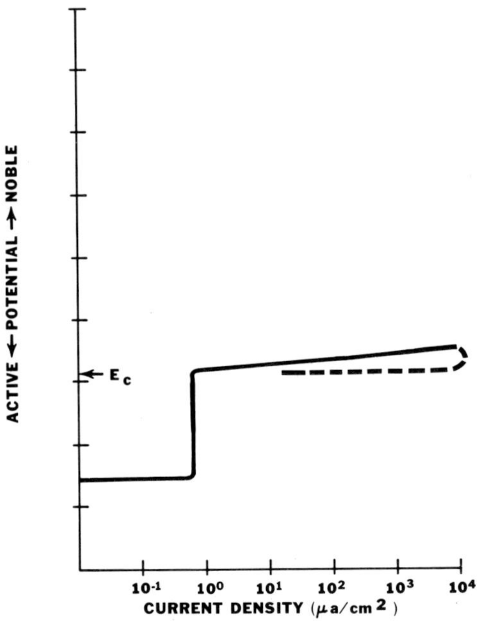

FIG. 5—*Schematic potentiodynamic polarization curve for aluminum and aluminum alloys in NaCl solution at room temperature.*

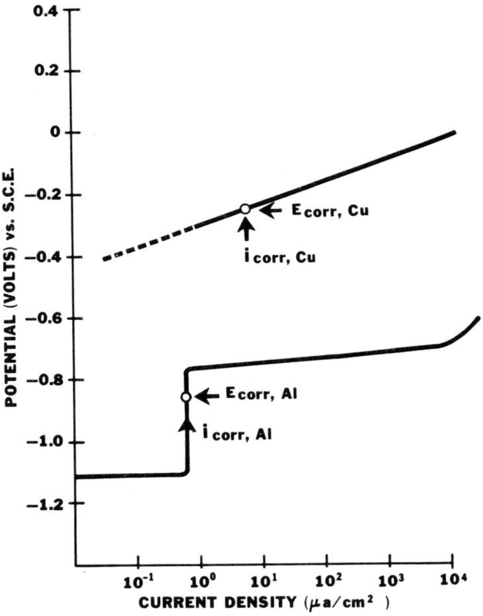

FIG. 6—*Potentiodynamic polarization curves for DHP copper and 3003 aluminum alloy in argon saturated 5 percent NaCl solution at 30 C (0.6 V/h scan rate).*

density in Fig. 5. Upon reversing the polarization direction, the current density decreases rapidly at the critical breakdown potential.

The polarization behavior of aluminum and aluminum alloy indicates that in 5 percent NaCl solution, initiation and propagation of pits and crevice corrosion occur at potentials above E_c, and protection from pit propagation and crevice corrosion occurs at potentials below E_c. The galvanic behavior of aluminum and aluminum alloys can be explained best by referring to this polarization information.

In Fig. 6, potentiodynamic polarization curves obtained in 5 percent NaCl solution for 3003 aluminum alloy and DHP copper are superimposed. Mixed potentials for galvanic couples of these metals in various area ratios are listed in Table 2.

TABLE 2—*Mixed potentials for copper/3003 aluminum galvanic couples in 5 percent NaCl.*

Cu/Al Area Ratio	Mixed Potential, V versus SCE
1/1	−0.755
150/1	−0.735
300/1	−0.704
1000/1	−0.686
2000/1	−0.620
4000/1	−0.558
8000/1	−0.548
12000/1	−0.484

Notice that polarization of 3003 aluminum alloys above E_c (—0.76 V versus the SCE) occurs in the galvanic couples. Extensive localized corrosion of the aluminum alloy was observed in the series of measurements.

Total immersion experiments in 5 percent NaCl solution were conducted with copper coated 3003 aluminum strip. The high copper/aluminum alloy area ratio at the edge produced galvanic tunneling as observed in the photomicrograph of a cross section in Fig. 7.

When coupled to stainless steels, galvanically induced localized corrosion of aluminum and aluminum alloys occurs. Specimens of Type 434 stainless steel and 3003 aluminum alloy (area ratio, 2/1) were coupled and immersed in 5 percent NaCl solution. Photomicrographs of the aluminum surfaces are shown in Figs. 8*a* and *b*. Although pitting occurs on the surface of electrically isolated 3003 aluminum, severe pitting is observed in the metal which was coupled galvanically.

Similar experiments were conducted with 3003 aluminum alloy coupled

FIG. 7—*Photomicrograph of a cross section* (×50) *of Cu/3003 Al/Cu strip showing galvanic tunneling after total immersion in aerated 5 percent NaCl solution at 30 C for 60 days.*

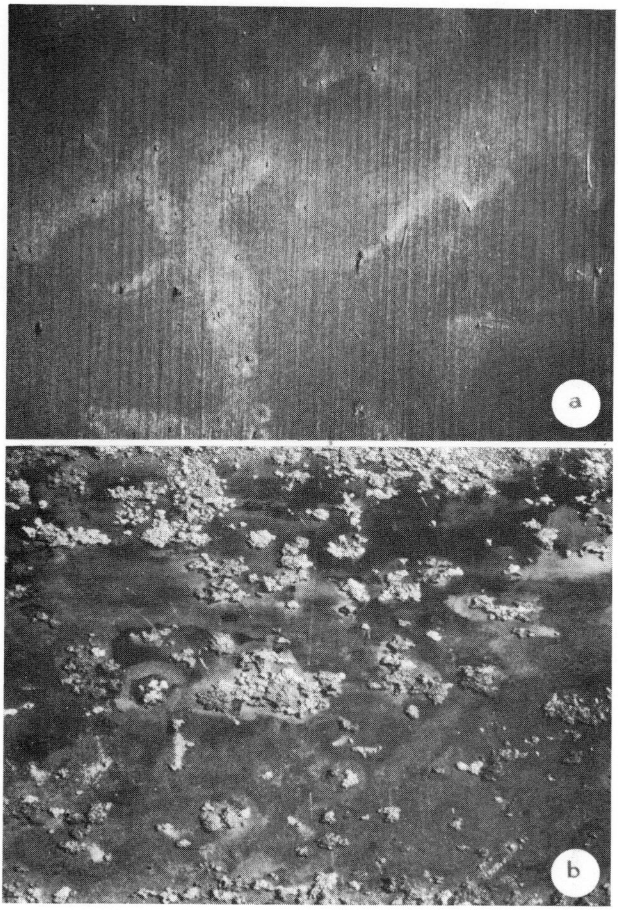

(*a*) Electrically isolated.
(*b*) Coupled to Type 434 stainless steel.

FIG. 8—*Photomicrographs* ($\times 10$) *of 3003 aluminum alloy surfaces showing galvanically accelerated pitting after total immersion for 40 days in aerated 5 percent NaCl solution at 30 C.*

to zinc. The mixed potential for this couple (−1.07 V versus the SCE) is below E_c. Localized corrosion was not observed on the surface of the aluminum alloy provided that alkaline buildup did not occur.

Galvanic behavior of stainless steels is difficult to predict because of the passive behavior exhibited by these alloys over a wide range of potentials. In addition, pitting and crevice corrosion due to the localized breakdown of the passive film occurs in aqueous chloride solutions.

Polarization techniques and critical breakdown potentials, E_c, have been widely used as a measure of the pitting susceptibility of alloys in chloride solutions [*9–12*]. However, E_c values are not useful for predicting the galvanic behavior of stainless steels. Measurement of critical protection potentials, E_p, is more valuable for this purpose [*12*].

Polarization behavior of stainless steel in 5 percent NaCl solution is shown in Fig. 9. The solid line represents anodic potentiodynamic polarization from active to noble potentials. The dashed line represents reverse polarization back to more active potentials.

Polarization above the critical breakdown potential, E_c, results in a marked increase in current density due to initiation of pitting. Reverse polarization below E_c does not reduce the current density appreciably,

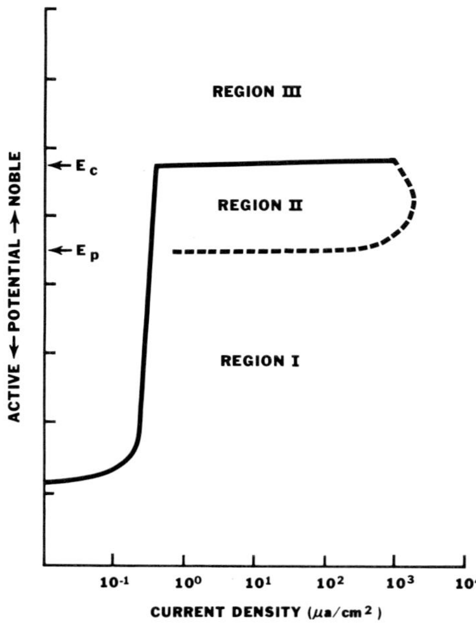

FIG. 9—*Schematic potentiodynamic polarization curve for stainless steel in NaCl solution at room temperature.*

indicating that propagation of existing pits and crevices corrosion occurs. At the critical protection potential, E_p, the current density approaches zero. Below E_p propagation of existing pits and crevice corrosion does not occur.

The corrosion behavior of stainless steels can be thus separated into three potential regions: Region I where protection from pit propagation and crevice corrosion is observed, Region II where propagation of existing pits and crevice corrosion occurs, and Region III where initiation of pitting and crevice corrosion occurs. This behavior is observed in oxygen containing and oxygen free electrolytes, as shown in the potentiodynamic polarization curves for Type 434 stainless steel in air saturated and argon saturated 5 percent NaCl solutions in Fig. 10. The curves are similar except that in the absence of oxygen, the full anodic polarization curve is observed. Controlled potential data, superimposed in Fig. 10, are in good agreement with the potentiodynamic polarization curves. At potentials above the critical protection potential, E_p, extremely high weight losses were observed due to crevice corrosion and pitting. Thus, by this treatment, the galvanic behavior of stainless steel can be predicted in oxygen free and oxygen containing electrolytes.

The relative resistance to pitting of stainless steels is measured frequently by the magnitude of the critical breakdown potential (E_c). In

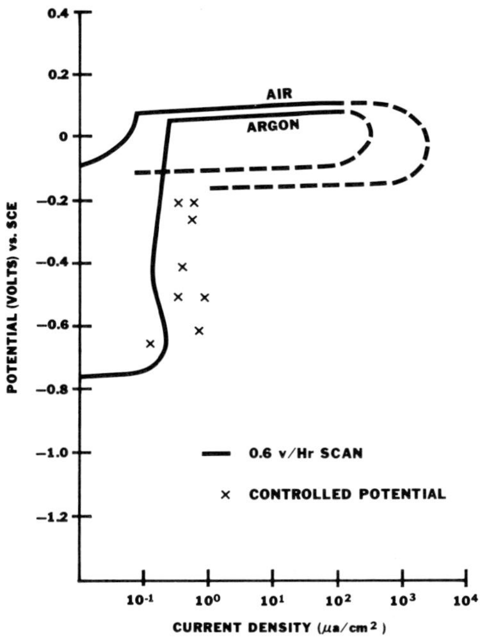

FIG. 10—*Polarization behavior of Type 434 stainless steel in 5 percent NaCl solution at 30 C (potentiodynamic polarization at 0.6 V/h scan rate; reverse polarization; X-controlled potential weight loss data).*

Fig. 11, polarization curves for Types 409, 430, and 434 stainless steels are shown. Both the critical breakdown potential (E_c) and the critical protection potential (E_p) become more noble in the order, Type 409, Type 430, and Type 434. The ranking of the pitting resistance of these metals is in the same order.

However, these metals behave similarly when galvanically coupled to copper because their mixed potentials lie in Region II (Fig. 9) where pit propagation and crevice corrosion occurs.

For example, galvanic pitting occurs at breaks in copper coatings on Type 409 stainless steel (Fig. 12*a*), and galvanic tunneling occurs at the

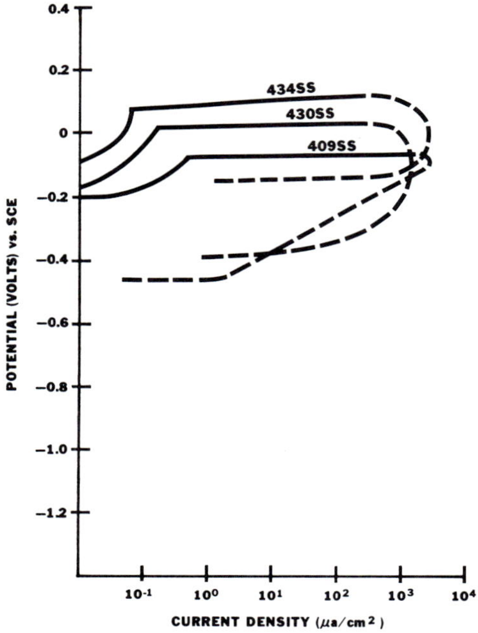

FIG. 11—*Potentiodynamic polarization curves for Types 409, 430, and 434 stainless steels in air saturated 5 percent NaCl solution at 30 C (0.6 V/h scan rate).*

exposed edges (Fig. 12*b*). Mixed potentials versus time are shown in Fig. 13 for DHP copper, Type 409 stainless steel, and the DHP copper/409 stainless steel couple. Notice that the potential of Type 409 stainless steel fluctuates over a wide range of potentials. The mixed potential for the DHP copper/409 stainless steel couple is more stable and after about 40 h remains more noble than the potential of monolithic Type 409 stainless steel. This potential is within Region II, the pit propagation and crevice corrosion region.

Galvanic couples of stainless steels with metals which exhibit mixed potentials within Region I (below E_p) yield protection from pit propagation and crevice corrosion. Mixed potentials for couples of Type 409 stainless steel with lead, 1006 low-carbon steel, and 3003 aluminum alloy in 5 percent NaCl solution are listed in Table 3. The potentials are below E_p and within Region I. Complete protection of Type 409 stainless steel

TABLE 3—*Mixed potentials of galvanic couples in 5 percent NaCl.*

Galvanic Couple	Area Ratio	Mixed Potential, V versus SCE
Cu/409 SS	100/1	−0.3
Pb/409 SS	100/1	−0.54
1006 LCS/409 SS	5/1	−0.6
3003 Al/409 SS	1/1	−0.76

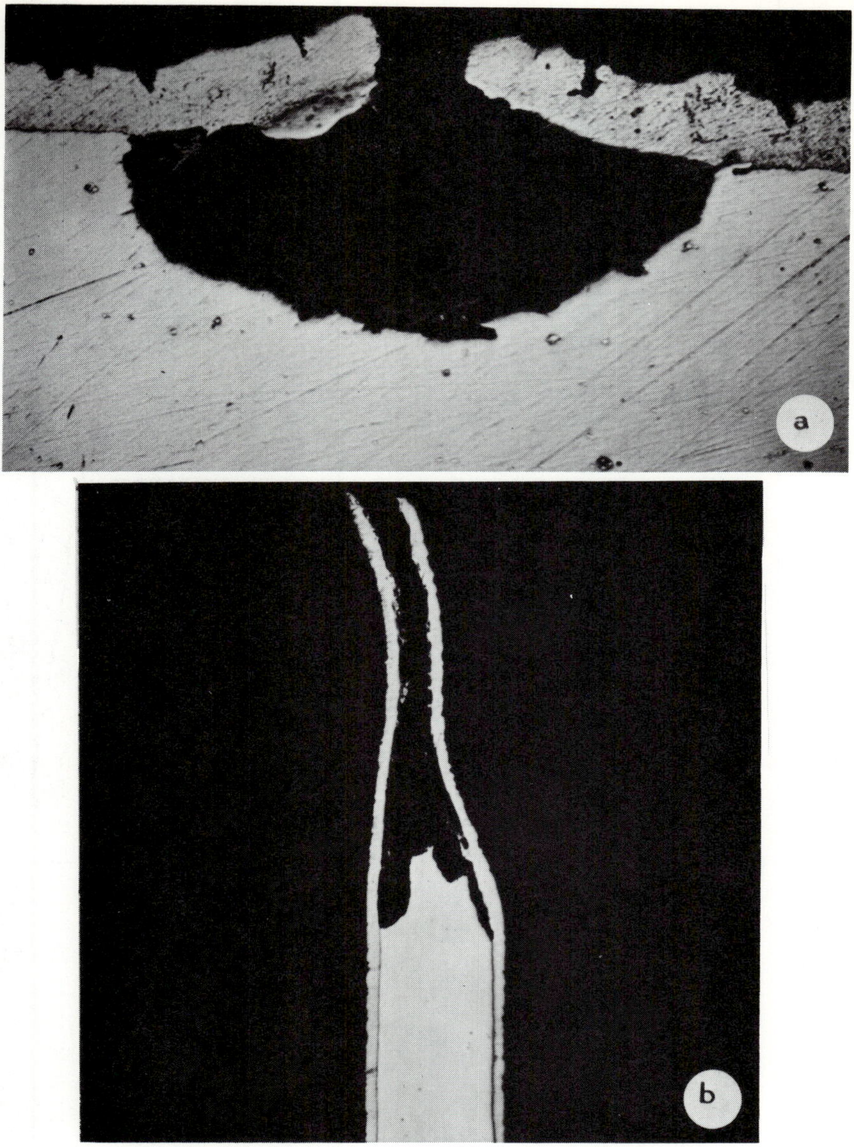

(*a*) Galvanic pitting at damage site in copper coating (×250).
(*b*) Galvanic tunneling at exposed 409 SS edge (×50).
FIG. 12—*Photomicrographs of cross sections of Cu/409 SS/Cu strip showing corrosion behavior after total immersion in aerated 5 percent NaCl solution at 30 C for 190 days.*

from pitting and crevice corrosion was observed in these couples. Similar results were obtained with Types 430 and 434 stainless steels.

The photomicrograph of a cross section in Fig. 14 of lead coated

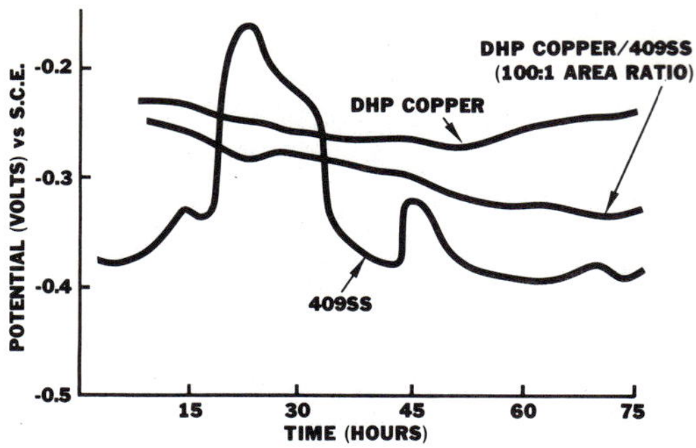

FIG. 13—*Mixed potential behavior of DHP copper and Type 409 stainless steel in aerated 5 percent NaCl solution at 30 C.*

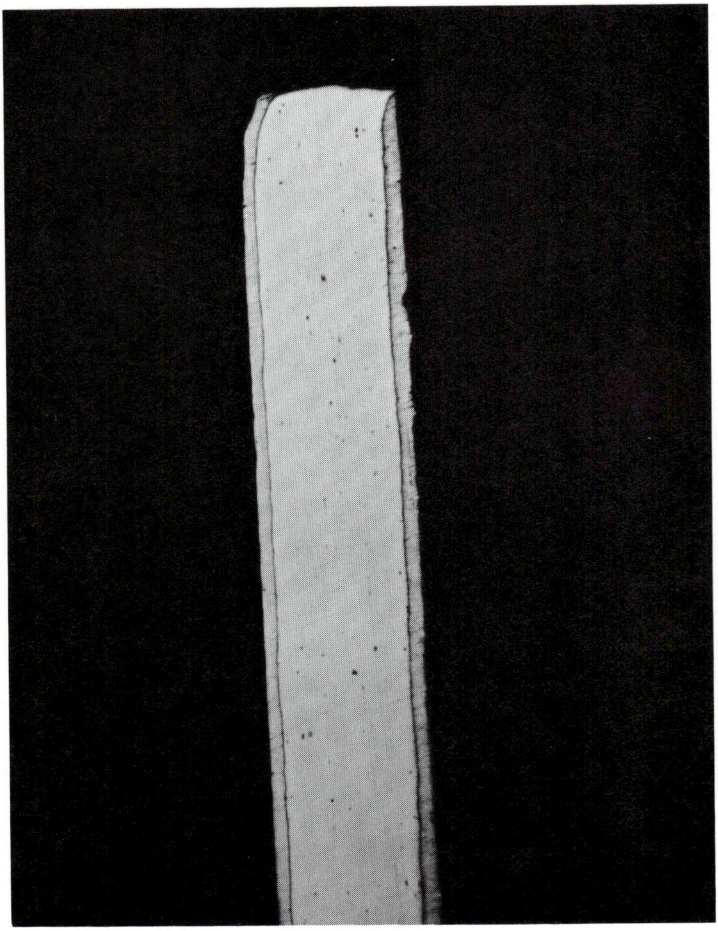

FIG. 14—*Photomicrograph of a cross section* (×50) *of Pb/Cu/409 SS/Cu/Pb strip showing galvanic protection of Type 409 stainless steel at exposed edge after total immersion in aerated 5 percent NaCl solution at 30 C for 190 days.*

Type 409 stainless steel shows that protection from tunneling occurs. This behavior should be compared to that of copper coated Type 409 stainless steel in Fig. 12*b*.

The photomicrographs of surfaces in Fig. 15*a* and *b* show pitting of Type 409 stainless steel occurs in 5 percent NaCl solution, while galvanic protection occurs in Type 409 stainless steel/1006 low-carbon steel couples.

Conclusions

Although the galvanic series of metals can be useful in predicting galvanic behavior of metals in sodium chloride solutions, use of the mixed potential theory and polarization of metals can lead to more precise information.

Galvanically induced localized corrosion occurs in chloride solutions under extreme area ratio conditions or due to breakdown of protective films.

Galvanic tunneling and pitting in metal coatings such as copper on 1006 LCS occur due to the high copper/1006 low-carbon steel area ratio.

When aluminum and aluminum alloys are coupled to metals such as copper, stainless steel, and 1006 low-carbon steel, galvanic polarization above E_c, the critical breakdown potential, occurs. Severe localized corrosion such as pitting and crevice corrosion of aluminum and aluminum alloys occurs under these conditions in NaCl solutions. Galvanic polarization below E_c yields protection from localized corrosion as long as alkaline buildup does not occur.

Galvanic behavior of stainless steels is explained best by referring to the polarization behavior of these metals. Above E_c, the critical breakdown potential, pit initiation, and crevice corrosion occur. This behavior occurs when stainless steels are galvanically polarized above E_c. Between E_p, the critical protection potential, and E_c, propagation of existing pits and crevices occurs. Galvanic couples of copper and stainless steel have mixed potentials in this region. Below E_p, protection from pitting and crevice corrosion is observed. Couples of stainless steel with lead, 1006 low-carbon steel, aluminum, and aluminum alloys have mixed potentials below E_p with the result that the stainless steel is protected from pitting and crevice corrosion.

Stainless steels are listed in the galvanic series of metals for a particular environment according to the steady-state potentials. The series is valuable in predicting the effect of stainless steels on other metals in galvanic couples; however, the opposite is not true. Stainless steels should be listed according to E_p, the critical protection potential, in order to predict accurately galvanic effects of other metals on them. This would also avoid the confusion resulting from the placement of stainless steels in two positions (active and passive) in the galvanic series.

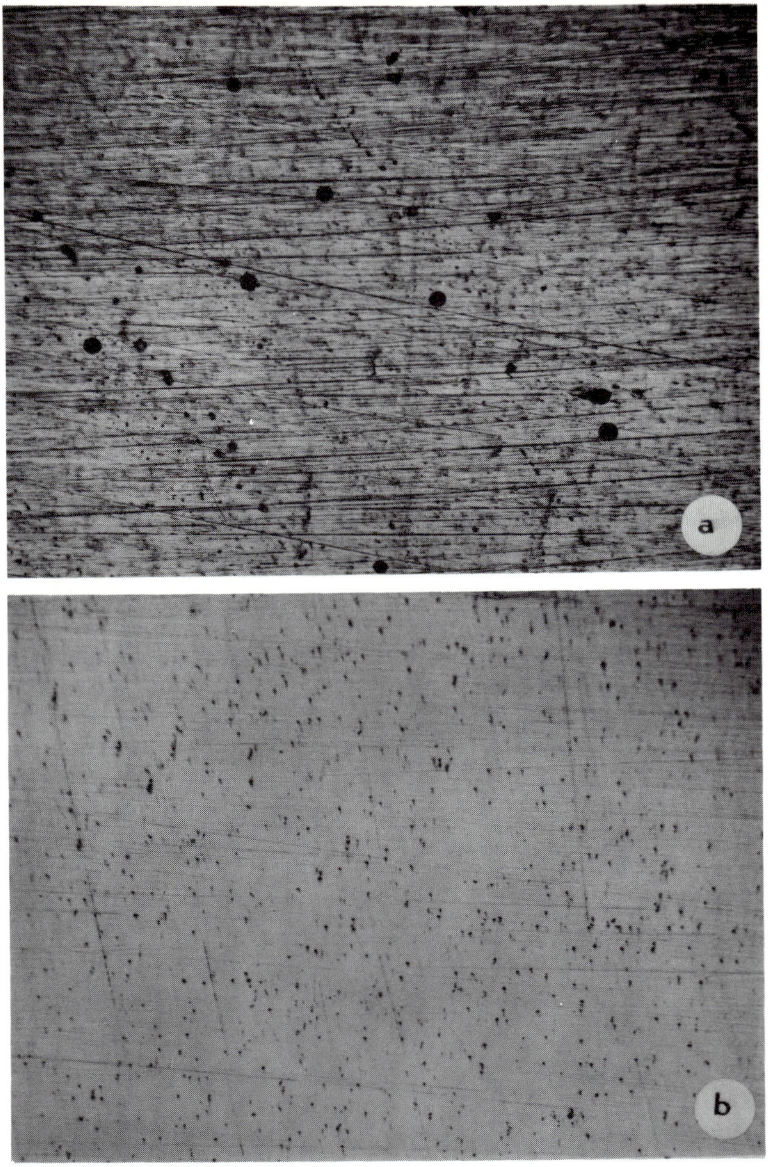

(*a*) Electrically isolated.
(*b*) Coupled to 1006 LCS.

FIG. 15—*Photomicrographs* (×*50*) *of Type 409 stainless steel surfaces showing corrosion behavior after total immersion in aerated 5 percent NaCl solution at 30 C for 219 days.*

Acknowledgments

The author wishes to acknowledge the assistance of Gardner Haynes, John Spadaro, and Daniel Jean in this work.

References

[*1*] Fontana, M. and Greene, N. in *Corrosion Engineering,* McGraw-Hill, New York.
[*2*] Wagner, C. and Traud, W., *Zeitschrift für Electrochemie,* ZEELA, Vol. 44, 1938, p. 391.
[*3*] Makrides, A. C., *Journal,* Electrochemical Society, JESOA, Vol. 107, 1960, p. 869.
[*4*] Greene, N. D., "Experimental Electrode Kinetics," Rensselaer Polytechnic Institute, Troy, N. Y., 1965.
[*5*] Littlewood, R., *Corrosion Science,* CRRSA, Vol. 3, 1963, p. 99.
[*6*] France, W. D. and Lietz, R. W., *Corrosion,* SBIIA, Vol. 24, 1968, p. 298.
[*7*] Myers, J. R., Gruewler, F. G., and Smulczenski, L. A., *Corrosion,* SBIIA, Vol. 24, 1968, p. 352.
[*8*] Bohni, H. and Uhlig, H. H., *Journal,* Electrochemical Society, JESOA, Vol. 116, 1969, p. 906.
[*9*] Brennert, S., *Journal,* Iron and Steel Institute, JISIA, Vol. 135, 1937, p. 101.
[*10*] Hospadaruk, V. and Petrocelli, J. V. *Journal,* Electrochemical Society, JESOA, Vol. 113, 1966, p. 878.
[*11*] Horvath, J. and Uhlig, H. H., *Journal,* Electrochemical Society, JESOA, Vol. 115, 1968, p. 791.
[*12*] Pourbaix, M., Klimzack-Mathieiu, L., Mertens, C., Meunier, J., Vanleugenhaghe, C., DeNuncky, L., Laureys, J., Neelemans, L., and Warzee, M., *Corrosion Science,* CRRSA, Vol. 3, 1963, p. 239.

W. D. France, Jr.[1]

Crevice Corrosion of Metals

REFERENCE: France, W. D., Jr., **"Crevice Corrosion of Metals,"** *Localized Corrosion—Cause of Metal Failure, ASTM STP 516,* American Society for Testing and Materials, 1972, pp. 164–200.

ABSTRACT: Shielded metal surfaces are susceptible to crevice corrosion when exposed to corrosive environments. Crevice corrosion commonly occurs at lap joints, threaded connections, and gasket fittings, in porous welds, and under debris deposits. This review considers the testing methods, the mechanisms of attack, and the procedures for controlling crevice corrosion of metals under a variety of corrosive conditions.

KEY WORDS: corrosion, crevice corrosion, poultice corrosion, concentration cell corrosion, corrosion tests, corrosion specimens, stress corrosion, anodic polarization, cathodic polarization, pitting, electrochemistry, inhibitors, pH, cathodic protection, seawater corrosion, electrolytes, atmospheric corrosion, polarization

Crevice corrosion[2] is a form of localized attack that occurs at shielded areas on metal surfaces exposed to certain environments. The sites for this type of corrosion often are unavoidable because of the structural or functional design and may even arise during the exposure period. Examples include spot-welded lap joints, threaded or riveted connections, gasket fittings, porous welds, valve seats, coiled or stacked sheet metal, marine or debris deposits, and the meniscus at a waterline.

The penetration of corrosive solutions into these relatively inaccessible areas, with widths that are typically a few thousandths of an inch, can result in various types of failures: the metal surface may become stained or perforated by the corrosive; the mechanical strength may be reduced below tolerance limits so that fracture occurs from the applied load or from the wedging action of the corrosion products; operating components may seize; or protective coatings may be disbonded from the metal surface. Some examples of crevice corrosion [*1–6*][3] are pictured in Fig. 1 to show the appearance of several metal surfaces after exposure to various

[1] Senior research chemist, Research Laboratories, General Motors Corp., Warren, Mich. 48090.

[2] Also known as concentration cell, contact, deposit, differential aeration, fissure, gasket, interface, poultice, waterline, and wedge corrosion.

[3] The italic numbers in brackets refer to the list of references appended to this paper.

(*a*) Iron diesel cylinder liner under rubber sealing gasket in contact with water [*1*].
(*b*) Breakdown of titanium above solution level (dashed) under wet salt crust from 5.3 *M* NaCl at 80 C [*2*].
(*c*) Type 316 stainless steel flange (after removal of rubber asbestos gasket) exposed to acetic and formic acid solutions at 120 F (49 C) [*3*].
(*d*) Type 304 stainless steel bolt and washers exposed for six months in quiescent seawater [*4*].
(*e*) Stainless steel Thornton nail orthopedic implant after an exposure of four years in a human body [*5*].
(*f*) Titanium pump shaft after service in chlorinated brine solution [*6*].

FIG. 1—*Examples of crevice corrosion.*

environments. These and other examples [*7–18*] indicate that crevice corrosion may be observed in some rather diverse industries, products, and applications. Undoubtedly, such case studies have led to the conclusions that crevices are design problems which cause most equipment corrosion failures [*19*] and that concentration cell (crevice) corrosion costs industry more than other types of corrosion [*20*].

Moody [*21*] recognized crevice corrosion over 60 years ago on iron specimens in contact with glass vessels, which he prevented by coating the ends of the specimens with paraffin. During the ensuing years, this type of corrosion has been the subject of a significant amount of work. Particular emphasis has been placed on developing mechanisms of crevice corrosion, determining factors affecting this localized attack, and establishing protection criteria.

This review considers the testing methods, the theoretical mechanisms of attack, and the procedures that have been investigated for preventing or controlling the crevice corrosion of several metals and alloys in a variety of corrosive environments.

Experimental Procedures

Various crevice corrosion tests ranging from the simple to the complex have been devised, but the procedures and the specimen designs generally are unique to the originating laboratory, with only a few examples gaining widespread application. Consequently, there are virtually no standard practices specifically for crevice corrosion, although the examination of all corrosion test specimens for localized attack, especially at specimen holders or spacers, is generally recommended in ASTM procedures for corrosion testing. ASTM methods [*22,23*] for establishing the effects of gasket materials on metals have been published, and these provide some guidelines for the development of standard tests for crevice corrosion.

Test Specimens and Assemblies

The development of specimens for the evaluation of crevice corrosion susceptibility has presented a challenge to the ingenuity of researchers. Among the necessary considerations are crevice dimensions, shapes, and relative areas. For example, Ellis and LaQue [*24*] demonstrated that the extent of crevice corrosion between overlapping stainless steel panels exposed to seawater for 87 days was proportional to the freely exposed area outside the crevice. Another factor is the mating surface, whether metal or nonmetal. Crevices formed with dissimilar metals will involve the added factor of galvanic corrosion [*25*]; those formed with nonmetal mating surfaces, such as plastics, may exhibit variable corrosion behavior that is related to ions leached from the material [*26*] or to the degree of material stiffness [*27*].

Various procedures have been used to obtain crevice specimens for exposure to corrosive environments with the objective of simulating actual components in service. Some typical sample configurations with metal/metal or metal/nonmetal combinations include sand or debris piled on sheet metal [*28,29*], rubber bands wrapped around pieces of metal [*29*], spot-welded lap joints [*28*], specimens partially immersed in solution [*2,30*], rubber grommets in sheet metal [*31*], stacked coupons with and without spacers [*32,33*], bolted or riveted joints [*3,8,34,35*], wire wrapped on a threaded bolt [*25,36*], and a hairpin-shaped tapered sheet [*37*].

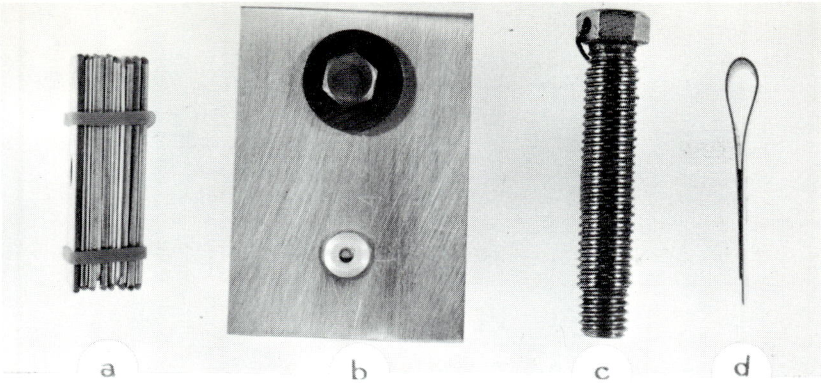

(*a*) Stacked metal and nonmetal coupons with rubber band straps.
(*b*) Sheet metal with gasketed bolt and rivet.
(*c*) Metal wire wrapped on nonmetal bolt.
(*d*) Hairpin tapered crevice in sheet metal.

FIG. 2—*Specimens for crevice corrosion tests.*

These last four examples are illustrated in Fig. 2. In each case the extent of corrosion is determined by interrupting the test for visual examination and the measurement of specimen weight loss.

The development of artificial crevice assemblies has been of particular importance for the accurate determination of crevice corrosion potentials and currents. Three recent designs are shown in Fig. 3 to illustrate different procedures for accurate control of crevice geometry and dimensions during electrochemical corrosion tests [*38–40*]. With the Rosenfeld-Marshakov device (*a*) the specimen is clamped in the plastic frame after a fixed crevice width is regulated by calibrated spacers;[4] the Lizlovs cell (*b*) allows reproducible changes in the crevice geometry and dimensions *during* the course of an experiment; and the France-Greene assembly (*c*) permits electrode potential measurements along the length of a crevice. Some applications of these assemblies for the interpretation and evaluation of crevice corrosion are discussed subsequently.

Evaluation of Crevice Corrosion Susceptibility

The evaluation of crevice corrosion susceptibility is often accomplished with weight loss measurements and relative rankings based on visual examination. Such procedures can require lengthy exposure times for the localized attack to develop unless the initiation and propagation processes are accelerated or evaluated in another way, as with electrochemical measurements. For example, Pourbaix [*41*] has characterized seven types of cyclic anodic polarization curves from which predictions can be made

[4] Rosenfeld and Marshakov also have described a unique sectional electrode for the determination of current distribution within a crevice [*38*].

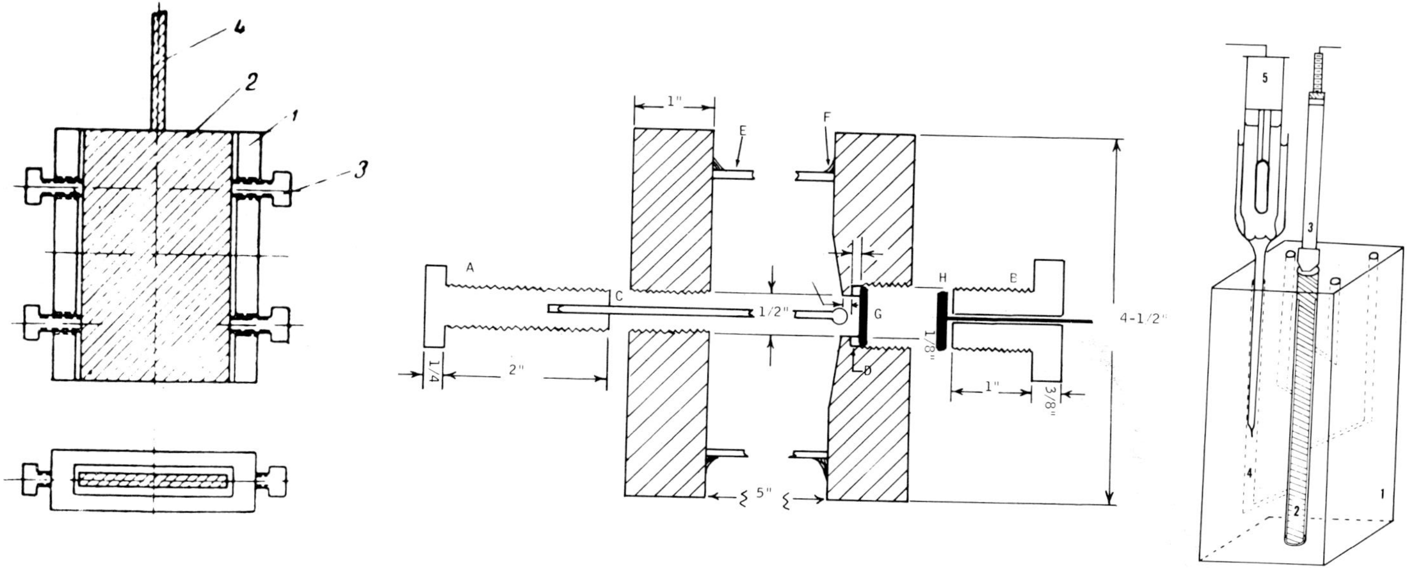

a b c

(*a*) Taken from Rosenfeld and Marshakov [*38*]
1. Glass cover plate.
2. Metal specimen.
3. Nonmetallic screw.
4. Electrical connection rod.

(*b*) Taken from Lizlovs [*39*]
A. Teflon bolt.
B. PVC bolt.
C. Glass rod with glass bead at the end.
D. Teflon gasket.
E. Glass tubing.
F. Silicone rubber mount.
G. Stainless steel working electrode.
H. Copper disk with copper lead wire.

(*c*) Taken from France and Greene [*40*]
1. Plexiglass block.
2. Specimen.
3. Electrode holder.
4. Potential probe.
5. Saturated calomel reference.

FIG. 3—*Artificial crevice assembly designs.*

regarding the possibility of general, pitting, and crevice corrosion. Differentiation is made between curves that indicate no occurrence of crevice corrosion and those that indicate crevice corrosion susceptibility under conditions of differential aeration.

One of the more significant advances in the development of evaluation procedures has been made by Wilde and Williams [*31*]. These authors conducted anodic polarization measurements with several alloys that had been previously evaluated in seawater for more than four years. Cyclic potentiodynamic anodic polarization curves for metals containing synthetic crevices exhibited two critical potentials—the pitting potential (E_c) and the protection or repassivation potential (E_p). The experimental procedure involved a potential scan in the noble direction from the corrosion potential to a point beyond the pitting potential. When the anodic current density reached 2000 $\mu A/cm^2$, the scanning direction was reversed, and the potential at which the downward scan intercepted the upward scan was designated the protection potential.[5] Three of their curves are reproduced in Fig. 4 to illustrate some possible variations. The curve for Hastelloy C, which is not susceptible to crevice corrosion, exhibits reversibility, whereas the curves for Incoloy 825 and Carpenter 20 Cb-3 exhibit slight and extensive hysteresis loops, respectively. Comparison of these data with seawater exposures indicated that the susceptibility to crevice corrosion is related to the presence of hysteresis during cyclic polarization. They also found that the difference between the pitting and protection potentials (difference potential) could be correlated directly with the extent of crevice corrosion after more than four years in seawater. This is shown by the relationship in Fig. 5 for several stainless steel alloys where pitting (rather than a synthetic crevice) was the precursor to crevice corrosion. Such a test that requires hours rather than years of service to provide information on the susceptibility and extent of crevice corrosion certainly deserves further consideration and interlaboratory testing in other environments.

In Situ *Detection of Crevice Corrosion*

The ability to detect and monitor the progress of crevice corrosion without disturbing the system is hindered by the fact that the evidence of attack usually is restricted to relatively inaccessible areas. Various approaches for circumventing this problem have been demonstrated, such as the visual examination of crevice corrosion in the progress through a transparent glass or plastic plate [*43,44*] that is held against a metal surface to

[5] This definition of the protection potential is different than that proposed by Pourbaix, who defines the protection potential, below which corrosion ceases, as the potential where the current density meets the ordinate axis (that is, a practically zero current density value on a linear scale) during the downward scan of a cyclic polarization experiment [*42*].

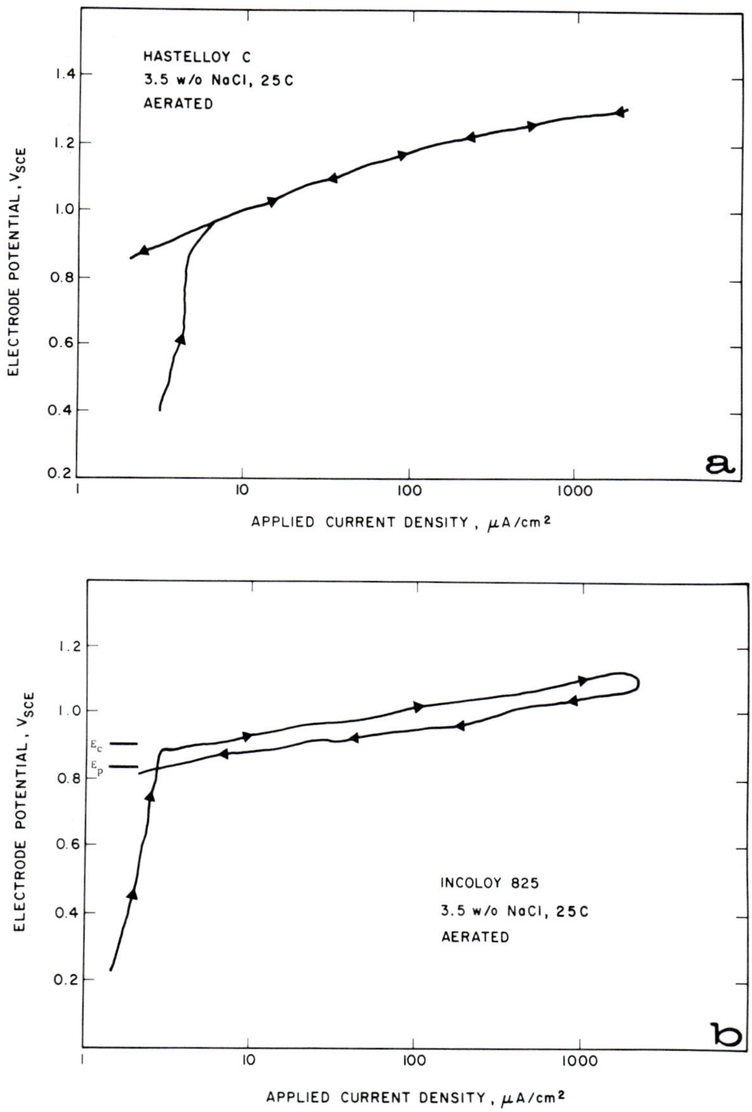

form the crevice.[6] A similar visual procedure in conjunction with electrochemical measurements has been used by Mueller [*45*] whose hemispherical test electrodes are placed in contact with hemispherical grooves in a plexiglass plate.

For quantitative measurements of localized corrosion, Jones and Greene

[6] An interesting technique for demonstrating the detection of differential aeration at a corrosive drop on a metal surface involves the incorporation of indicators into the drop to differentiate between anodes and cathodes within this unique crevice [*29*].

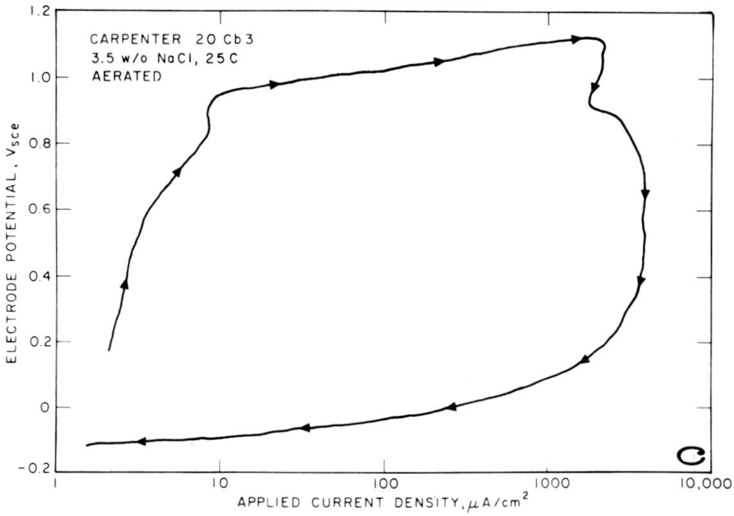

(*a*) Hastelloy C.
(*b*) Incoloy 825.
(*c*) Carpenter 20Cb-3.

FIG. 4—*Cyclic potentiodynamic anodic polarization curves for metals containing synthetic crevices* [31].

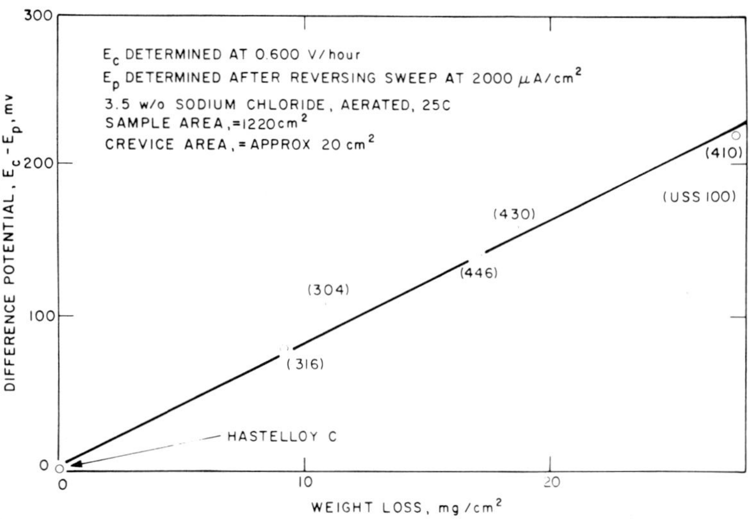

FIG. 5—*Correlation between difference potential and corrosion weight loss of stainless steel alloys exposed in seawater for 4.25 years* [31].

[*46*] have demonstrated an electrochemical technique suitable for detecting crevice attack before any visible evidence could be observed. Corrosion potentials and corrosion rates determined by linear polarization were found to provide the necessary data. The combination of a marked increase in corrosion current, a significant (30 to 50 mV) shift in the corrosion potential to active values, and the initial occurrence of erratic corrosion potential fluctuations revealed the existence of localized corrosion. The spontaneous initiation of crevice corrosion resulted in as much as a hundredfold increase in the measured overall corrosion currents.

For both laboratory and process control applications, the development of experimental resistance probes for continuous detection and measurement of crevice corrosion is particularly interesting [*47*]. These probes were made from 6061 aluminum alloy wire fitted with Teflon disks to form a crevice region. Experience had shown that this combination of materials was susceptible to severe crevice corrosion. With this system the resistance of the aluminum electrical conductor increases as the cross sectional area is decreased by corrosion. Thus, the resistance is proportional to the corrosion rate, and this special probe exhibited a dramatic increase in the resistance reading when crevice corrosion was initiated.

Crevice Corrosion Mechanisms

Significant contributions to the understanding of crevice corrosion began some 50 years ago with research by McKay [*48*] on metal-ion concentration cells and by Evans [*44*] on differential aeration. Through the years, a number of mechanisms have been proposed, modified, invalidated, rejected, or accepted, often with exceptions.[7] There is little doubt that corrosion in crevices is a complex process that involves several competitive and synergistic factors, which have been investigated and discussed in some detail by several researchers, particularly Evans [*49*], Rosenfeld and Marshakov [*38*], and their colleagues.[8] Both the classical and modern experimental work are contributing to the development of some generalized, and possible unifying, concepts of mechanisms that deserve continued consideration and refinement.

Concentration Cells

Much of the theoretical interpretation of crevice corrosion has been associated with some sort of concentration cell. This concept is based on the fact that the initiation and propagation of localized attack on metal surfaces is influenced by nonuniformity of the corrosive solution. For example, variations in the concentration of ions or dissolved gases may

[7] There are some other mechanisms that are primarily of historical interest (that is, the hydrogen bubble egress and rival interface theories) [*43,50,51*].

[8] Additional publications are cited in these two references.

arise along the same metal surface because of the existence of crevices that are relatively inaccessible to fresh electrolyte compared with freely exposed surfaces. Such evironmental conditions cause potential differences and influence electrode kinetics to the extent that an electrochemical cell is established, wherein an oxidation process (that is, corrosion) occurs at the anodic sites, and some reduction process (for example, oxygen reduction) occurs at the cathodic sites.

A description and critique of several concentration cell mechanisms that have contributed to the understanding of crevice corrosion are discussed in the following paragraphs.

Metal-Ion Cell—This mechanism [*48*] is based on the development of differences in metal-ion concentration between the inside and outside of a crevice. Laboratory experiments with an electrochemical cell consisting of two short-circuited copper electrodes, each immersed in a different concentration of copper sulfate solution in electrical contact through a porous barrier, have shown that corrosion occurs predominantly in the solution with the low metal-ion concentration. From the Nernst equation, the electromotive force (emf) of such a divided cell with a 0.0001 normal solution in the anode chamber and a 1.0 normal solution in the cathode chamber is 0.13 V. By anology, such a cell can be formed in practice after the metal ions formed by the initial corrosion reactions accumulate within the crevice where solution flow or diffusion is restricted. Once a difference in metal-ion concentration is established, accelerated dissolution (that is, $M \rightarrow M^+ + e^-$) is anticipated at anodes just outside the crevice, whereas the reduction of metal ions (that is, $M^+ + e^- \rightarrow M$) is anticipated at the cathodes within the crevice. A schematic representation of the operation of this crevice cell is illustrated in Fig. 6*a*.

Differential Aeration Cell—This mechanism [*44,49*] depends on the occurrence of different dissolved oxygen concentrations in the solution adjacent to a metal surface.[9] Electrochemical cell experiments, where one half-cell consists of a metal in a deaerated solution and the other half-cell consists of a metal in an aerated solution, indicate that accelerated dissolution occurs at low oxygen concentrations while the reduction of oxygen occurs where the O_2 levels are high. This type of cell is illustrated in Fig. 6*b*. In a crevice situation the oxygen is replenished more easily on the exterior surfaces than within the shielded crevice, where corrosion is accelerated as the result of oxygen depletion.

Critique—Schafer and Foster [*32*] have considered the metal-ion concept invalid because it depends on the M^+ concentration at the anode source being lower than elsewhere on the metal surface and because the reaction is not self sustaining (that is, continued operation of the cell even-

[9] This discussion of differential aeration is limited to crevice corrosion, but there are more extensive applications of this general mechanism [*49*].

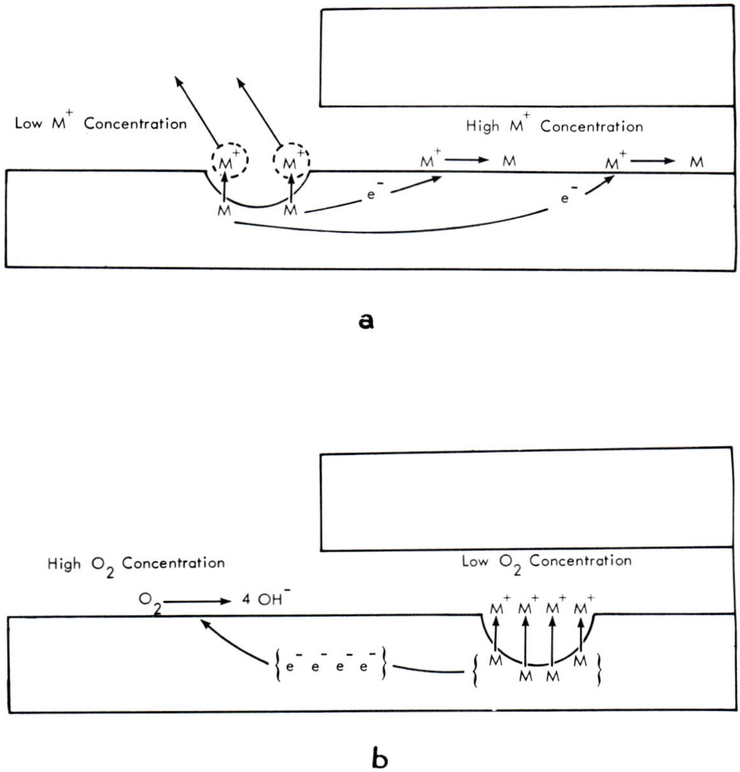

(*a*) Metal-ion cell.
(*b*) Differential aeration.
FIG. 6—*Schematic representation of crevice corrosion mechanisms.*

tually should eliminate the metal-ion concentration differences, and the reaction would cease). These authors suggest that the actual cathodic reaction is oxygen reduction rather than metal-ion reduction and that the corrosion observed at the mouth of a crevice is simply a special case of the differential aeration cell. They support this viewpoint with the fact that corrosion at the crevice opening is stopped by removal of dissolved oxygen.

In a related example Evans [*49*] has shown that although iron and zinc demonstrate the differential aeration mechanism, copper does not. When a copper electrode in an aerated solution is coupled with one in a deaerated solution, corrosion occurs in the aerated environment. Evans has attributed this reverse effect to the operation of a metal-ion cell. The bubbling action during aeration stirs the solution enough to remove the cuprous corrosion products from the freely exposed surface where the attack is then accelerated.

The reconciliation of these observations depends on some flexibility in the interpretation of the metal-ion mechanism. Although the metal-ion cell cannot be self sustaining without some additional cathodic process, the data on copper are difficult to explain without this concept. It appears probable that initially metal ions concentrate within the crevice during corrosion which is supported by oxygen reduction. Subsequently, the concentration of M^+ and the *combination* of M^+ and O_2 reduction contribute to the location of corrosion just outside the crevice.

Tomashov [*53*] has questioned the validity of the differential aeration mechanism under conditions such as metal in chloride environments and suggested that conclusions based on the external current of a laboratory macrocell can be misleading. This is because the aerated specimen (so-called cathode) of the electrochemical cell sometimes exhibits a greater weight loss than the deaerated specimen to which it is coupled. He explains that the activity of microcells on the cathode is enhanced by aeration. Although Evans [*49*] admitted that several investigators have encountered this apparently contradictory situation, he notes that this is an experimental artifact. It is traceable to the high resistance of the electrochemical cell or circuitry, which can accentuate the action of local anodes on the macrocathode, thereby causing a greater weight loss at the aerated specimen.

Another observation has been that the metal-ion and the differential aeration concentration cells are competitive [*54*]. A low oxygen concentration favors accelerated attack which generates a high metal-ion concentration in stagnant areas, thereby stifling further anodic reaction. If these two mechanisms were the only ones operating, the final location of attack would depend upon which one became dominant. However, some of the other factors that must be considered are discussed in the following paragraphs.

Active-Passive Cell—This cell is sometimes considered to be another example of differential aeration, but it deserves special attention; the depletion of any oxidizer in a crevice can initiate localized attack on active-passive metals because the establishment and maintenance of passivity depends upon the availability of sufficient oxidizer concentrations. Therefore, a deficiency of oxygen can prevent the reformation of protective oxide films at damaged areas on a metal surface, which become active and corrode at an accelerated rate. Other oxidizers (for example, Cu^{+2} and Fe^{+3}) can have similar effects, as indicated by the depletion of ferric ions in a stainless steel crevice. After 9 h in a hot sulfuric acid solution which contained 10 g/liter of ferric sulfate, the stainless steel crevice potential changed from a passive value of 0.53 V versus saturated calomel electrode (SCE), indicative of a 0.3 mils per year (mpy) corrosion rate, to an active value of —0.36 V versus SCE, indicative of 6000-mpy corrosion rate [*55*].

Rosenfeld and Marshakov [*38*] have measured similar potential changes with time for several metals in a neutral electrolyte and have studied the changes in electrolyte composition and characteristics during crevice corrosion. With adjustable crevice assemblies (Fig. 3*a*) these investigators determined anodic and cathodic polarization curves which demonstrated the acceleration of anodic corrosion reactions and the retardation of cathodic reactions within crevices. These data indicated that the extent of anodic and cathodic control determines whether corrosion is accelerated inside or outside the crevice. As shown schematically in Fig. 7*a,* with the reduced effectiveness of the cathodic process, the negative shift in electrode potentials should tend to reduce corrosion at active potentials (*A*) and either increase (*B*) or unaffect (*C*) corrosion at passive potentials, depending on the proximity of the reduction curves to the anodic current density maximum.

Critique—Such a graphical presentation facilitates the understanding of crevice corrosion, especially for active-passive cells, but the conclusion related to corrosion at active potentials may be misleading if the anodic interior and cathodic exterior areas of the crevice are considered a corrosion couple. For example, with the oxidation-reduction reactions shown in Fig. 7*b*,[10] the corrosion current inside the crevice (i'_{in}) is greater than that outside the crevice (i'_{out}) at the corrosion potential of the couple ($E'_{couple\ (in/out)}$), and crevice corrosion would be predicted. However, it is evident that in the uncoupled state, the exterior is corroding at a faster rate (i_{out}) than the interior (i_{in}) of the crevice. Thus, some flexibility is important in the theoretical interpretation of polarization curves for crevice corrosion.

Other Concentration Cells—In addition to the concentration cells discussed above, Myers and Obrecht [*12*] have suggested that there are at least three additional types, namely, hydrogen-ion, neutral salt, and inhibitor cells. For example, high chloride ion concentrations or low inhibitor concentrations generally are associated with anodic rather than cathodic regions on a metal surface. Furthermore, Rosenfeld and Marshakov [*38*] showed that acidification (low pH) of the solution within the crevice was a major contribution to accelerated and autocatalytic crevice corrosion, especially in near-neutral solutions.[11] Subsequently, Korovin and Ulanovskii [*58*] found that pH changes had a far greater influence on corrosion rates and potentials than did changes in oxygen concentration.

[10] Other publications [*56,57*] provide details on the manipulation and theoretical analysis of such polarization curves for various combinations of exchange current densities, reversible potentials, corrosion potentials, Tafel slopes, and limiting diffusion currents.

[11] Acidification is the usual, but apparently not the only pH change that can occur within crevices. Rosenfeld and Marshakov also discuss "linear-selective dissolution" and "differential concentration cells" in which alkalization is responsible for accelerated corrosion. Their results with a divided cell and three phase boundaries (for example, iron/dielectric/acid) indicated that the corrosion at the anode along the dielectric perimeter was associated with a pH increase due to acid depletion. Thus, depending on the bulk electrolyte pH, the anode electrolyte may be of higher or lower pH than the bulk solution.

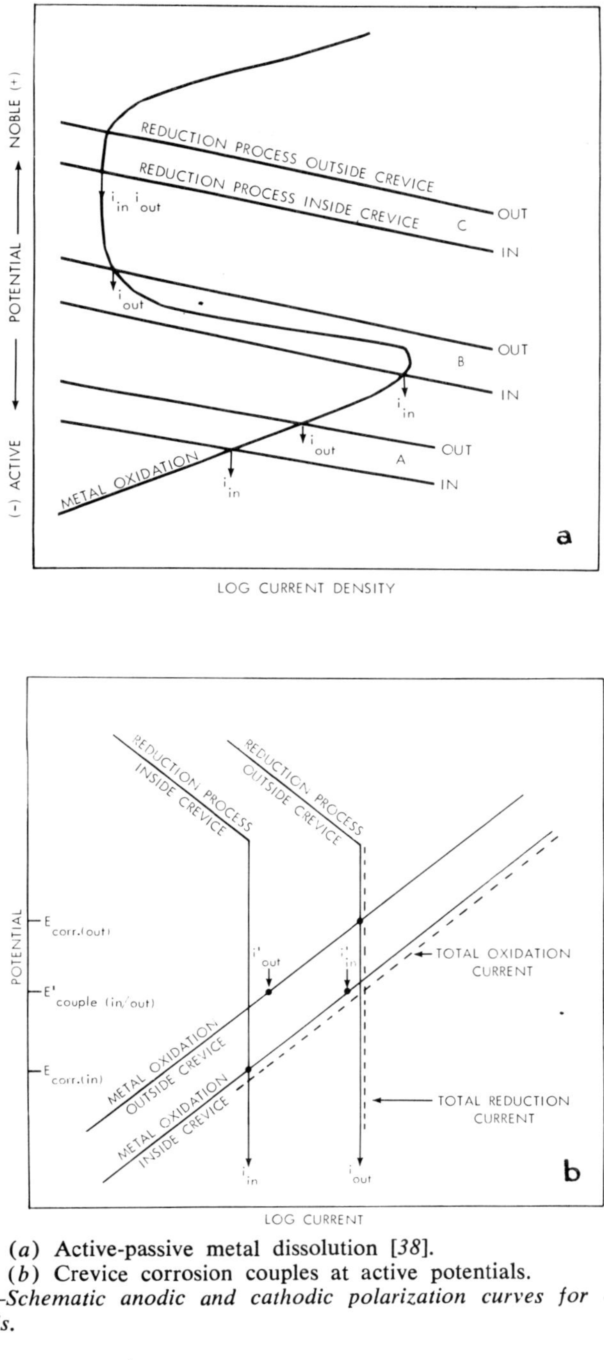

(a) Active-passive metal dissolution [*38*].
(b) Crevice corrosion couples at active potentials.

FIG. 7—*Schematic anodic and cathodic polarization curves for crevice corrosion analysis.*

For example, with a combined reduction in oxygen concentration (9 to 1 mg/liter) and change in pH (8.3 to 2.7), the potential shift in the active direction was 372 mV, of which 355 mV was associated with the pH decrease and 17 mV with the reduction in oxygen concentration. For a bulk pH of 6 to 7 in halide solutions at 150 C, Griess [*34*] found that the pH within a titanium crevice where corrosion occurred was about 1, whereas the pH was 4 to 5 when significant corrosion was absent.

Besides acidification effects, Griess showed that the severity of titanium crevice corrosion increased with the solution salt concentration. Although his experiments indicated that the anion was relatively unimportant, other investigators have demonstrated the significance of anions. For example, Strocchi et al [*59*] examined the crevice corrosion behavior of several activated and passivated austenitic stainless steels (Types 302, 305, 316, and 347) in contact with nonconducting materials (for example nitrile rubber O-rings) during immersion in acid-chloride solutions. These steels exhibited different susceptibility to crevice corrosion, but a common feature was that oxygen was necessary for the initiation of crevice corrosion at three phase contacts (metal/nonmetal/solution) and that aggressive anions (for example, chloride) contributed significantly to the breakdown of passivity and corrosion.

Unified Crevice Corrosion Mechanism

Based on the type of information developed in the preceding sections, Fontana and Greene [*57*] have proposed a basic mechanism of crevice corrosion. Initially the anodic dissolution (that is, $M \rightarrow M^+ + e^-$) and cathodic reduction (for example, $O_2 + 2H_2O + 4e^- \rightarrow 4OH^-$) processes occur uniformly over the entire metal surface, including the crevice interior. The oxygen in the shielded crevice area is consumed after some incubation period, but the decrease in the cathodic reaction rate is negligible because of the small area involved. Consequently, the corrosion of the metal inside and outside the crevice continues at the same rate.[12] With the cessation of the cathodic hydroxide-producing reaction, however, the migration of mobile negative ions (for example, chlorides) into the crevice area is required to maintain charge balance. These processes are shown schematically in Fig. 8. The resulting metal chlorides hydrolyze in water to insoluble metal hydroxides and free acid. Both the chloride anions and low pH accelerate crevice corrosion in a manner similar to autocatalytic pitting, while the reduction reaction cathodically protects the exterior surfaces. Also, for active-passive metals that depend on protective oxide films for their corrosion resistance, the breakdown of passivity and active

[12] Equal corrosion rates inside and outside the crevice would be anticipated from Fig. 7*b* if the metal oxidation reactions were considered to be represented by a single line.

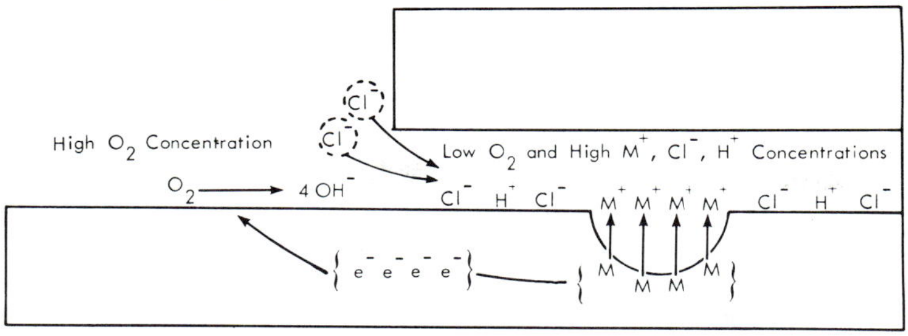

FIG. 8—*Fontana-Greene mechanism of crevice corrosion* [57].

corrosion in crevices are favored by the increased concentration of chlorides and hydrogen ions.

This mechanism represents a rather inclusive and unifying description of crevice corrosion. It takes into account metal-ion concentration, oxygen depletion, hydrolysis and acidification, aggressive anion migration, and both active and active-passive dissolution behavior. Undoubtedly, there are exceptions to the strict interpretation of this mechanism, for example the observed occurrence of crevice corrosion when aggressive anions are absent from the solution. However, this mechanism, which might be appropriately named "differential concentration cell",[13] represents a significant contribution and starting point for specific refinements.

Crevice-Related Localized Corrosion Phenomena

The phenomenological and mechanistic relationship between various forms of localized attack have been under increased scrutiny lately. With an emphasis on similarities, Brown [*60*] has proposed an occluded corrosion cell concept in which local acidification by hydrolysis is a common feature of localized attack, such as crevice corrosion, filiform corrosion, pitting, stress-corrosion cracking, intergranular attack, exfoliation, and tuberculation. Such an approach facilitates the classification of corrosion phenomena. In the following paragraphs the specific role of crevice corrosion in the initiation or propagation of several types of localized corrosion is considered.

Filiform (Underfilm) Corrosion

Filiform corrosion has been described by Uhlig [*61*] as an example of a differential aeration cell and by Fontana and Greene [*57*] as a special type of crevice attack in which the growing filament becomes a "selfpropagating crevice." The characteristic long and narrow filaments with active heads and colored corrosion product tails have been observed on steel, aluminum, and magnesium surfaces under coatings that are permeable to

[13] This term has been used before, although not as a description of this particular mechanism [*38*].

moisture. At 65 to 95 percent relative humidities, the attack is particularly severe, and the result is unacceptable appearance, especially on the exteriors of canned products. In a recent study with aluminum, Hoch and Tobias [*62*] found that the heads had a pH of less than 1 compared with a range of 7 to 8.5 for the cathodic areas, which were approximately 100 times greater than the anodic areas. When these conditions were simulated electrochemically with differential aeration cells, 60-μA/cm^2 corrosion currents and 20-mV potential differences between anodic and cathodic areas were measured. This corrosion rate corresponds to a growth rate of 2 mm/week for a filament with a 0.5 mm width and a 0.1 mm depth, which approximates observed rates. Kane [*63*] has reported the concentration of chloride by electrochemical transport at the filament head on magnesium. Thus, these data on differential aeration, acidic anodes, chloride concentration, and large cathode-to-anode ratios support the similarities between filiform and crevice corrosion.

Pitting Corrosion

Pitting has been recognized often as a result of corrosive attack initiated at crevices [*20,31,61*], and the extent and depth of pitting has been recorded as a measure of crevice corrosion [*24*]. A consideration of the similarities between pitting and crevice corrosion led Schaffer et al [*64*] to suggest that they are identical processes. In both cases the locations for corrosion initiation were considered dependent on differential aeration or inhibitor depletion, and the autocatalytic propagation was associated with an acid mechanism. Their conclusion was that pitting at microscopic sites is a limiting case of crevice corrosion at macrosites.

Actual similarities between the propagation of pits and crevices are evident from the microscopic and electrochemical investigations of Rosenfeld and Danilov [*65*]. They considered that nearly closed pits on stainless steels severely retarded the diffusion of passivators and demonstrated the inhibition of pit growth by destroying the shield layer to permit fresh solution access to the pit interior.

Smialowska et al [*66*] recently examined the pitting of stainless steels and found that predominant initiation sites were complex sulfide-oxide inclusions in the metal. Dissolution of the sulfide shell around an oxide particle resulted in a narrow crevice between the oxide and the metal. From this active site, the propagation of pits can be considered one kind of crevice corrosion.

Stress-Corrosion Cracking

For metals under tensile stress, solution stagnation in crevices can create an environment in which stress-corrosion cracking occurs; tests designed to evaluate both effects simultaneously with specimens like a stressed sheet containing a tapered crevice [*67*] can provide valuable information on synergistic interactions. Examples have been reported where stainless and carbon steel specimens failed during operation of pilot plant loops at elevated temperature because of previously entrapped caustic solutions [*68*] in crevice areas. Such failures did not occur with simple crevice geometries that could be rinsed effectively. When annealed and sensitized Inconel 600, Inconel 625, Incoloy 800, and stainless steels (Types 304, 304L, and 347) were stress corrosion tested in autoclaves with high pressure water at 316 C, Copson and Economy [*69*] found that a combination of high stress, high oxygen concentration, and crevices where oxidizing cations (for example Cr^{+6}, Fe^{+3}, and Cu^{+2}) could accumulate, promoted intergranular cracking. No attack was observed without crevices, but Armijo [*70*] has reported subsequently that crevices were not necessary for the initiation of cracking in sensitized Incoloy 800 and Type 304 stainless steel. Besides the initiation of stress-corrosion cracking from existing crevices, Sylwestrowicz [*71*] has proposed a mechanism for stress-corrosion cracking of a copper-beryllium alloy where preferential attack at grain boundaries of slip planes forms crevices with sharp apexes from which cracking develops. Stress concentrations at a crevice tip should be enhanced by the wedging action of solid corrosion products: Pickering et al [*72*] have measured wedging pressure (7 ksi) which resulted in crack tip stresses on the order of the metal yield strength.

Bombara [*73*] has suggested an electrochemical and geometric analogy between stress-corrosion cracking and crevice corrosion. His model is based on the initiation of stress-corrosion cracks from deep, narrow pits that appear to be similar to crevices. A further generalization is that once a crack is initiated from this site, the geometry of the advancing crack resembles a tapered crevice; crack propagation then depends on factors such as those outlined in Staehle's mechanistic interpretation of processes operating during stress-corrosion cracking [*74*].

Other Localized Corrosion Phenomena

Localized corrosion, such as exfoliation, fatigue, and intergranular corrosion, also has been related to crevice effects. For example, exfoliation corrosion of aluminum aircraft skin at countersunk fastener holes resulted from corrosion when solution penetrated the crevice area between the fastener and the unprotected periphery of the hole [*75,76*]. Also, the occurrence of crevice corrosion in fatigue cracks has resulted in the interesting development of an electrochemical crack detection system [*77*]. When cracked specimens are exposed to a chloride-ferrocyanide solution, a visible precipitate is formed at the crack because of the initiation of crevice corrosion. The technique has been applied to high-temperature and high-strength alloys to detect cracks smaller than those detectable with conventional liquid penetrants. Also, there is an inhibitive, nondestructive aspect to this test because the corrosion action evidently blunts the crack tip.

Prevention and Control of Crevice Corrosion

The primary method for reducing crevice corrosion susceptibility is effective design of components and structures. However, when crevices are unavoidable, cathodic protection and inhibitors may provide adequate corrosion control. When these techniques are unacceptable or unsuitable, a material change would be another possible solution. Each of these prevention and control methods are discussed next along with some examples to illustrate suitable applications and limitations.

Design Factors

The recognition and, of course, the elimination of crevices are the most direct approaches for solving crevice corrosion problems. Numerous specific recommendations for minimizing crevice corrosion through proper design have been developed, and experience has shown that the following procedures should be advantageous [*19,20,57,78,79*]: seal existing crevices with noncorrosive caulking compounds or by welding; weld joints rather than bolt or rivet them; butt-weld joints instead of spot-welding lap joints; use nonporous materials for gaskets or insulators to avoid moisture absorption and retention; eliminate marine or debris deposits by periodic cleaning; avoid designs that have sharp corners or areas where moisture and debris can collect; paint or coat faying surfaces before assembly; and inspect welds for porosity and material components for fabrication or workmanship defects.

Although the intensity of crevice corrosion is considered greater with metal/metal mating surfaces than with metal/nonmetal combinations [*43*], the use of noncorrosive and nonporous materials for sealants and gaskets is also important. As demonstrated by Wyche et al [*80*] the amount of corrosion protection depends on the characteristics of the material, and attack can be actually stimulated by substances within organic compounds, such as sulfur-bearing rubbers that were more harmful than rubbers without sulfur. Low moisture permeability is an important materials property during immersion conditions and especially during wet and dry atmospheric exposures where corrosive moisture from condensation, leakage, or precipitation can be absorbed, retained, and concentrated in contact with metal surfaces.

Evans [*49*] has suggested zinc coatings on steel windows to prevent the seizing caused by voluminous ferrous corrosion products, and he also has referenced a number of papers related to crevice corrosion in buildings.

Among the more recent innovations are clad metals that can serve as effective substitutes for mechanical connections [*81*]. Procedures like explosion cladding offer solid metallurgical bonds that can provide galvanic cathodic protection. One example is an explosion-formed transfer joint between the aluminum superstructure and steel hull of a marine vessel, which apparently provides resistance to crevice and galvanic corrosion [*82*]. Another interesting case study has been reported by Burton [*79*] who observed that the penetration of chloride-containing water into crevices between stainless steel heat exchanger tubes and tube sheets can result in corrosion and stress cracking, as depicted in Fig. 9*a*. One clever remedy for this problem involves material and design changes (Fig. 9*b*) that result in cathodic protection of the stainless steel when the steel corrodes in the crevice area. Other cathodic protection applications are considered in the next section.

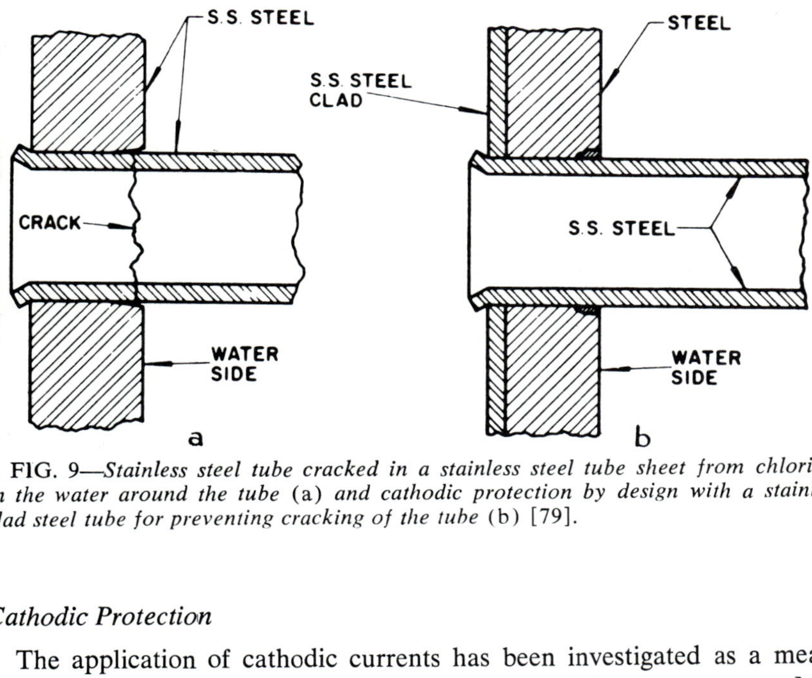

FIG. 9—*Stainless steel tube cracked in a stainless steel tube sheet from chlorides in the water around the tube* (a) *and cathodic protection by design with a stainless clad steel tube for preventing cracking of the tube* (b) [79].

Cathodic Protection

The application of cathodic currents has been investigated as a means of controlling crevice corrosion of metals, especially in seawater. May and Humble [*83*] exposed nickel, Monel, and stainless steels (Types 302, 316, 410, and 430) to quiet seawater at Kure Beach, and their results showed that crevice corrosion and pitting were reduced effectively when the specimens were coupled galvanically to magnesium sacrifical anodes. Later work by Vreeland and Bedford [*84*] with some of these alloys (Types 304, 316, and Monel) under similar seawater conditions confirmed these findings and provided statistical information on the erratic occurrence of crevice corrosion. Various steel anode combinations were tested to determine their performance in reducing or eliminating the extent and depth of attack.

Lennox et al [*83*] conducted experiments in quiescent seawater at Key West with stainless steels (Types 304, 316, 17–4 PH, and 20Cb-3) having crevices at insulators or rubber grommets. The results of exposures for several hundred days demonstrated that crevice corrosion was the predominant type of attack when compared with general or pitting corrosion. None of the alloys tested were immune to the attack in the absence of cathodic protection, but 20Cb-3 was the most resistant. Additional data for stainless steels Types 205, 304, 410, and 430 at 5600 ft depths exhibited similar crevice corrosion characteristics. The usefulness of these alloys for seawater service depended upon cathodic protection which was applied with either stainless steel or aluminum anodes. Subsequent work by these authors [*86*] demonstrated the application of cathodic protection to Type 304 stainless steel O-ring seals in quiescent and flowing seawater. Although flow rates of 0.5 ft/s (0.15m/s) were found to increase the crevice corrosion damage, both steel and zinc anodes still provided effective cathodic protection. For the seawater with pH 7.8 to 8.2, the pH in the active crevice was 1.2 to 2.0. However, with steel and zinc sacrificial anodes, the pH in the protected crevice was 5.5 to 6.5 and above 10, respectively.

For some alloys, the elimination of crevice corrosion by cathodic currents can cause other problems. When aluminum anodes were used to protect the 17–4 PH alloy, hydrogen embrittlement and cracking resulted [*85*], and severe hydrogen blistering was observed on Types 410 and 430 stainless steels at the required current levels [*83*]. In this regard, Vreeland's experiments [*87*] have indicated the careful potential control that is necessary for successful cathodic protection of precipitation hardened stainless steel in seawater. For a steel with a 175 ksi yield strength, the required cathodic potential was between —0.6 and —0.75 V versus SCE. Potentials more noble than 0.6 V versus SCE could result in crevice corrosion and pitting, whereas potentials more active than —0.75 V versus SCE could result in hydrogen embrittlement. The effective range was extended to —1.3 V versus SCE for a 154-ksi precipitation hardened steel.

Cathodic protection for controlling crevice corrosion in nonmarine environments has not been applied extensively. However, Mueller's laboratory experiments [*45*] with several steels in solutions of sodium chloride with chlorine dioxide have demonstrated a decrease in general and crevice corrosion with cathodic protection. For example, the crevice corrosion rate of Type 316 stainless steel was reduced from an average of 3.9 mpy to less than 1 mpy with a 0.05-mA/cm^2 cathodic current at —0.06 V versus SCE. It was concluded that cathodic protection appears promising for bleached pulp washers, assuming that the necessary cathodic current can be supplied to the very inaccessible crevices.

Inhibitors

The use of inhibitors for controlling crevice corrosion presents definite problems because the relative inaccessibility of the crevice site makes replenishment of depleted inhibitors difficult or impossible. For example, Peers and Evans [*88*] found intense waterline attack of iron that was partially immersed in chloride solution containing sufficient anodic inhibitors (for example, phosphates and carbonates) to passivate the immersed area. Their data supported the concept that this type of crevice suffered attack because of the exhaustion and slow replenishment of inhibitors to the meniscus. Rosenfeld and Marshakov [*38*] also found that increased inhibitor concentrations were necessary to provide protection for metals containing crevices. For iron corroding in a near-neutral solution, it was necessary to increase the inhibitor concentration tenfold over that required for a freely corroding surface to protect a 0.05-mm crevice.

Accessibility to mating surfaces prior to assembly facilitates the application of an inhibitor system. One approach is to cover these surfaces with inhibited paints, such as red lead for steel and zinc chromate for aluminum [*89*]. Other investigators [*90*] have wrapped metal foils in a paper that was saturated with inhibitor solutions before exposing a stack of these specimens to a humid sulfur dioxide atmosphere. They found that some of the solutions under test actually enhanced corrosion and that the addition of surface active agents improved the performance.

The chances for complete inhibition of crevice corrosion are improved when only short-term service is required. In their laboratory evaluation of Type 410 stainless steel in inhibited refinery cleaning solution at 140 F (60 C), Freedman and Dravnieks [*91*] observed that the exposed surfaces were protected adequately, but that the crevice areas were attacked. The addition of 0.3 percent lead acetate or stannous chloride eliminated crevice corrosion, but tests performed in the presence of hydrogen sulfide showed no beneficial effects from these salts.

Hatch [*92*] has investigated the inhibition of a differential aeration cell composed of a bare steel panel in one chamber and a steel panel wrapped with filter paper to restrict oxygen availability in the other chamber. He conducted electrochemical measurements of potential and the short-circuit current, which is a measure of attack on the wrapped panel. A variety of inhibitors (for example, sodium nitrite, sodium benzoate, and zinc sulfate) were evaluated, and the results indicated the concentration ranges where the treated solution provided either inhibition or stimulation of attack.

Materials Selection

The selection of materials for service where crevice corrosion is a potential problem involves a comparison of their performance under crevice conditions. However, with the possible exception of data from seawater tests, the lack of standard tests has contributed to some erratic and nonreproducible results that thwart attempts to establish meaningful ratings of materials by a comparison of data from several sources. A ranking by Tuthill and Schillmoller [*57,93*] of metals and alloys in terms of crevice corrosion resistance in quiet seawater is given in Table 1. An

TABLE 1—*Relative crevice corrosion resistance of metals and alloys in quiet seawater* [93,57].

Metal or Alloy	Resistance
Hastelloy C Titanium	inert
90Cu-10Ni-1.5Fe 70Cu-30Ni-0.5Fe Bronze Brass	best
Austenitic nickel cast iron Cast iron Carbon steel	neutral
Incoloy 825 Carpenter 20 Ni-Cu alloy Copper	less
316 stainless steel Ni-Cr alloys 304 stainless steel Series 400 stainless steels	pit initiation at crevices

additional factor that must be considered is the effect of seawater flow rate, and these same authors [*4,93*] have provided guidelines for materials subjected to seawater which is stagnant, quiescent (quiet), flowing moderately, or flowing at high speed. For example, Hastelloy C and titanium exhibit no attack over this flow rate range; aluminum alloys under stagnant or quiescent conditions exhibit little, moderate, or considerable attack depending on the alloy; 70Cu-30Ni alloy exhibits little to no attack over the entire flow rate range; and stainless steels exhibit little to considerable attack over the flow rate range.

Although such lists provide useful guidelines, the order of merit is not necessarily applicable to other media or even to the specified media under varied test conditions. Consequently, actual materials evaluation tests under the actual conditions of exposure are essential for reasonable success in the prediction of crevice corrosion behavior. Some of the more significant investigations of several materials in a variety of environments are discussed in the following paragraphs.

Titanium—A rather extensive amount of information is available on the crevice corrosion behavior of titanium, particularly in halide environments [*28,32,94–102*]. Generally, this metal is considered quite resistant to halide environments [*28*], especially seawater. However, Kane [*63*] has indicated that the severity of attack increases with temperature, especially above 250 F (121 C), halide concentration, and hydrogen ion concentration. Some of these data have been represented in the form of a corrosion diagram that Feige and Kane [*96*] prepared to show the pitting and crevice corrosion of Ti-50A in brine solutions.

Ruskol and Klinov [*97*] found that cracks and fissures in titanium corrode hundreds of times faster than the exposed surface in 0.75 to 2.0 *N* H_2SO_4. Schlain and Kenahan [*98*] reported that for other acid solutions, including hydrochloric, oxalic, and formic, corrosion initiated at crevice sites rapidly becomes general corrosion over the external surface that was previously passive.

Bomberger [*32*] has demonstrated that wet chlorine attacks titanium in deep crevices. He attributes this to the dehydration of the gas under the stagnant crevice conditions. Once the moisture content is below the amount required for inhibition, acidic and hygroscopic corrosion products are produced and accelerate the attack. Reaction with available water yields hydrochloric acid and titanium oxides.

Griess [*34*] has conducted anodic and cathodic polarization measurements in addition to conventional corrosion tests with tapered crevices, some of which were exposed in autoclaves. His results in high-temperature halide and sulfate solutions indicated that alloying with sufficient concentrations of molybdenum, nickel, or palladium should improve the crevice corrosion resistance of titanium. Several investigators [*99,100*] have noted the significantly improved crevice corrosion resistance of a Ti-0.2Pd alloy compared with titanium. Its performance is associated with the surface enrichment of palladium and the resulting improvement in the ease of passivation [*101*]. However, tests with this alloy in hot seawater loops have shown that it too is attacked after prolonged exposure to chlorides at elevated temperatures [*102*].

Aluminum—In a recent review Godard [*63*] concluded that crevice attack is not generally responsible for significant deterioration of aluminum. He associated practical problems with thin sections (for example, less than 0.040 in. (0.1 cm)), metal appearance, and effects of bulky corrosion products. Examples included the staining of stacked or coiled aluminum sheet by moisture. Crevice corrosion in the atmosphere and fresh water was considered negligible. However, such attack in the form of pitting has been noted in seawater, where the extent is related to the general corrosion resistance of the alloy, in soil, and in marine atmospheres where appreciable corrosion rates are probably proportional to the salt concentration in the atmosphere. It has been shown that when aluminum was exposed to sodium chloride solutions the corrosion rate in crevices was ten times faster than that on the exterior surface [*103*].

Other investigators [*104*] have conducted experiments in natural and simulated seawater with sintered aluminum powder (SAP) sheets clad with aluminum and aluminum alloys. They found that certain clad alloys were more resistant to crevice attack than others.

Stainless Steels—Although this series of alloys has been discussed previously in this paper, some additional data are worth mentioning. With regard to relative rankings, the Wilde and Williams [*31*] data for stainless steels exposed for more than four years in seawater showed that the crevice corrosion resistance decreased in the following order, Types 316, 304, 446, 430, and 410. Although improvement in crevice corrosion resistance of stainless steels often has been associated with the addition of molybdenum as an alloying element, other investigators [*85*] concluded that Type 316 is not necessarily superior to Type 304 stainless steel for applications in quiescent seawater. Such discrepancies undoubtedly arise from the fact that the same variables are not always controlled, or controllable, to the same extent in different experiments.

Data also have been reported for the crevice corrosion resistance of stainless steel in contact with various sealing compounds [*80*] that provide a considerable range of protection, with nitrile rubber O-rings [*59*], and with powdered metal and oxide gaskets [*105*]. Pray and Berry [*106*] found that seizing, as a result of ferric oxide formation at the periphery of a journal-sleeve arrangement, was more prevalent with Types 410 and 430 stainless steels than with Types 347 and 17–4.

Wyche et al [*80*] did not observe crevice corrosion on specimens of stainless steels (Types 301, 302, and 316) with spot-welded lap joints that were exposed for several years in a marine atmosphere. Evidently, the conditions necessary for atmospheric crevice corrosion of these alloys are more severe than those for crevice corrosion during immersion in seawater.

Miscellaneous Materials—The crevice corrosion performance of various other materials has been studied, although the information in the literature is somewhat sparse. However, in addition to some of the practical examples noted in the introduction, the following references should be mentioned: crevice corrosion of zinc, cadmium, and lead in seawater [*107*]; the dealuminization of aluminum bronze by chloride crevice attack [*108*]; the dezincification of brass under deposits [*105*]; the filiform and crevice corrosion of magnesium [*63*]; the waterline attack of zinc in sodium chloride solutions [*109*]; the crevice corrosion and pitting of nickel alloys [*35*] and a nickel-copper alloy [*84*] in seawater.

Electrolytic Crevice Corrosion

The phenomenon of electrolytic crevice corrosion was recognized in 1965 during a study of electrode mounting procedures [*110*]. In contrast to electrochemical crevice corrosion, which generally becomes initiated after some incubation period because of changes in the stagnant crevice solution, electrolytically induced crevice corrosion can be initiated along a narrow electrolyte path immediately when current is applied to a metal-electrolyte system. Besides their contribution to the development of this mechanism, France and Greene [*40*] have demonstrated the passivation of crevices during anodic protection. A significant amount of data on electrolytic crevice corrosion has been published recently by a group of researchers [*27,59,73,111–114*], including Bombara, Sinigaglia, Strocchi, Taccani, and Vicentini.

Phenomenological and Mechanistic Aspects

The early work [*110*] on materials for mounting electrodes revealed that that presence of microscopic crevices (for example, 0.001 in. (0.0025 cm)) between the metal electrode and some encapsulating resins caused a pronounced effect on the active-passive anodic polarization curves. At passive potentials the measured current densities were as much as 100 times greater than the expected values, as shown in Fig. 10 [*115*]. Since these observations were disclosed, several authors have reported similar behavior for various metal-electrolyte systems [*116–118*]. These examples demonstrate clearly that the avoidance of crevices in and adjacent to

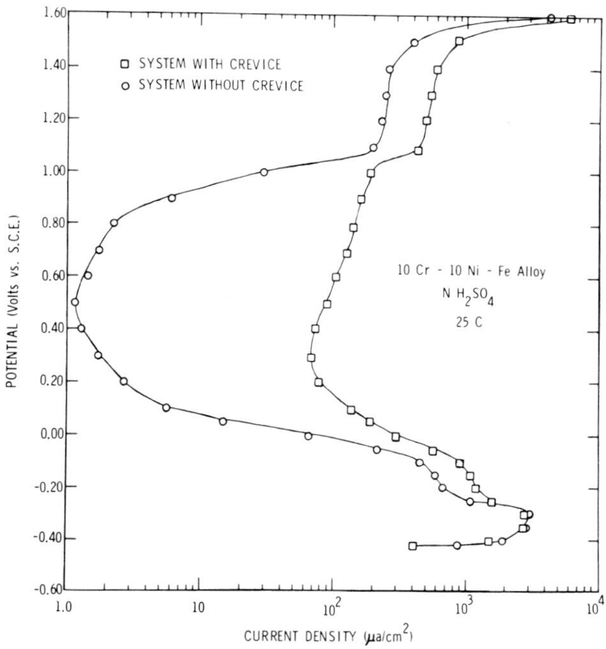

FIG. 10—*Comparison of potentiostatic anodic polarization curves for systems with and without crevices* [115].

electrodes is necessary for the determination of accurate active-passive polarization curves.

The electrochemical similarity between porous powder metal compacts and crevices is illustrated in Fig. 11 with Tikkanen's data [*119*] for sintered stainless steel. In this case the voids associated with incomplete sintering apparently act as microcrevices giving a characteristic increase in passive current density with increased compact porosity. The near coincidence of the critical maximum current densities indicates that changes in real surface area are insignificant compared with active crevice effects.

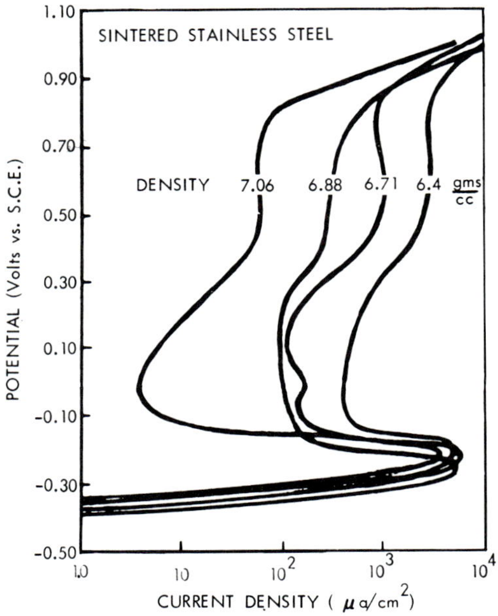

FIG. 11—*Influence of powder compact density on stainless steel anodic polarization characteristics* [119].

Besides recommendations for suitable electrode mounting procedures, the initial study [*110*] established the basis for an electrolytic crevice corrosion mechanism. Measurements showed that the crevice interior can remain active and corrode rapidly (for example, 1200 $\mu A/cm^2$ at —0.42 V versus SCE) even when the exterior metal surface is maintained at passive potentials characteristic of low-corrosion rates (for example, 2 $\mu A/cm^2$ at 0.25 V versus SCE). Thus, the high passive current densities represent the large current density contribution from active crevice sites where the large potential (IR) drop between the passive and active sites is a result of the high resistance along the narrow electrolyte path within the crevice.

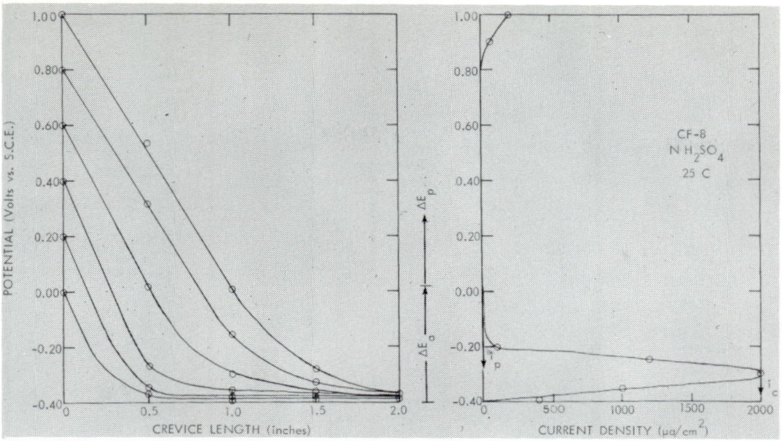

FIG. 12—*Crevice potential variation of CF-8 stainless steel together with its anodic polarization curve* [40].

Other studies [*40,55*] with a special crevice assembly (Fig. 3*c*) demonstrated that the crevice potential is not an absolute value, but varies with the crevice length. Such an electrode potential distribution along the length of a 0.009-in. (0.023-cm) wide crevice is shown in Fig. 12 along with an anodic polarization curve for this CF-8 stainless steel in *N* H_2SO_4. These data can be compared to data for Type 304 stainless steel, as shown in Fig. 13. Although the exterior metal surface was controlled at similar potentials in both experiments, the potential gradients are significantly different. For example, at a control potential of 0.6 V versus SCE, the crevice potential for the CF-8 alloy decreased to active values within the first 0.5 in. (1.3 cm), whereas the crevice potential of the

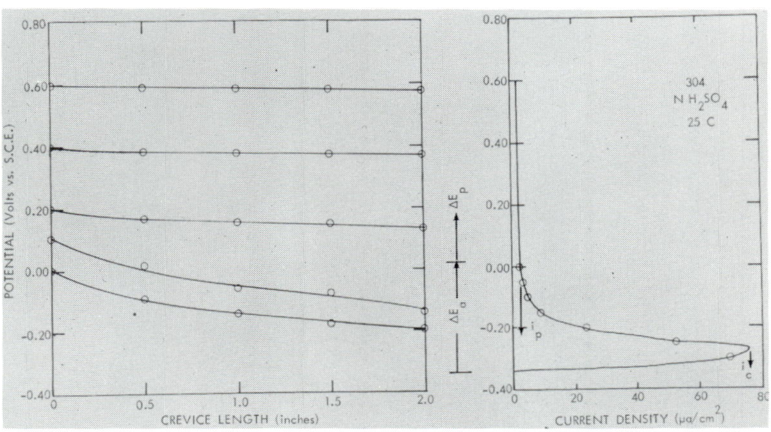

FIG. 13—*Crevice potential variation of Type 304 stainless steel together with its anodic polarization curve* [40].

Type 304 alloy remained at passive potentials along the entire 2-in. (5.1-cm) crevice length.

The appearance of specimens after extended testing was consistent with the data in Figs. 12 and 13. For example, a CF-8 specimen that was exposed in the crevice assembly at a control potential of 0.6 V versus SCE for 75 h is shown in Fig. 14. The unattacked or passive region is evident on the left, while the position of the current density maximum is indicated by the area of severe attack. Corrosion actually diminished to the right of this point (slight grain faceting) as expected from the data in Fig. 12.

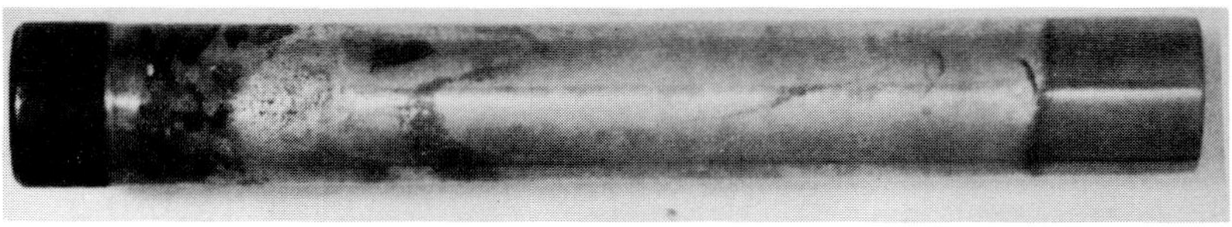

FIG. 14—*Crevice specimen of CF-8 maintained at a control potential of 0.6 V versus SCE for 75 h in* N H_2SO_4 *at 25 C. Crevice face* (bottom) *on left* [40].

Based on Ohm's Law, the resistivity equation, and the Fokin-Timonin analysis of potential distributions along thin wires polarized in relatively large diameter tubes [*120*], two equations have been derived for predicting the length of crevice that could be passivated with an applied anodic potential. The first equation defines the length of an active-passive crevice that can be passivated, and the second equation defines the length of a passivated crevice that can be maintained passive [*40*].

$$L_p = \Delta E_p \sqrt{\frac{W}{\rho i_c \, \Delta E_a}} \text{ (active-passive crevices)} \quad (1)$$

$$L_p = \sqrt{\frac{\Delta E_p W}{\rho i_p}} \text{ (passive crevices)} \quad (2)$$

where:

L_p = passive length,
ΔE_p = passive potential range,
W = crevice width,
ρ = solution resistivity,
i_c = critical anodic current density,
i_p = passive current density, and
ΔE_a = active potential range.

The electrochemical terms are designated in Figs. 12 and 13, and for the data shown, ΔE_p = control potential −0.02 V.

A comparison of experimental and calculated passive crevice lengths showed that there was good correlation, as indicated in Table 2.

Solovena et al [*101*] have made this same analysis for the electrolytic crevice corrosion of titanium in an acid-chloride solution at 98 C. Their calculated and experimental values for the passive crevice lengths showed the same good correspondence as the data in Table 2.

In another investigation of geometric factors associated with electrolytic crevice corrosion, Bombara et al [*27*] derived equations for the location of the maximum crevice attack within a crevice for any value of the anodic control potential. The solution to their final equation gave two possible values, one related to a previously activated surface and the other related to a previously passivated surface. Thus these two calculations can provide supplemental information to Eqs 1 and 2.

TABLE 2—*Comparison of calculated and experimental passive crevice lengths* [40].

Stainless Steel	Control Potential (V versus SCE)	L_p (in.) Experimental	L_p (in.) Calculated[a]
CF-8	0.00	0	0
	0.20	0.13	0.17
	0.60	0.50	0.55
304	0.02	0	0
	0.10	0.37	0.42
	0.20	8.0	8.3[b]

[a] Calculated from Eq 1.
[b] Calculated from Eq 2.

Prevention and Evaluation of Electrolytic Crevice Corrosion

Although anodic protection has been applied successfully as an industrial corrosion control method, the preceding data indicate that the protection of crevices in complex plant structures can pose a problem. However, the methods [*40*] for improving the feasibility of anodically protecting structures containing crevices follow directly from Eq 1. Because localized attack is avoidable only when the passive length equals or exceeds the crevice length, variables that increase the numerator and decrease the denominator of this equation are advantageous. However, there are some practical limitations, such as the fact that environments in industrial applications are usually fixed (that is, ρ is constant) and engineering designs seldom permit the widening of the crevice to increase the (W) term. Obviously, any attempt to eliminate crevices with sealants must be complete because very narrow crevices are the worst cases.

The most significant factor in Eq 1 was found to be the critical anodic current density which can vary by orders of magnitude for different metals and alloys. The importance of this finding is that a relative ranking of electrolytic crevice corrosion resistance for metals and alloys in a given environment can be established on the basis of the critical current density (i_c). Such a ranking is presented in Table 3. Because standard procedures were not used to determine each of the polarization curves referenced and because there even can be variations in the critical current density for specimens within a given alloy classification [*129*], this list should be considered a guideline based on the referenced data rather than an inflexible ranking.

TABLE 3—*Predicted electrolytic crevice corrosion resistance*[a] *(H_2 saturated, N H_2SO_4, 25 C).*

Metal	Representative i_c ($\mu A/cm^2$)	Reference (for i_c value)
Cr-10Ni	1	121
Ti-6Al-4V	2	122
Hastelloy C	9	123
Titanium	15	123
304-L stainless steel	35	123
347 stainless steel	40	124
316 stainless steel	45	123
304 stainless steel	75	40
201 stainless steel	125	125
20Cr-80Ni	300	121
Cu-45.4Ni-2.35Al	350	126
CF-8 stainless steel	2 000	40
Chromium	4 000	121
430 stainless steel	9 000	118
Nickel	12 000	121
Fe-16Cr	25 000	124
Iron	120 000	127
Fe-15Co	200 000	128

[a] Ranked in the order of decreasing resistance.

Sinigaglia et al [*112*] also have investigated the electrolytic crevice corrosion susceptibility of several austenitic stainless steels in acid and acid-chloride solutions. In a series of constant potential polarization experiments with specimens having a neoprene ring to form a crevice site, they determined a critical potential below which crevice corrosion occurred and above which the specimen completely passivated. On the basis of the critical potential, the crevice corrosion resistance of these steels in sulfuric acid solutions decreased in the following order: $321 > 316 \geqslant 304 > 302$, and in the presence of chloride ions the ranking was: $316 > 321 \geqslant 302 > 304$. Also, it was found that the intensity of attack decreased as the crevice width increased, a confirmation of Eqs 1 and 2.

In other constant potential experiments Lizlovs [*39*] has evaluated the electrolytic crevice corrosion resistance of several ferrous alloys in sodium chloride solutions with a unique cell (Fig. 3*b*). The initiation of crevice corrosion was marked by an increase in polarization current when the glass bead was moved against the specimen electrode to create a crevice at the interface. Comparison of the current-time curves at constant potential provided the following ranking in order of decreasing resistance to crevice corrosion: 25Cr-3.5Mo-Fe, 17Cr-3Mo, 316, 304 = 25Cr-Fe, 17Cr-Fe. Besides the beneficial effects of molybdenum as an alloying element, sulfate ions in acid-chloride solutions have been shown to improve the electrolytic crevice corrosion resistance of ferritic stainless steels [*117*].

Critique

The theoretical analysis of electrolytic crevice corrosion demonstrated that the ability to passivate crevices is controlled by electrolyte resistivity, crevice dimensions, and the electrochemical behavior of the metal-electrolyte system. However, recent work has indicated that other factors, such as solution composition and concentration can contribute to this phenomenon. Such effects have been noted by Mazza [*2*] in his study of the preferential corrosion of titanium in crevices and under wet salt deposits on partially immersed specimens (Fig. 1*b*). Although short-term electrolytic tests showed breakdown voltages of about 9 V for specimens in hot, concentrated chloride solutions at 176 F (80 C), the presence of crevices or salt crusts reduced this value to about 6 V during 24-h exposures. His polarization curves showed passive rather than active-passive behavior which was characteristic of the metals described in the previous section, and probably as a result of this fact the measured potential difference between the inside and outside of the crevice was only about 250 mV after 15 min and then became nearly zero after a short time. Because of this small potential difference, Mazza concluded that crevice corrosion of titanium in this environment occurred because of changes in the crevice solution rather than potential (IR) drops. The polarization current was assumed to cause polarization concentration and a reduction in the stability of passivity due to accumulation of corrosion products and chloride ions, hydrolysis, interactions of complex ions, reactions of chlorine with water, and oxygen and water depletion. Thus, there are other factors besides potential differences that can contribute to electrolytic crevice corrosion, and these must be considered before accurate predictions of passive crevice lengths can be made.

Vermilyea and Tedmon [*130*] have derived mathematical equations for predicting the variations in concentration and potential in a crevice with inert sides. They found that small potential differences caused relatively large concentration changes and correspondingly large corrosion currents at constant applied potential. During constant potential polarization at 0.00 V versus SCE the current increased to a maximum value of 25 μA. Calculations indicated that the solution concentration changed from 10^{-3} *M* Na_2SO_4 in the bulk solution to 7.8×10^{-3} *M* at the mouth of the crevice to 1.0 *M* at the bottom of the crevice. The potentials associated with these three positions were 0.00, —0.028, and —0.105 V versus SCE, respectively.

In summary, there are some unique aspects to electrolytic crevice corrosion, such as calculable potential drops within a crevice and the immediate initiation of attack when current is applied. Except in the original research [*40,55*], investigators have not tended to differentiate between electrolytic and natural corrosion, and its appears that further research that would establish the extent of correlation between the two phenomena would be worthwhile. Based on the success of electrochemical techniques for the detection and quantitative evaluation of crevice corrosion, continued applications and further advances in this field can be anticipated.

Conclusions

Based on the literature reviewed for this paper on crevice corrosion, the following conclusions are pertinent: (1) numerous test assemblies and procedures are available, but standard methods for crevice corrosion testing are definitely required; (2) significant progress is being made in the development of unifying concepts of mechanisms for electrochemical and electrolytically induced crevice corrosion; and (3) appropriate designs are among the more effective ways of avoiding crevice corrosion, but consideration of materials selection, anodic and cathodic protection, and inhibitors is worthwhile when crevices are unavoidable.

Acknowledgment

The author wishes to thank the publishers and the following individuals and their colleagues for permission to use material from their publications: B. F. Brown and the Marine Corrosion Section, Naval Research Laboratory; W. H. Burton, Allied Chemical Corp.; L. W. Gleekman, Materials and Corrosion Engineering Services; N. D. Greene, University of Connecticut; E. V. Kunkel, Celanese Chemical Co.; F. L. LaQue, International Nickel Co.; E. A. Lizlovs, Climax Molybdenum Co.; F. Mazza, University of Milano (Italy); I. L. Rosenfeld, Institute of Physical Chemistry of the Academy of Sciences (USSR); M. H. Tikkanen, Institute of Technology (Finland); and particularly B. E. Wilde, U. S. Steel Corp. for prepublished data. Thanks are also given to M. D. Cooper, L. L. Lewis, L. C. Rowe, and R. L. Chance, General Motors Research Laboratories, for their helpful editorial comments.

References

[*1*] Speller, F. N. and LaQue, F. L., *Corrosion*, SBIIA, Vol. 6, 1950, p. 209.
[*2*] Mazza, F., *Corrosion*, SBIIA, Vol. 23, 1967, p. 223.
[*3*] Kunkel, E. V., *Corrosion*, SBIIA, Vol. 10, 1954, p. 260.
[*4*] Brown, B. F., *Machine Design*, MBUIA, Vol. 40, 18 Jan. 1968, p. 165.

[*5*] Greene, N. D. and Jones, D. A., unpublished data (see Fontana, M. G. and Greene, N. D., *Corrosion Engineering,* McGraw-Hill, New York, 1967, p. 288).
[*6*] Gleekman, L. W., panel discussion moderator, *Materials Protection,* MAPRA, Vol. 6, No. 10, 1967, p. 22.
[*7*] Muroaka, J. S., *Materials Protection,* MAPRA, Vol. 9, No. 3, 1970, p. 23; Rigo, J. H., *Materials Protection,* MAPRA, Vol. 5, No. 4, 1966, p. 54.
[*8*] Gray, K. O., *Materials Protection,* MAPRA, Vol. 3, No. 7, 1964, p. 46.
[*9*] Kirchner, O. E. and Morris, F. M., *Corrosion,* SBIIA, Vol. 7, 1951, p. 161; staff feature, *Corrosion,* SBIIA, Vol. 17, No. 2, 1961, p. 14.
[*10*] Lechnick, W., Brown, P. E., and Schultz, B. G., "Induction of Crevice Corrosion in Reactor Materials by Concentration of Coolant Additives," *High Purity Water Corrosion of Metals,* National Association of Corrosion Engineers, Houston, Tex., 1968.
[*11*] Moehrl, K. E., *Corrosion,* SBIIA, Vol. 17, No. 2, 1961, p. 26; DePaul, D. J., *Corrosion,* SBIIA, Vol. 13, 1957, p. 75t.
[*12*] Myers, J. R. and Obrecht, M. F., "Deposit and Corrosion Problems in Potable Water Systems of Buildings," Paper No. 90, 27th Annual Conference, National Association of Corrosion Engineers, Chicago, 1971.
[*13*] Turnbull, J. M., *Corrosion,* SBIIA, Vol. 16, 1960, p. 11.
[*14*] Moskvin, V. M., Alekseev, S. N., and Novgorodskii, V. I., *Protection of Metals,* PTMNA, Vol. 1, 1965, p. 499; Boyd, W. K. and Tripler, A. B., Jr., *Materials Protection,* MAPRA, Vol. 7, No. 10, 1968, p. 40.
[*15*] Suzuki, T., Saito, H., Yoshioka, K., and Kitamura, Y., *Corrosion Engineering,* BOGJA, Vol. 19, No. 3, 1970.
[*16*] Smith, L. L., *Tappi,* TAPPA, Vol. 47, 1964, p. 118A.
[*17*] Hutcherson, C. H., *Materials Protection,* MAPRA, Vol. 7, No. 6, 1968, p. 27.
[*18*] Hudson, J. C. and Wormwell, F., *Chemistry and Industry,* CHINA, Aug. 1957, p. 1078.
[*19*] Landrum, R. J., *Chemical Engineering,* CHEEA, Vol. 118, 24 Feb. 1969.
[*20*] Coburn, S. K., *Materials Protection,* MAPRA, Vol. 6, No. 2, 1967, p. 33.
[*21*] Moody, G. T., *Journal,* Chemical Society, JCSOA, Vol. 89, 1906, p. 720.
[*22*] "Standard Method of Test for Corrosion Produced by Leather in Contact with Metal," Designation: D 1611–60, *Book of Standards,* American Society for Testing and Materials, 1969.
[*23*] "Standard Method of Test for Corrosive and Adhesive Effects of Gasket Materials on Metal Surfaces," Designation: F 64–69, *Book of Standards,* American Society for Testing and Materials, 1969.
[*24*] Ellis, O. B. and LaQue, F. L., *Corrosion,* SBIIA, Vol. 7, No. 11, 1951, p. 362.
[*25*] Godard, H. P., *Materials Protection,* MAPRA, Vol. 2, No. 6, 1963, p. 38.
[*26*] Rasmussen, L. M., *Corrosion,* SBIIA, Vol. 11, 1955, p. 155t.
[*27*] Bombara, G., Sinigaglia, D., and Taccani, G., *Electrochimica Metallorum,* ELMTB, Vol. 3, 1968, p. 81.
[*28*] Renshaw, W. G. and Bish, P. R., *Corrosion,* SBIIA, Vol. 11, 1955, p. 41t.
[*29*] *Corrosion in Action,* The International Nickel Company, Inc., New York, 1955.
[*30*] Peers, A. M., *Transactions,* Faraday Society, TFSOA, Vol. 51, 1955, p. 1748.
[*31*] Wilde, B. E. and Williams, E., "The Use of Current/Voltage Curves for the Study of Localized Corrosion and Passivity Breakdown on Stainless Steels in Chloride Media," submitted for publication in *Electrochemica Acta,* as part of the Proceedings of the 3rd International Conference on Passivity, Cambridge, England, July 1970.
[*32*] Bomberger, H. B., *Titanium Review,* Vol. 10, No. 1, Aug. 1962.
[*33*] Rice, J. K. and Loucks, C. M., *Materials Protection,* MAPRA, Vol. 4, No. 2, 1965, p. 14.
[*34*] Griess, J. C., Jr., *Corrosion,* SBIIA, Vol. 24, 1968, p. 96.

[*35*] Niederberger, R. B., Ferrara, R. J., and Plummer, F. A., *Materials Protection and Performance,* Vol. 9, No. 8, 1970, p. 18.

[*36*] Compton, K. G., Mendizza, A., and Bradley, W. W., *Corrosion,* SBIIA, Vol. 11, 1955, p. 383t.

[*37*] Fontana, M. G., *Industrial and Engineering Chemistry,* IECHA, Vol. 47, Nov. 1955, p. 81A.

[*38*] Rosenfeld, I. L. and Marshakov, I. K., *Corrosion,* SBIIA, Vol. 20, 1964, p. 115t.

[*39*] Lizlovs, E. A., *Journal,* Electrochemical Society, JESOA, Vol. 117, 1970, p. 1335.

[*40*] France, W. D., Jr., and Greene, N. D., Jr., *Corrosion,* SBIIA, Vol. 24, 1968, p. 247.

[*41*] Pourbaix, M., "Theoretical and Experimental Considerations in Corrosion Testing." presented at the *Symposium on State of the Art in Corrosion Testing,* Annual Meeting, American Society for Testing and Materials, Toronto, Canada, 22 June 1970.

[*42*] Pourbaix, M., Klimzack-Mathieiu, L., Mertens, Ch., Meunier, J., Vanleugenhaghe, Cl., deMunck, L., Laureys, J., Neelemans, L., and Warzee, M., *Corrosion Science,* CRRSA, Vol. 3, 1963, p. 239.

[*43*] Mears, R. B. and Evans, U. R., *Transactions,* Faraday Society, TESOA, Vol. 30, 1934, p. 417.

[*44*] Evans, U. R., *Journal,* Institute of Metals, JIMEA, Vol. 30, 1923, p. 239.

[*45*] Mueller, W. A., "Electrochemical Study of Crevice Corrosion of Steels in Solutions of Sodium Chloride and Chlorine Dioxide," Paper No. 78, 26th Annual Conference, National Association of Corrosion Engineers, Philadelphia, 1970; Mueller, W. A., *Corrosion,* SBIIA, Vol. 25, 1969, p. 473.

[*46*] Jones, D. A. and Greene, N. D., *Corrosion,* SBIIA, Vol. 25, 1969, p. 367.

[*47*] FitzPatrick, V. F. and Ayres, J. A., *Materials Protection,* MAPRA, Vol. 9, No. 2, 1970, p. 27.

[*48*] McKay, R. J., *Transactions,* Electrochemical Society, TESOA, Vol. 41, 1922, p. 201.

[*49*] Evans, U. R., *The Corrosion and Oxidation of Metals,* Edward Arnold, Ltd., London, 1960.

[*50*] McCulloch, L., *Journal,* American Chemical Society, JACSA, Vol. 47, 1925, p. 1940.

[*51*] Evans, U. R., *Journal,* American Chemical Society, JACSA, Vol. 48, 1926, p. 1601.

[*52*] Schafer, G. J. and Foster, P. K., *Journal,* Electrochemical Society, JESOA, Vol. 106, 1959, p. 468.

[*53*] Tomashov, N. D., *Theory of Corrosion and Protection of Metals,* Macmillan, New York, 1966, p. 470.

[*54*] Hibert, C. L., *Product Engineering,* PRENA, Vol. 27, No. 10, 1956, p. c6.

[*55*] France, W. D., Jr., "Controlled Potential Investigations of Localized Corrosion Phenomena," Ph.D. thesis, Rensselaer Polytechnic Institute, Troy, N. Y., 1966.

[*56*] Stern, M., *Journal,* Electrochemical Society, JESOA, Vol. 105, 1958, p. 638.

[*57*] Fontana, M. G. and Greene, N. D., *Corrosion Engineering,* McGraw-Hill, New York, 1967.

[*58*] Korovin, Y. M. and Ulanovskii, I. B., *Corrosion,* SBIIA, Vol. 22, 1966, p. 16.

[*59*] Strocchi, P. M., Vicentini, B., and Sinigaglia, D., *Electrochimica Metallorum,* ELMTB, Vol. 1, 1966, p. 239.

[*60*] Brown, B. F., *Corrosion,* SBIIA, Vol. 26, 1970, p. 249.

[*61*] Uhlig, H. H., *Corrosion and Corrosion Control,* Wiley, New York, 1963.

[*62*] Hoch, G. M. and Tobias, R. F., "Mechanisms of Filiform Corrosion," Paper No. 19, 27th Annual Conference, National Association of Corrosion Engineers, Chicago, 1971.

[*63*] Godard, H. P., Jepson, W. B., Bothwell, M. R., and Kane, R. L., *The Corrosion of Light Metals,* Wiley, New York, 1967.

[64] Schafer, G. J., Gabriel, J. R., and Foster, P. K., *Journal,* Electrochemical Society, JESOA, Vol. 107, 1960, p. 1002.
[65] Rosenfeld, I. L. and Danilov, I. S., *Corrosion Science,* CRRSA, Vol. 7, 1967, p. 129.
[66] Szklarska-Smialowska, Z., Szummer, A., and Janik-Czachor, M., *British Corrosion Journal,* BCRJA, Vol. 5, 1970, p. 159.
[67] Copson, H. R. and Berry, W. E., *Corrosion,* SBIIA, Vol. 16, 1960, p. 79t.
[68] Larrick, A. P. and Weed, R. D., *Materials Protection,* MAPRA, Vol. 7, No. 4, 1968, p. 34.
[69] Copson, H. R. and Economy, G., *Corrosion,* SBIIA, Vol. 24, 1968, p. 55.
[70] Armijo, J. S., *Corrosion,* SBIIA, Vol. 24, 1968, p. 319.
[71] Sylwestrowicz, W. D., *Corrosion,* SBIIA, Vol. 25, 1969, p. 405.
[72] Pickering, H. W., Beck, F. H., and Fontana, M. G., *Corrosion,* SBIIA, Vol. 18, 1962, p. 230t.
[73] Bombara, G., *Corrosion Science,* CRRSA, Vol. 9, 1969, p. 519.
[74] Staehle, R. W., *Corrosion,* SBIIA, Vol. 23, 1967, p. 202.
[75] Erben, A. R., *Materials Protection,* MAPRA, Vol. 4, No. 8, 1965, p. 18.
[76] Miller, R. N., *Materials Protection,* MAPRA, Vol. 6, No. 2, 1967, p. 55.
[77] *Industrial Research,* Dec. 1970, p. 58 and letter from Griessel, R. W., Research and Development Center, General Electric Co., Schenectady, N. Y., 1971.
[78] Mears, R. B., *Australasian Corrosion Engineering,* ACSEA, Vol. 6, No. 12, 1962.
[79] Burton, W. H., *Materials Protection,* MAPRA, Vol. 6, No. 2, 1967, p. 22.
[80] Wyche, E. H., Voigt, L. R., and LaQue, F. L., *Transactions,* Electrochemical Society, TESOA, Vol. 89, 1946, p. 149.
[81] Pocalyko, A., *Materials Protection,* MAPRA, Vol. 4, No. 6, 1965, p. 10.
[82] *Iron Age,* IRAGA, 4 Feb. 1971, p. 56.
[83] May, T. P. and Humble, H. A., *Corrosion,* SBIIA, Vol. 8, No. 2, 1952, p. 50.
[84] Vreeland, D. C. and Bedford, G. T., *Materials Protection and Performance,* Vol. 9, No. 8, 1970, p. 31.
[85] Lennox, T. J., Jr., Groover, R. E., and Peterson, M. H., *Materials Protection,* MAPRA, Vol. 8, No. 5, 1969, p. 41.
[86] Peterson, M. H., Lennox, T. J., Jr., and Groover, R. E., *Materials Protection,* MAPRA, Vol. 9, No. 1, 1970, p. 23.
[87] Vreeland, D. C., *Materials Protection,* MAPRA, Vol. 9, No. 1, 1970, p. 32.
[88] Peers, A. M. and Evans, U. R., *Journal,* Chemical Society, JCSOA, 1953, p. 1093.
[89] Shreir, L. L., *Corrosion,* Wiley, New York, 1963.
[90] Shreider, A. V. and Gintsberg, S. A., *Zhurnal Prikladnoi Khimii,* ZORKA, Vol. 33, 1960, p. 2541.
[91] Freedman, A. J. and Dravnieks, A., *Corrosion,* SBIIA, Vol. 14, 1958, p. 567t.
[92] Hatch, G. B., *Corrosion,* SBIIA, Vol. 21, 1965, p. 179.
[93] Tuthill, A. H. and Schillmoller, C. M., *Transactions,* Marine Technology Society, MTSTA, Vol. 3, 1965.
[94] Awaya, H., Yoshimatsu, H., Kajiyama, Y., and Tomoguchi, S., *Corrosion Engineering,* BOGJA, Vol. 16, No. 3, 1967.
[95] Jackson, J. D. and Boyd, W. K., *Applications Related Phenomena in Titanium Alloys, ASTM STP 432,* American Society for Testing and Materials, 1968, p. 218.
[96] Feige, N. G. and Kane, R. L., *Materials Protection and Performance,* Vol. 9, No. 8, 1970, p. 13.
[97] Ruskol, Y. S. and Klinov, I. Y., *Protection of Metals,* PTNMA, Vol. 2, 1966, p. 544.
[98] Schlain, D. and Kenahan, C. B., *Corrosion,* SBIIA, Vol. 12, 1956, p. 422t.
[99] Shimose, T. and Takamura, A., *Journal,* Japan Institute of Metals, NIKGA, Vol. 29, 1965, p. 416.

[*100*] Takamura, Akira, *Corrosion,* SBIIA, Vol. 23, 1967, p. 306.
[*101*] Solovena, Z. F., Fokin, M. N., and Timonin, V. A., *Protection of Metals,* PTNMA, Vol. 6, 1970, p. 187.
[*102*] *Battelle Technical Review,* BATRA, Vol. 16, No. 5, 1967.
[*103*] Rosenfeld, I. L. and Marshakov, I. K., *Zhurnal Fizicheskoi Khimii,* ZORKA, Vol. 31, 1957, p. 2328.
[*104*] Tolstaya, M. A., Flegontova, L. N., and Dmitriev, Y. V., *Protection of Metals,* PTNMA, Vol. 6, 1970, p. 514.
[*105*] Uhlig, H. H., *The Corrosion Handbook,* Wiley, New York, 1955.
[*106*] Pray, H. A. and Berry, W. E., Atomic Energy Commission Publication, BMI-886, 24 Nov. 1953.
[*107*] Ulanovskii, I. B., *Korroziya Metallov i Splavov,* KMSPA, No. 2, 1965, p. 359.
[*108*] Gleekman, L. W. and Swandby, R. K., *Corrosion,* SBIIA, Vol. 17, 1961, p. 144t.
[*109*] Bianchi, G., *Corrosion,* SBIIA, Vol. 14, 1958, p. 245t.
[*110*] Greene, N. D., France, W. D., Jr., and Wilde, B. E., *Corrosion,* SBIIA, Vol. 21, 1965, p. 275.
[*111*] Strocchi, P. M., Sinigaglia, D., Vicentini, B., *Electrochimica Metallorum,* ELMTB, Vol. 2, 1967, p. 38.
[*112*] Sinigaglia, D., Taccani, G., and Bombara, G., *Electrochimica Metallorum,* ELMTB, Vol. 3, 1968, p. 297.
[*113*] Strocchi, P. M., Sinigaglia, D., and Vicentini, B. in *Proceedings,* Third International Congress on Metallic Corrosion, Vol. 1, Moscow, 1969, p. 281.
[*114*] Strocchi, P. M., Taccani, G., and Sinigaglia, D., *Electrochimica Metallorum,* ELMTB, Vol. 4, 1969, p. 49.
[*115*] France, W. D., Jr., *Materials Research and Standards,* MTRSA, Vol. 9, No. 8, 1969, p. 21.
[*116*] Bond, A. P. and Lizlovs, E. A., *Journal,* Electrochemical Society, JESOA, Vol. 115, 1968, p. 1130.
[*117*] Lizlovs, E. A. and Bond, A. P., *Journal,* Electrochemical Society, JESOA, Vol. 116, 1969, p. 574.
[*118*] Recommended Practice for Standard Reference Method for Making Potentiostatic and Potentiodynamic Anodic Polarization Measurements," Designation: G 5–71, *Book of Standards,* American Society for Testing and Materials, 1971.
[*119*] Tikkanen, M. H. A., "Modern Methods on Studying Corrosion Resistance of Sintered Stainless Steels," Hoeganaes Sponge Iron Corp., *Stainless Steel Powders Seminar,* Detroit, 25 Feb. 1965, p. 41.
[*120*] Fokin, M. N. and Timonin, V. A., *Doklady Akademii Nauk SSSR,* DANKA, Vol. 164, No. 1, 1965, p. 150.
[*121*] Hodge, F. G. and Wilde, B. E., *Corrosion,* SBIIA, Vol. 26, 1970, p. 146.
[*122*] Peters, J. M. and Myers, J. R., *Corrosion,* SBIIA, Vol. 23, 1967, p. 326.
[*123*] Greene, N. D. and Judd, G., *Corrosion,* SBIIA, Vol. 21, 1965, p. 15.
[*124*] France, W. D., Jr., and Greene, N. D., *Corrosion,* SBIIA, Vol. 24, 1968, p. 403.
[*125*] Walker, M. S. and Rowe, L. C., *Corrosion,* SBIIA, Vol. 25, 1969, p. 47.
[*126*] Mansfeld, F. and Uhlig, H. H., *Journal,* Electrochemical Society, JESOA, Vol. 115, 1968, p. 900.
[*127*] Steigerwald, R. F. and Greene, N. D., *Journal,* Electrochemical Society, JESOA, Vol. 109, 1962, p. 1026.
[*128*] Mottern, M. M. and Myers, J. R., *Corrosion,* SBIIA, Vol. 24, 1968, p. 197.
[*129*] Wilde, B. E. and Greene, N. D., Jr., *Corrosion,* SBIIA, Vol. 25, 1969, p. 300.
[*130*] Vermilyea, D. A. and Tedmon, C. S., Jr., *Journal,* Electrochemical Society, JESOA, Vol. 117, 1970, p. 437.

E. A. Lizlovs[1]

Crevice Corrosion of Some High-Purity Ferritic Stainless Steels

REFERENCE: Lizlovs, E. A., **"Crevice Corrosion of Some High-Purity Ferritic Stainless Steels,"** *Localized Corrosion—Cause of Metal Failure, ASTM STP 516,* American Society for Testing and Materials, 1972, pp. 201–209.

ABSTRACT: Crevice corrosion of some ferritic stainless steels of low-interstitial content containing 2 percent molybdenum and chromium from 20 to 28 percent, and of 25Cr steels containing up to 5 percent molybdenum was studied using the potentiostatic polarization technique and a special cell developed at the laboratory. The crevice corrosion resistance in chloride containing media increased with increased chromium content at the same molybdenum level and with increased molybdenum content at the same chromium level. The 26Cr-1Mo steel was immune to crevice corrosion at room temperature. As the molybdenum content was increased, the immunity to crevice corrosion was extended to higher temperatures (up to 60 C), and also, as the chromium content was increased from 25 to 28 percent for the 2 percent molybdenum level, the immunity to crevice corrosion extended from room temperature to higher temperatures (up to 50 C).

KEY WORDS: corrosion, crevice corrosion, polarization, ferritic stainless steel, molybdenum, chromium, corrosion tests

High-purity ferritic stainless steels are involved in growing commercial interest. The excellent corrosion resistance of this class of alloys has been demonstrated in work at the Climax Research Laboratory and by other investigators [*1–5*].[2] In particular, the earlier work in this laboratory indicated that molybdenum-containing ferritic stainless steels of low-interstitial content possess a remarkable resistance to crevice corrosion in chloride media. In an extension of the previous work, the crevice corrosion resistance of several high-chromium ferritic alloys was investigated.

Materials

The experimental alloys were produced by induction melting and casting in vacuum. The chemical compositions are given in Table 1. The ingots

[1] Senior research associate, Climax Molybdenum Company of Michigan (a subsidiary of American Metal Climax, Inc.) Ann Arbor, Mich. 48106.

[2] The italic numbers in brackets refer to the list of references appended to this paper.

TABLE 1—*Chemical compositions of the alloys studied.*

Steel Type	Composition, %								
	Cr	Mo	Ni	Si	P	S	N	C	Al
26Cr-1Mo	26.15	0.99	...	...	...	...	0.016	0.003	...
20Cr-2Mo	19.70	2.01	...	...	...	...	0.010	0.003	...
22.5Cr-2Mo	22.52	2.00	...	...	...	...	0.011	0.003	...
25Cr	25.66	0.002	0.13	...	...	...	...	0.002	...
25Cr-2Mo	25.02	1.99	...	...	...	...	0.013	0.003	...
28Cr-2Mo	27.86	2.00	...	...	...	...	0.021	0.007	...
25Cr-3.5Mo	25.58	3.49	0.13	...	...	...	...	0.002	...
25Cr-5Mo	25.25	5.24	0.13	0.019	0.012	0.002	0.013	0.002	0.006

were hot forged at 2000 F (1090 C) to ¾-in. (20-mm)-thick plate, then hot rolled to 0.30 in. (7.6 mm) at 1900 F (1040 C) and finally cold rolled to 0.150-in. (3.8-mm)-thick strip. All materials were annealed at 1800 F (980 C) for 1 h and water quenched.

Experimental Procedures

Potentiostatic Crevice Corrosion Tests

Potentiostatic crevice corrosions tests were conducted in a cell developed in the Climax Laboratory [*6*]. A crevice in this cell is generated by pressing a glass bead against the disk-shaped electrode. The unwanted crevices are eliminated by pressing the electrode against a Teflon gasket by means of a polyvinyl chloride bolt. The crevice corrosion cell was adopted for use above room temperature by wrapping heating tape around the cell. The applied heat was controlled by a thermistor immersed in the electrolyte. Temperature control was ± 0.2 C.

All potentiostatic crevice corrosion experiments were performed in 1 *N* sodium chloride (NaCl) solution at room temperature (21 to 23 C), 30, 40, 50, and 60 C. The electrolyte was purged continuously with prepurified nitrogen. Experiments usually were conducted at 0.600 and 0.400 V and occasionally at 0.000 V with reference to saturated calomel electrode (SCE). Polarization current and potential were recorded as functions of time.

The electrolyte was allowed first to reach the desired temperature. Because of the cell arrangement, the electrode was in contact with the solution during this period. Then the potentiostat with a voltage preset at 0.0 V or less was switched on. After the initial current surge subsided, the electrode potential was adjusted manually to the desired value; and, after the polarization current stabilized, the crevice was applied to the electrode surface, that is, the glass bead was advanced until it pressed against the electrode surface. The electrode with the crevice applied was usually kept at a given potential for 2 to 4 h, and then the crevice was removed from the electrode surface. Finally, the electrode was removed from the assembly and inspected for corrosion damage. The results were considered acceptable only when the corrosion damage occurred under the bead and in no other place.

Crevice Corrosion Tests in Ferric Chloride Solutions

Crevice corrosion tests with most of the materials were also carried out in 10 percent ferric chloride ($FeCl_3$) solutions (pH1). Specimens for these tests consisted of 1 by ¾ by 0.15 in. thick (25 by 20 by 3.8 mm) surface ground and polished through No. 600 wet silicon carbide paper, and finally through 0.3μm alumina-distilled water slurry (same polishing procedure as for the disk electrodes for potentiostatic tests). Crevices were formed by stretching a rubber band around the specimen with two Teflon cylinders placed on the major faces of the specimen. Thus, four crevices were generated, two where the rubber band contacted the edges and two where the Teflon cylinders contacted the faces of the specimen. Crevice corrosion damage was estimated by visual inspection and weight loss determination.

Results

Potentiostatic Crevice Corrosion Tests

Polarization currents as a function of time for various stainless steel electrodes at different temperatures in 1 *N* NaCl are shown in Figs. 1 through 5. Since the electrode area for all specimens was the same (1.26

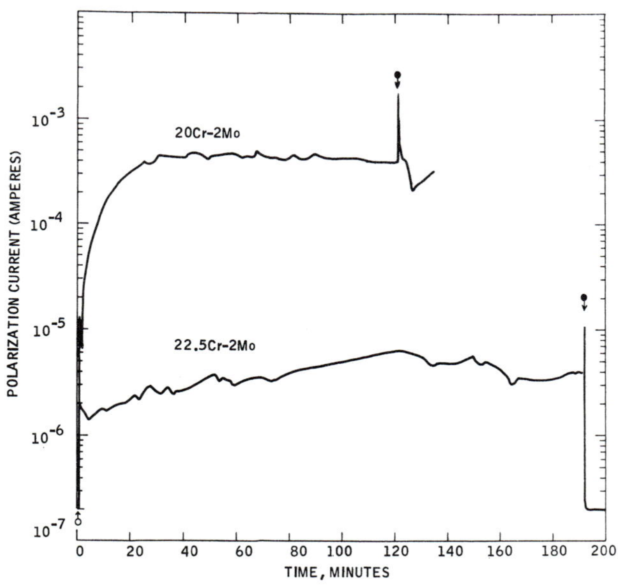

○↑-Crevice applied to electrode surface.
●↓-Crevice removed from electrode surface.

FIG. 1—*Variations in polarization current with time at 0.00 V (SCE) in 1* N *NaCl at 22 ± 1 C.*

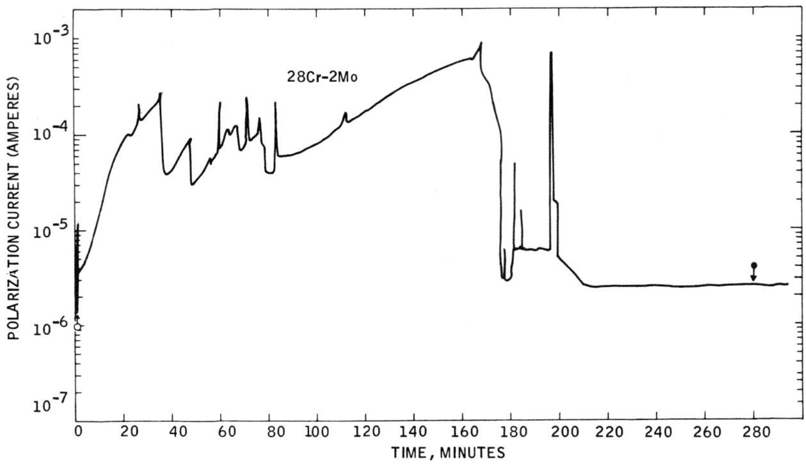

-Crevice applied to electrode surface.
-Crevice removed from electrode surface.

FIG. 2—*Variations in polarization current with time at +0.600 V (SCE) in 1* N *NaCl at 50 ± 0.2 C.*

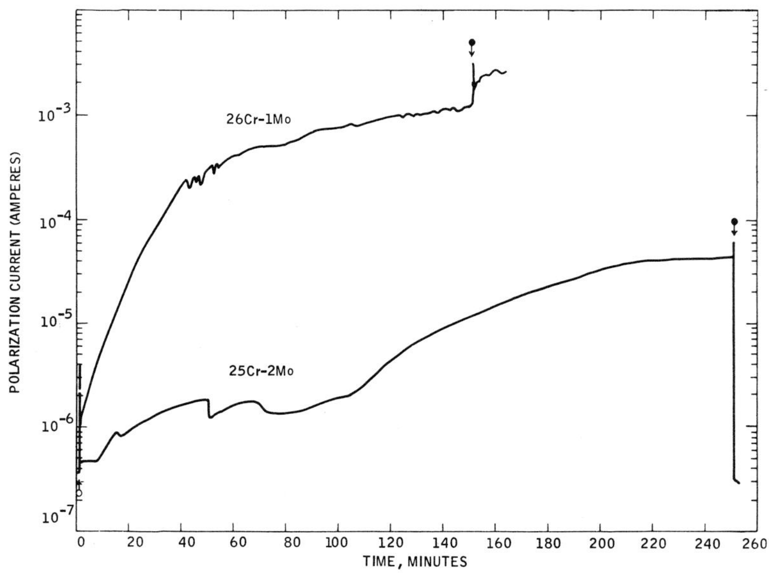

-Crevice applied to electrode surface.
-Crevice removed from electrode surface.

FIG. 3—*Variations in polarization currents with time at +0.400 V (SCE) in 1* N *NaCl at 30 ± 0.2 C.*

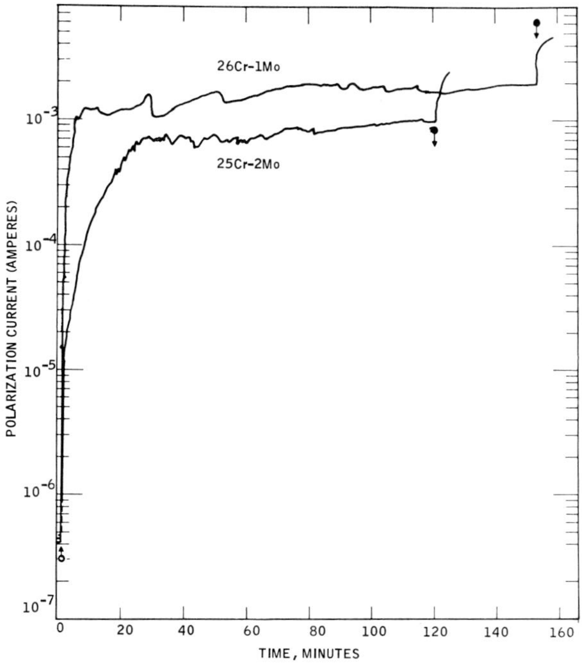

○↑-Crevice applied to electrode surface.
●↓-Crevice removed from electrode surface.

FIG. 4—*Variations in polarization currents with time at +0.400 V (SCE) in 1* N *NaCl at 40 ± 0.2 C.*

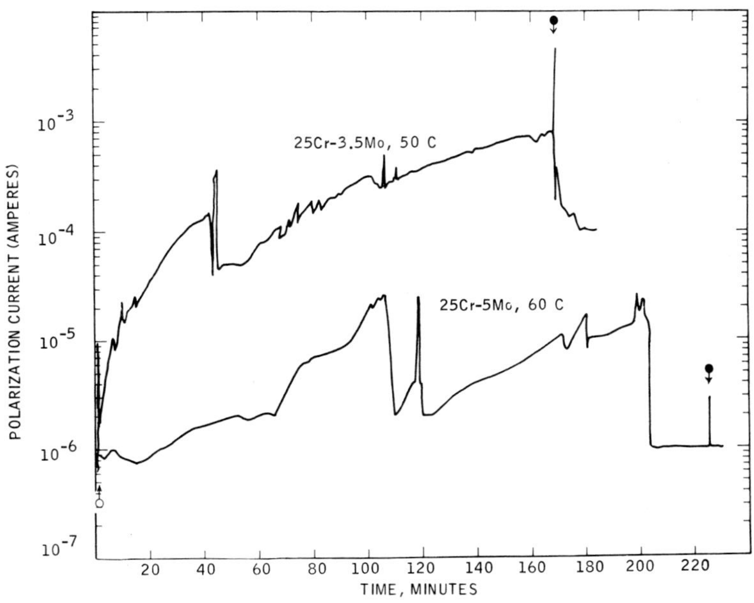

○↑-Crevice applied to electrode surface.
●↓-Crevice removed from electrode surface.

FIG. 5—*Variation in polarization currents with time at +0.600 V (SCE) in 1* N *NaCl.*

cm^2) and the same glass bead was applied to the surface, the polarization current itself rather than current density may be used to compare the relative rates of crevice corrosion. The results from the potentiostatic crevice corrosion testing are summarized in Table 2.

Polarization currents as a function of time for various stainless steel electrodes at different temperatures in 1 *N* NaCl are shown in Figs. 1 through 5. Only a small polarization current, 10^{-7} to 10^{-6} A, is recorded prior to the application of the glass bead to electrode surface, characteristic of stainless steel in a passive state with virtually no corrosion. After the glass bead is pressed against the electrode surface, the passive polarization current either remains unchanged, if material is resistant to crevice

TABLE 2—*Results of controlled potential crevice corrosion test in 1* N *NaCl cell on ferritic stainless steels.*

	Presence of Crevice Corrosion at Indicated Test Temperature				
Steel Type	21 to 22 C	30 C	40 C	50 C	60 C
20Cr-2Mo	yes	...	...	...	...
22.5Cr-2Mo	yes	...	...	...	...
25Cr	yes	...	...	...	...
26Cr-1Mo	no	yes	yes	...	...
25Cr-2Mo	no	yes	yes	...	...
28Cr-2Mo	no	no	no	yes	...
25Cr-3.5Mo	no	no	no	yes	...
25Cr-5Mo	no	no	no	no	yes

corrosion, or rises sharply if crevice corrosion is initiated under given experimental conditions. Since the only anodic reaction is the dissolution of metal under the glass bead, the anodic current is directly proportional to the amount of metal dissolved under the glass bead, and the time-polarization current diagram shows the initiation and progress of crevice corrosion. However, since corrosion rates are potential dependent, only corrosion rates obtained at the same potential can be compared with each other. The results from potentiostatic crevice corrosion testing are summarized in Table 2.

In general, as the chromium content was increased, the resistance to crevice corrosion was also increased. The lowest resistance to crevice corrosion was shown by the 20Cr-2Mo alloy. The crevice corrosion for this material was initiated readily at a rather low temperature (22 ± 1 C) at 0.00 V. Upon removal of the crevice from the electrode surface, a stable pit developed at the site of the crevice damage, as evidenced by continuing high polarization current (Fig. 1). Under the same experimental conditions, crevice corrosion was also initiated on the 22.5Cr-2Mo alloy, but the crevice corrosion current in this case was two orders of magnitude less than for the 20Cr-2Mo alloy, and the electrode repassivated upon removal of the crevice (Fig. 1). No crevice corrosion at room temperature could be initiated on the 25Cr-2Mo alloy. For the 28Cr-2Mo alloy some crevice corrosion could be initiated at 50 C; however, this material repassivated after some initial crevice corrosion before the bead was removed from the surface even at +0.60 V (Fig. 2).

Molybdenum content plays an even more striking role in increasing resistance to crevice corrosion. High-purity 25Cr steel suffered crevice corrosion at 0.00 V and room temperature [*6*]. Crevice corrosion current for 25Cr alloy was approximately the same as for 20Cr-2Mo alloy except the 25Cr steel repassivated after the bead was removed from the surface, while 20Cr-2Mo continued to corrode. Both the 26Cr-1Mo and 25Cr-2Mo alloys were resistant to crevice corrosion at room temperature (up to approximately 24 C). Crevice corrosion could be initiated on both materials at 30 C, although the crevice corrosion current for the 25Cr-2Mo alloy was about two orders of magnitude lower than for the 26Cr-1Mo steel (Fig. 3). On removal of the crevice, the 25Cr-2Mo electrode repassivated while the 26Cr-1Mo alloy continued to corrode at +0.40 V, that is, a stable pit developed at the site of the crevice corrosion damage. At 40 C both materials suffered severe crevice corrosion and stable pits developed at +0.40 V. The crevice corrosion current for the 26Cr-1Mo alloy was about twice that for the 25Cr-2Mo alloy (Fig. 4); thus, the 25Cr-2Mo steel is more resistant to crevice corrosion than the 26Cr-1Mo steel. The 25Cr-3.5Mo and 25Cr-5Mo steels were completely resistant to crevice attack at temperatures of 40 C or lower.

Crevice corrosion on 25Cr-3.5Mo steel was initiated at 50 C and on 25Cr-5Mo at 60 C (Fig. 5). The rate of crevice corrosion for 25Cr-3.5Mo steel at +0.60 V was approximately the same as for 28Cr-2Mo, except that 25Cr-3.5Mo steel did not repassivate but continued to corrode, indicating that a stable pit was formed at the site of crevice damage.

Crevice corrosion on the 25Cr-5Mo at +0.60 V and 60 C could be initiated only after repeated application of the bead to the surface. The corrosion rate was quite low (not exceeding 30 μA), and repassivation occurred while the bead was still applied to the electrode surface (Fig. 5).

Crevice Corrosion Tests in Ferric Chloride

The results from crevice corrosion tests in acidified $FeCl_3$ solution are presented in Table 3. Comparison of Table 2 with Table 3 shows that the $FeCl_3$ test was slightly less severe than the polarization test. Nevertheless, both tests ranked the alloys in the same order of resistance. It should be noted that the $FeCl_3$ test, even at room temperature, is a very severe test, and its severity increases greatly with the temperature.

Discussion

Comparison of the results from potentiostatic crevice corrosion tests gave the following order of merit for resistance to crevice corrosion:

TABLE 3—*Crevice corrosion tests in 10FeCl$_3$·6H$_2$O + 0.1* N *HCl.* Weight Loss in mg/dm^2 per Day Computed over Entire Coupon Area (test duration in days given in parentheses)

	Test Temperature			
Steel Type	25 C	30 C	40 C	50 C
20Cr-2Mo	64 (1)	197 (1)	...	...
22.5Cr-2Mo	0 (7)	3 (1)	...	...
25Cr	670 (2.5)	...	...	...
26Cr-1Mo	0 (23)	6.2 (4)	229 (1)	764 (1)
25Cr-2Mo	0 (7)	0 (1)	0 (1)	210 (1)
28Cr-2Mo	0 (7)	0 (4)	0 (1)	13 (1)
25Cr-3.5Mo	0 (24)	...	...	7 (5)
25Cr-5Mo	0 (63)	...	...	0 (11)

20Cr-2Mo (least resistant) <25Cr <22.5Cr-2Mo <26Cr-1Mo <25Cr-2Mo <28Cr-2Mo <25Cr-3.5Mo <25Cr-5Mo. This order of merit is also supported by the crevice corrosion tests in ferric chloride. Thus, an increase in chromium content at the same molybdenum level, or an increase of molybdenum content at the same chromium level, increases the resistance to crevice corrosion and the temperature region over which materials are immune to crevice corrosion. In other words, when the ferritic stainless steels have a single phase microstructure, higher chromium or molybdenum contents stabilize the passive film against localized breakdown that causes crevice and pitting corrosion. While pitting corrosion was not studied in detail, it can be generalized that stainless steels which are immune to crevice corrosion are also immune to pitting corrosion at the same chloride concentration and temperature. No contradiction has been found for this statement either in the work of this laboratory or in the literature.

The materials in potentiostatic crevice corrosion tests showed three types of behavior (excluding the case of immunity to crevice corrosion): repassivation with the bead still applied to the electrode surface, repassivation after the bead is removed from the surface, and continuation of corrosion after the bead is removed from the surface.

Materials having the ability to repassivate with the bead still on the surface could be highly resistant to crevice corrosion, provided the repassivation under the bead occurs at sufficiently high potentials (+0.60 V (SCE) or higher). It is reasonable to assume that materials that continue to corrode after the removal of the crevice may pit at a given chloride ion concentration and temperature; it is possible that these materials may not pit under some conditions in ferric chloride or in potentiostatic tests. If their passive films were damaged mechanically or a temporary crevice generated on a surface of the material, a growing pit could be expected at the site of the damage.

The good agreement between the electrochemical crevice corrosion tests and the ferric chloride exposure tests is considered especially important because of the difficulty that has been experienced in evaluating crevice corrosion resistance by laboratory procedures. Although further studies, especially long-time exposure tests, will be still required to confirm the results, it now appears that by electrochemical testing, conditions leading to immunity to crevice corrosion can be defined, that is, a temperature and chloride concentration can be established below which crevice corrosion would not occur under any circumstances for a given condition within a reasonable time period.

Conclusions

1. Increase in chromium or molybdenum content or both for high-purity single-phase ferritic stainless steels increased their resistance to crevice corrosion in chloride environment.

2. Ferritic stainless steels containing 25 to 26 percent chromium were immune to crevice corrosion at room temperature and 1 *N* NaCl concentration provided they contained at least 1 percent molybdenum. Increase in molybdenum content resulted in increase in temperature at which 25Cr steels were immune to crevice corrosion.

3. The relative order of merit of resistance to crevice corrosion was: 25Cr-2Mo (least resistant <25Cr <22.5Cr-2Mo <25Cr-1Mo <25Cr-2Mo <25Cr-3.5Mo <28Cr-2Mo <25Cr-5Mo.

References

[1] Chernova, G. P. and Tomashov, N. D., *Korroziya Metallov i Splavov,* KMSPA, Sb. 2, Moscow, 1965, pp. 7–13.

[2] Schwartz, C. D., Franson, I. A., and Hodges, R. J., *Chemical Engineering,* CHEEA, 20 April 1970, pp. 164–169.

[3] Lizlovs, E. A. and Bond, A. P., *Journal,* Electrochemical Society, JESOA, Vol. 116, No. 5, May 1969, pp. 574–579.

[4] Bond, A. P. and Lizlovs, E. A., *Werkstoffe und Korrosion,* WSKRA, Vol. 21, No. 5, May 1970, pp. 336–342.

[5] Lizlovs, E. A. and Bond, A. P., *Journal,* Electrochemical Society, JESOA, Vol. 118, No. 1, Jan. 1971, pp. 22–28.

[6] Lizlovs, E. A., *Journal,* Electrochemical Society, JESOA, Vol. 117, No. 10, Oct. 1970, pp. 1335–1337.

R. P. Jackson[1] *and D. van Rooyen*[1]

Crevice Corrosion of Some Ni-Cr-Mo-Fe Alloys in Laboratory Tests

REFERENCE: Jackson, R. P. and van Rooyen, D., **"Crevice Corrosion of Some Ni-Cr-Mo-Fe Alloys in Laboratory Tests,"** *Localized Corrosion—Cause of Metal Failure, ASTM STP 516,* American Society for Testing and Materials, 1972, pp. 210–221.

ABSTRACT: Results were obtained for six alloys, evaluated in the laboratory for crevice corrosion resistance. Two methods were used: (1) direct observation of corrosion damage in several electrolytes and (2) electrochemical determination of polarization curves. Results showed the ratings of the alloys in the order of decreasing resistance using both methods. In the direct crevice corrosion tests, it was found that the crevice size was important, especially for the more resistant alloys. Some scatter was found in those results. The polarization data provided more detail of corrosion behavior over a wide range of potential; therefore, the electrochemical test was considered a superior tool in alloy screening and development. Neither method, at present, can predict exact field performance.

KEY WORDS: corrosion, alloys, concentration cell corrosion, electrochemistry, electrolytes, polarization (charge separation), anodic polarization, pitting

Electrochemical methods have been used in corrosion studies for many years. Potentiostatic steps or sweeps are now used frequently in the course of obtaining polarization data, in addition to the older procedure based on controlled impressed current. Both techniques are generally accepted, although finer procedural details have varied quite widely. However, a correlation of the electrochemical data with direct immersion corrosion results has not been usually made in the literature.

The work in this paper was aimed at adding to the understanding of some of the test variables in potentiostatic polarization and in the direct immersion crevice corrosion experiments. The results obtained by these two laboratory methods were found useful for alloy screening and alloy development, and they provided some information about the mechanism of crevice corrosion.

[1] Research investigator and section supervisor, respectively, The International Nickel Co., Inc., Paul D. Merica Research Laboratory, Sterling Forest, Suffern, N. Y. 10901.

Experimental

Materials

Table 1 lists the compositions of the alloys used; heat treatments of the alloys are listed in Table 2.

Procedure for Direct Immersion

Figure 1 shows a picture of a crevice corrosion coupon including nuts and bolts made of Teflon[2]; these were screwed onto the specimen with a 0.25-mm Teflon washer in between, providing a "wide" crevice. In some alloys a tighter crevice was examined, formed by placing six very small, square pieces of thin foil of Teflon (0.0125 mm thick) under each of the outside corners of the bolt, which was then hand tightened as much as possible; the thicker washer was eliminated. In all tests the specimen size was 5 by 5 cm, and the volume of test solution was about one liter.

Immersion tests were made for 72 h in 5 test solutions using wide crev-

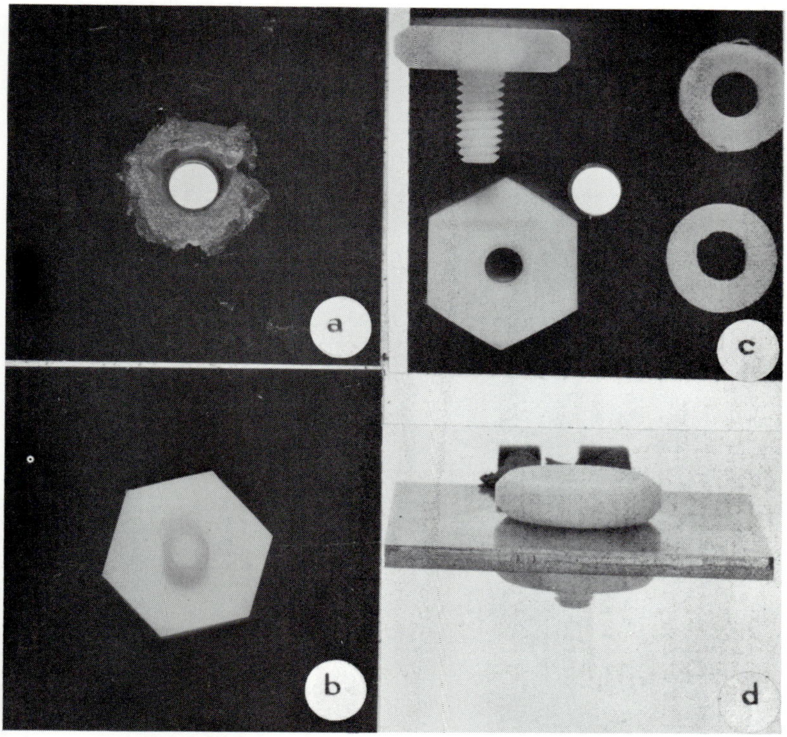

FIG. 1—*Makeup of the crevice corrosion specimens used in the experiments:* (a) *as crevice corroded,* (b) *front view,* (c) *bolt, nut, and washers used, and* (d) *side view.*

[2] Registered trademark of E. I. duPont de Nemours and Co., Inc.

TABLE 1—*Composition of alloys.*

Alloy	Cr	Mo	Ni	Fe	Cb	C	Al	Ti	Mg	Other
Exp Ni-base	17.9	12.3	balance	...	...	0.038	0.13	0.27	0.034	...
Exp Fe-base	19.7	8.4	26.8	balance	0.39	0.024	0.43	0.46	0.005	...
Hastelloy alloy C[a]	16.0	16.0	balance	5.0	...	0.04	0.08	...	...	W 3.8, Co 2.5 max
Incoloy alloy 825[a]	21.5	3.0	42.0	30.0	...	0.03	0.1	0.9	...	Cu 2.2
Inconel alloy 625[a]	21.5	9.0	61.0	2.5	3.6	0.05	0.2	0.2	...	...
AISI 316[a]	17.0	2 to 3	12.0	balance	...	0.05	...	...	...	...

[a] Approximate nominal.

TABLE 2—*Heat treatment of the alloys used.*

Hastelloy alloy C	commercial-mill annealed as received
Incoloy alloy 825	
Inconel alloy 625	
AISI 316	
Exp Fe-base—1 h/2200 F/OQ[a]	
Exp Ni-base—1 h/2000 F/WQ[b]	

[a] OQ = oil quenched.
[b] WQ = water quenched.

ices. These test solutions were 5 percent $FeCl_3$ at room temperature, 5 percent $FeCl_3$ + 5 percent HCl at room temperature, 5 percent $FeCl_3$ at 50 C, 5 percent $CuCl_2$ at 50 C, and 5 percent NaOCl at 50 C. The tighter crevice specimens were exposed only to 5 percent ferric chloride ($FeCl_3$) at 50 C, and Type 326 stainless and alloys 825 were omitted from this test. Chemically pure materials and deionized water were used for making the solutions.

The surfaces of the specimens were machine ground to about a 30-μin. finish. Before testing, they were degreased in acetone, followed by absolute alcohol, and rinsed in distilled water.

Electrochemical Procedure

Specimens with an exposed surface of 10 cm^2 for the stepwise potentiostatic polarization tests were prepared from sheet. All nonexposed surfaces were masked with two coats of a single component epoxy that cured at around 160 C. Just prior to testing, the specimens were abraded with No. 240 silicon carbide paper, degreased in acetone, and washed in distilled water.

The (anodic) polarization measurements were conducted with a Wenking potentiostat, and voltage increases were programmed on a Wenking stepping potentiometer. Potentials, relative to a saturated calomel electrode, and the logarithms of the current densities were recorded continuously on an *X-Y* recorder. Tests were done in acids under several conditions of temperature, pH, and chloride concentration, and hydrogen was bubbled through the solutions during the total time of the exposure. No potential corrections were made for IR drop in the solution. A fine probe was used for the potential measurements, with its tip at a distance of between two and three times its outside diameter away from the specimen surface. Under these conditions, IR drop corrections would be less than 2 mV at 10^3 $\mu A/cm^2$.

Since pH values approaching 1 have been reported by others for the liquid inside crevices in stainless steel,[3] it was felt desirable to do the initial electrochemical tests at this pH level; the four most resistant alloys, however, were found to show no active peaks, and the breakout potentials for

[3] Greene, N. D. et al, *Corrosion,* SBIIA, Vol. 21, 1965, p. 275.

these test materials were high and not well separated. Raising the test temperature from 50 to 65 C did not affect these results significantly. Consequently, more concentrated electrolytes were chosen.

Results

Penetration Measurements

Using the wider of the two crevices (0.25-mm spacer), the tests were severe enough to cause attack of Type 316 stainless steel and alloy 825 in all five test media, as shown in Table 3. In the same tests, crevice corrosion was not observed on Inconel[4] alloy 625 or Hastelloy[5] alloy C, while the two experimental alloys showed attack only in some instances. Crevice corrosion was considerably more severe in very confined areas, and undercutting of spacers by corrosion was prevalent.

TABLE 3—*Results in several solutions with a "wide" crevice (72-h exposures).*

	Average Penetration in Attacked Area, mmpy[a]					
Test	Type 316	Alloy 825	Exp Fe-base	Exp Ni-base	Alloy 625	Hastelloy Alloy C
5% $FeCl_3$, 50 C	410	200	35	0.01[b], 83	0.04[b]	0.02[b]
5% $CuCl_2$, 50 C	23 950	200	5	0.003[b], 15	0.003[b]	0.01[b]
5% NaOCl, 50 C	73	46	0.02[b]	0.03[b], 10	0.003[b]	0.006[b]
5% $FeCl_3$ + 5% HCl, RT[c]	250	360	21	0.006[b]	0.004[b]	0.02[b]
5% $FeCl_3$, RT[c]	320	32	0.14[b]	0.004[b]	0.002[b]	0.02[b]

[a] Millimeters per year. Where large disagreement was evident between replicates of the experimental Ni-base alloy, individual penetrations are listed for each specimen.
[b] General corrosion only; no visible local attack.
[c] RT = room temperature.

Table 4 lists the results obtained in hot $FeCl_3$ with a more restricted crevice (0.0125-mm spacers). Very severe attack was observed on most specimens of all four alloys tested. Hastelloy alloy C, with the lowest rate, showed individual penetration rates of up to 10 mm per year, and an overall average of about 5.2 mm per year. Other overall average rates in millimeters per year increased as follows: Inconel alloy 625, 9.5; experimental

TABLE 4—*Results in 5 percent $FeCl_3$ at 50 C with "tight" crevice (72-h exposures).*

Alloy	Average Penetration in Attacked Area, mmpy[a]	Avg
Hastelloy alloy C	0.02[b], 0.04[b], 4, 7, 10, 10	5.2
Inconel alloy 625	5, 7, 8, 12, 12, 13	9.5
Exp Ni-base	0.006[b], 9, 13, 17, 18, 21	13.0
Exp Fe-base	8, 10, 12, 20, 25, 25	16.7

[a] Millimeters per year.
[b] General corrosion only; no visible local attack.

[4] Registered trademark of The International Nickel Co., Inc.
[5] Registered trademark of Cabot Corp.

Ni-base alloy, 13; and experimental Fe-base alloy, 16.7. Once again, it was immediately noticed that much deeper penetrations had occurred underneath the Teflon spacers that were used, although the crevice corosion was not confined to the undercut regions.

Polarization Behavior

A solution of 0.5 hydrogen chloride (HCl) + 1 *N* sodium chloride (NaCl) was found to provide a reasonable separation of the polarization curves of the four most resistant alloys, as shown in Fig. 2. The order of increasing resistance to pitting, judged by the potentials at which passivity

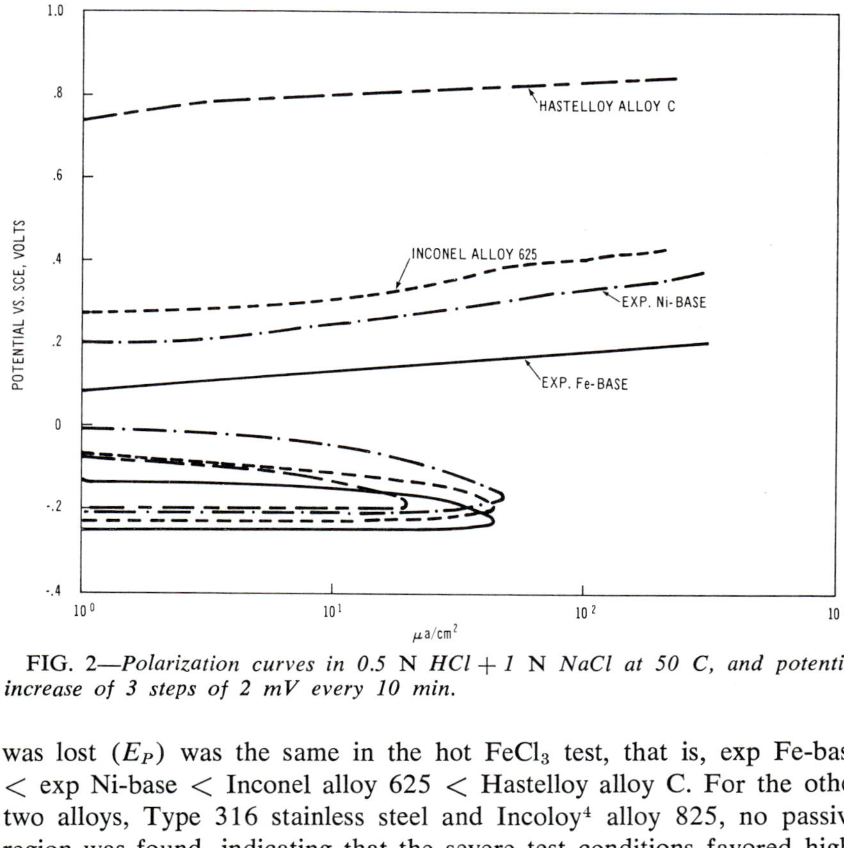

FIG. 2—*Polarization curves in 0.5* N *HCl + 1* N *NaCl at 50 C, and potential increase of 3 steps of 2 mV every 10 min.*

was lost (E_P) was the same in the hot $FeCl_3$ test, that is, exp Fe-base < exp Ni-base < Inconel alloy 625 < Hastelloy alloy C. For the other two alloys, Type 316 stainless steel and Incoloy[4] alloy 825, no passive region was found, indicating that the severe test conditions favored high-corrosion rates throughout for the less-resistant alloys. The two curves are plotted in Fig. 3. The critical current density, I_C (Fig. 2), of alloy C was the lowest in this series, but the other three were not much different from one another. In this instance, therefore, the E_P-values were more important in rating the alloys for crevice corrosion resistance.

After the tests were completed, the various specimens of the four resis-

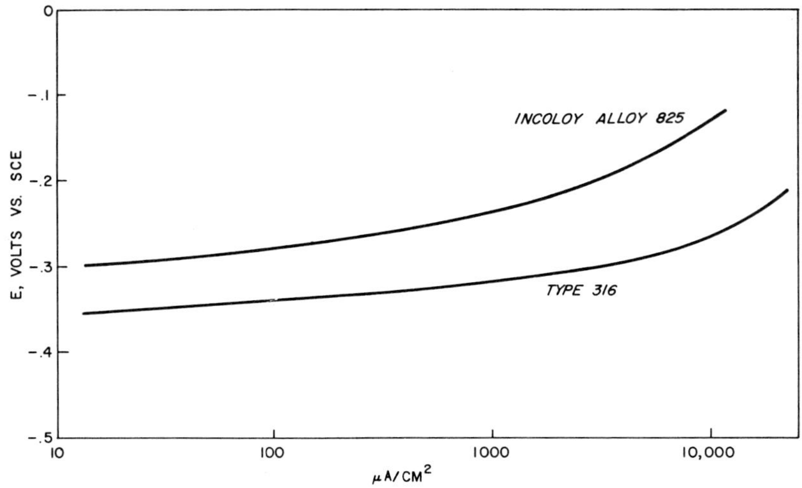

FIG. 3—*Polarization curves in 0.5* N *HCl +1* N *NaCl, and potential increase of 3 steps of 2 mV every 10 min.*

tant alloys were examined for pitting. The only one without localized attack was Hastelloy alloy C.

The effect of the rate of potential increase can be seen by comparing Figs. 2 and 4. Although the relative ratings of the alloys remained unchanged, the I_C-values tended to increase in the slower tests (done at 2-mV step every 10 min, compared to 2-mV step every 3.3 min) while the E_P-values decreased. Apparently equilibrium is not achieved in a few hours and may require several days. It is interesting to note that Hastelloy alloy C was not affected significantly by the change in stepping rate.

A similar effect of rate of potential change is shown for the experimental Fe-base alloy in Fig. 5, which contains data for a lower pH and higher Cl^- concentration (total $Cl^- = 3.5\ N$). In addition, the change in electrolyte caused an increase in anodic current densities, and passivity was obviously difficult to achieve at the test temperature, which was 50 C. The experimental Ni-base alloy also showed the effects of Cl^- concentration at two different Cl^- levels, as plotted in Fig. 6 for tests at 40 C.

Increasing test temperatures caused higher anodic current densities, as shown in Fig. 7 for the Fe-base alloy in 1 *N* HCl + 2.5 *N* NaCl. At 40 C there was reasonably good passivity, none at 60 C, and 50 C was intermediate between these two.

The usefulness of the polarization technique for alloy development and processing optimization is demonstrated in Fig. 8. The experimental Fe-base alloy was examined after various heat treatments, indicating an optimum solution annealing temperature in the vicinity of 2150 F (1177 C). Below this temperature, pitting resistance was indicated to be impaired. Results of exposures to $FeCl_3$ solutions agreed with the curves in Fig. 8.

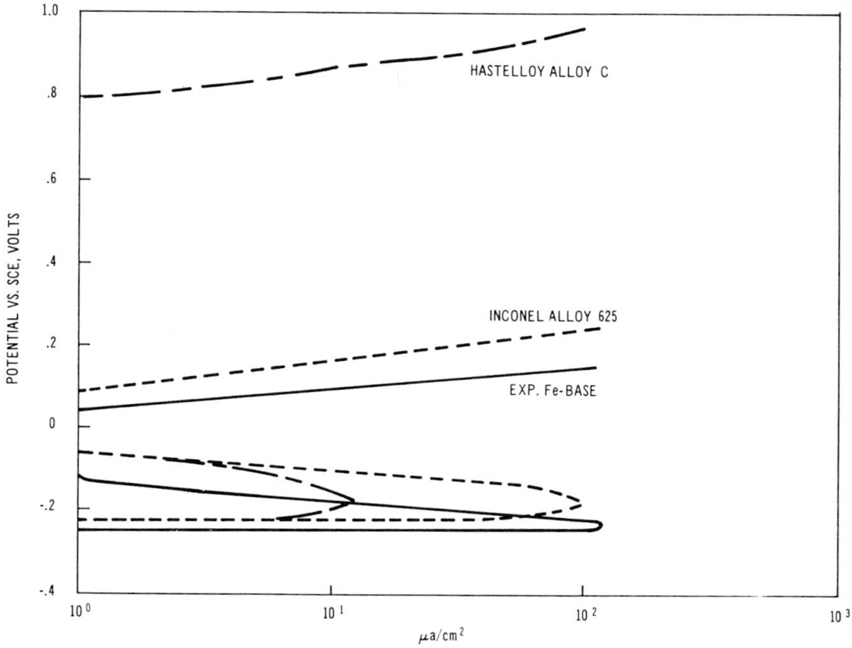

FIG. 4—*Polarization curves in 0.5* N *HCl + 1* N *NaCl, and potential increase of one 2-mV step every 10 min.*

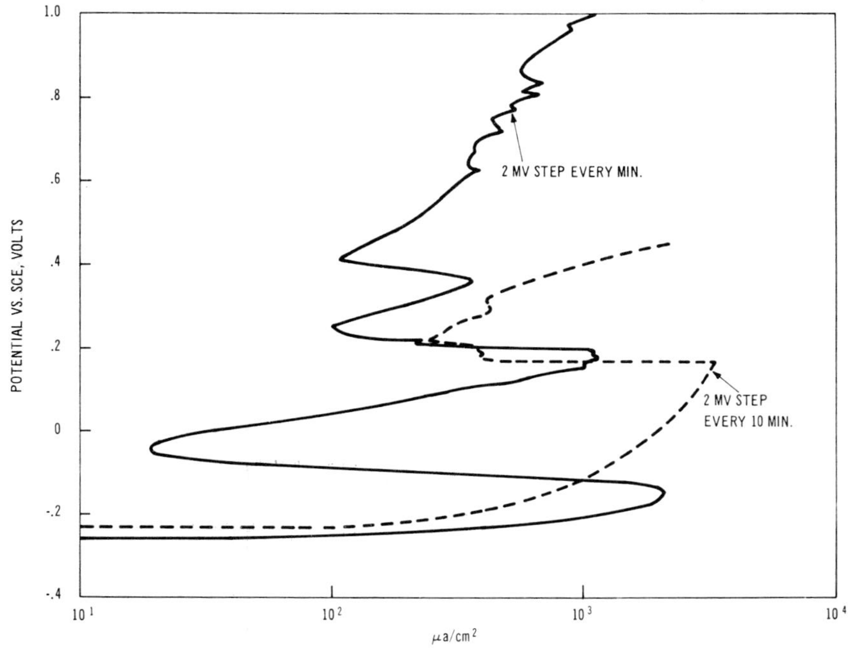

FIG. 5—*Effect of potential increase rate on polarization curves for exp Fe-base alloy in 1* N *HCl + 2.5* N *NaCl at 50 C.*

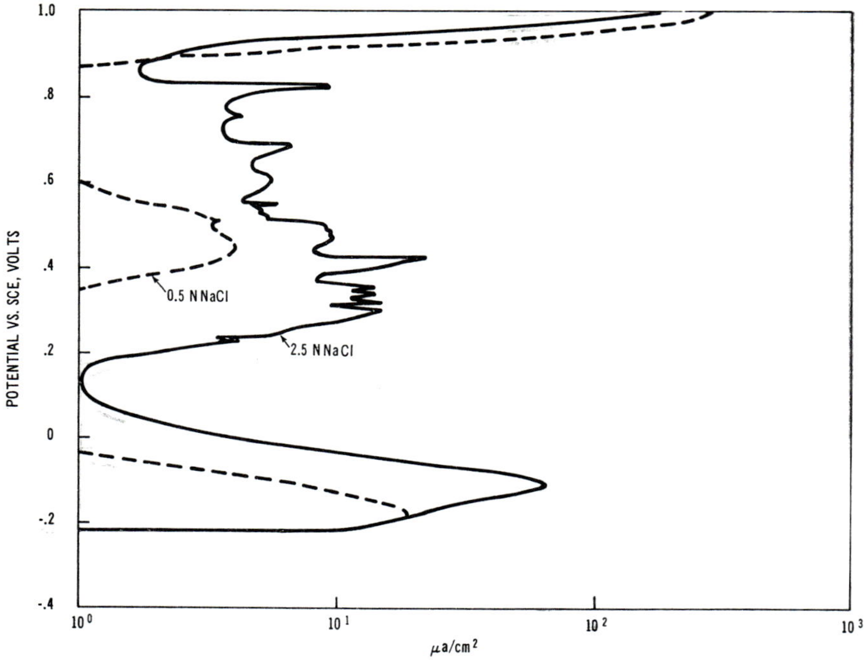

FIG. 6—*Effect of chloride ion concentration in exp Ni-base alloy polarization curves in 1* N *HCl at 40 C, with potential increased at 2-mV step every min.*

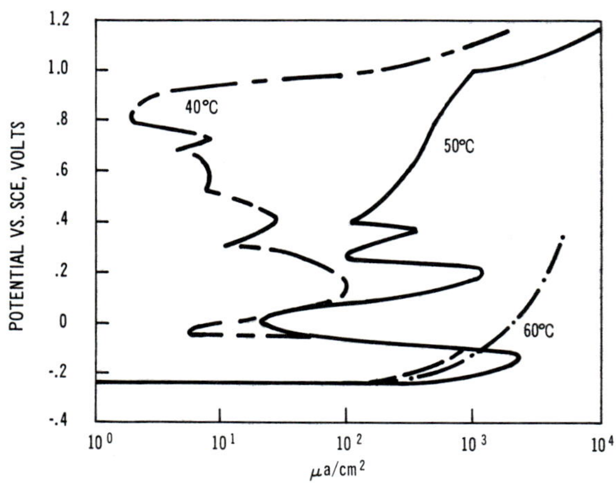

FIG. 7—*Effect of temperature on polarization curves for exp Fe-base alloy in 1* N *HCl + 2.5* N *NaCl, with potential increased at a 2-mV step every min.*

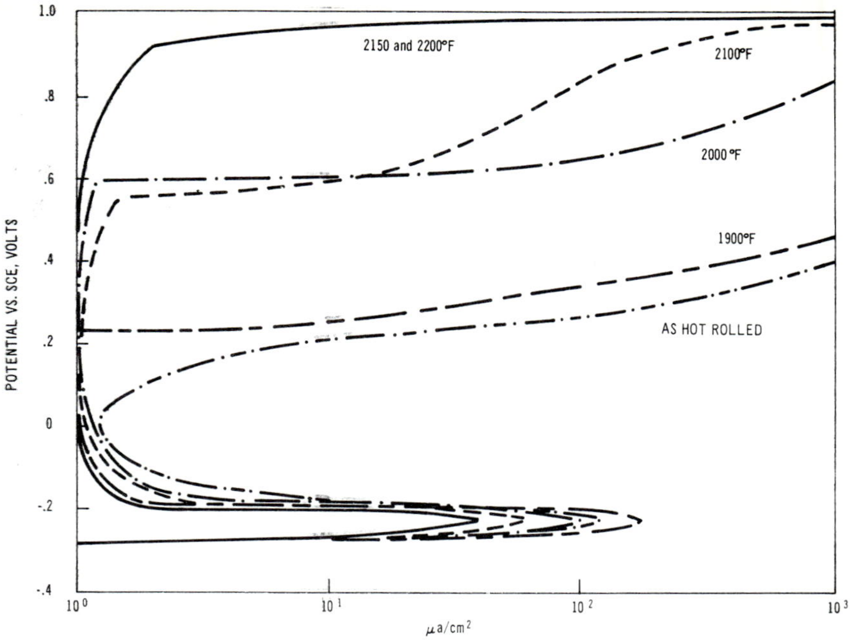

FIG. 8—*Effect of annealing temperature of exp Fe-base alloy on polarization curves in 1* N *H_2SO_4 + 1* N *NaCl, with potential increased at a 2-mV step every min.*

Discussion

There was a dependence of crevice corrosion on crevice geometry, and attack increased in severity as the crevice became tighter. This observation has practical implications and warns that widely scattered results can be expected if strict experimental control is not exercised. Even when it is, and assuming that laboratory results are reproducible, it would remain exceedingly difficult to translate them into practical terms (field performance).

In terms of mechanism, the direct immersion tests agree with suggestions[3] that corrosion is accelerated by the accumulation of Cl^- in the anodic area. The greatest and earliest accumulation of corrosion products (including H^+ and Cl^-) would be expected in the more confined areas, where the greatest penetration would then result.

Anodic polarization data were found more reproducible than the penetration determinations, and they provided some understanding of corrosion rates over a range of potential. The potential of the Fe-base experimental alloy was measured in hot 5 percent $FeCl_3$, by means of a probe through the Teflon used to make a tight crevice, while active corrosion was occurring. It indicated that the reaction was taking place in the region of the breakout (E_P) point of the anodic polarization curve, and not in the active region, as found for a different system by Greene and Judd and Jones and Greene.[6]

It seems that the environments in the polarization work were less severe than the one inside the "tight" crevice ($FeCl_3$ at 50 C), where alloy C corroded rapidly, but more severe than those inside wide crevices, where the experimental alloys and alloy 625 escaped crevice attack. The high-corrosion rate of alloy C in the tight crevice, for instance, corresponds to a calculated current density roughly of the order of 500 to 1000 $\mu A/cm^2$. and it occurred at a potential in the neighborhood of 0.35 V versus saturated calomel electrode (SCE). None of the curves showed such a high-current density for this alloy. Instances where polarization data would have predicted high rates of crevice attack for the experimental alloys and alloy 625, but where very low weight losses were found, suggest that the crevices were too wide to allow the buildup of corrosion products to a dangerous level.

In pursuing this point further, an experiment was done to find out if time was involved. An electrochemical test was carried out with alloy C at 50C, with the potential held at +0.4 V (SCE) in 0.5 *N* HCl + 1.5 NaCl. After several days, a breakout was evident, but the attack was located under the masking material and about an inch away from its edge; slight seepage had occurred. This result tends to show that the electrolyte conditions required for attacking Hastelloy alloy C at 50 C and 0.4 V (SCE) are more severe than the starting solution in the cell.

The electrochemical method ranked the alloys in the same order as the exposures in chemical environments, but provided considerably more detail, since a wide range of electrochemical potentials was covered. Useful anodic polarization data were obtained in NaCl solutions acidified by means of sulfuric acid (H_2SO_4) or HCl, at several concentrations and temperatures. Furthermore, this method rated the alloys reproducibly and appeared more reliable for screening purposes than weight loss experiments.

Although they appeared superior, the electrochemical results still could not provide exact details of how alloys would perform in the field.

Conclusions

1. In direct immersion tests, penetration measurements were influenced by crevice geometry. Relatively less resistant alloys (for example, Type 316 stainless steel) were readily shown to be susceptible. More highly resistant alloys (for example, Inconel alloy 625, Hastelloy alloy C)

[6] Greene, N. D. and Judd, G., *Corrosion,* SBIIA, Vol. 21, 1965, p. 15; also Jones, D. A. and Greene, N. D., Paper No. 67, Annual National Association of Corrosion Engineers Conference, Houston, Tex., March 1969.

appear to require large changes in crevice electrolyte for attack, associated with very tight crevices.

2. Polarization data, obtained potentiostatically, are very useful in the screening of alloys for crevice corrosion resistance. There is a need to exercise care in the choice of electrolyte and testing conditions in order to obtain a spread in results for a given series of materials.

3. The electrochemical tests rated the alloys in the same order as the immersion tests in various chemicals. However, the polarization data were more reproducible and provided more detail.

4. The work indicates that a reproducible laboratory crevice test should be developed for the more highly resistant alloys because of their sensitivity to variations in crevice geometry.

L. C. Covington[1] *and N. G. Feige*[1]

A Study of Factors Affecting the Hydrogen Uptake Efficiency of Titanium in Sodium Hydroxide Solutions

REFERENCE: Covington, L. C. and Feige, N. G., **"A Study of Factors Affecting the Hydrogen Uptake Efficiency of Titanium in Sodium Hydroxide Solutions,"** *Localized Corrosion—Cause of Metal Failure, ASTM STP 516,* American Society for Testing and Materials, 1972, pp. 222–235.

ABSTRACT: An electrochemical technique is described for estimating the tendency of titanium to absorb hydrogen on prolonged exposure to caustic solutions. Unalloyed titanium appears to be quite resistant. A possibility of localized hydrogen absorption and corrosion due to iron contamination is described. Ways of avoiding this problem are suggested.

KEY WORDS: corrosion, corrosion tests, titanium, hydrogen, sodium hydroxide, electrochemistry, polarization (charge separation)

Nickel has long been the standard material of construction for equipment handling hot caustic soda in the Chlor-Alkali industry. Due to the increasing price of nickel an interest is developing in alternate materials of construction. Titanium is a leading candidate.

Although no large scale service experience is presently available for titanium in caustic solutions, corrosion tests [*1,2*][2] of titanium coupons in chemical plant exposures have shown it to be resistant up to 37 percent sodium hydroxide (NaOH) at 230 F.

Some metals with good general corrosion resistance, however, are found to be embrittled severely by hydrogen on prolonged exposure to caustic solutions. Conventional corrosion testing techniques require an impractical amount of time to acquire the data needed to evaluate a metal's susceptibility to hydrogen embrittlement under such conditions. This paper describes an electrochemical technique which has been developed to predict the rate of hydrogen pickup with variations in time and temperature.

[1] Senior research engineer and manager, Market Development, respectively, Titanium Metals Corporation of America, West Caldwell, N. J. 07006.

[2] The italic numbers in brackets refer to the list of references appended to this paper.

Knowing the corrosion current and the hydrogen uptake efficiency of a metal in a particular environment, it is possible to calculate the rate at which hydrogen will be absorbed by the metal. The theoretical background for determining corrosion currents from electrochemical polarization curves has been well established by other investigators [*3,4*], so that will not be discussed in this paper. This will simply be a description of the techniques employed and the results obtained.

The relationship between hydrogen content and embrittlement of titanium has been determined by previous investigators [*5,6*] so no attempt is made here to relate hydrogen content to physical properties. This study was aimed at developing a method of estimating the rate at which hydrogen would be absorbed in specific enviroments.

Experimental Techniques

Corrosion currents (I_c) and hydrogen uptake efficiency (HUE) were measured for titanium in NaOH solutions covering the concentration range from 10 to 60 percent and temperatures from 70 to 220 F.

Corrosion currents were determined from polarization curves. The specimens for polarization measurements were ½-in. disks punched from sheet material. They were electropolished for 1 min at 20 V in a solution consisting of 60 ml of 70 percent perchloric acid, 590 ml of menthanol, and 350 ml of butyl cellusolve. Experience has shown that hydrogen pickup using this procedure on bulk samples at room temperature is virtually nil. The surface area exposed for polarization was 1 cm^2.

The NaOH solutions used were prepared from reagent grade NaOH and distilled water. A portion of each solution was titrated with standard hydrogen chloride (HCl) solution to determine the exact concentration.

Electrochemical polarization data were obtained using a Wenking 61RS potentiostat. A polarization cell was constructed using a cylindrical Vycor glass body and employing two electrode holders of the design of France, Jr. [*7*]. A Luggin probe connected to a saturated Calomel reference electrode was fused into the glass body. This cell is shown in Fig. 1. Platinum-clad titanium was used as the auxiliary electrode. The solution was circulated through the cell at the rate of 1700 ml/min from a 4000-ml reservoir. This resulted in a complete change of solution in the cell about every 20 s. The solution was purged with argon to remove all the air by bubbling the gas through the solution in the reservoir and venting to the atmosphere.

Electrochemical polarization experiments were carried out potentiostatically. The procedure followed was to start at the open circuit potential and proceed in either the anodic or cathodic direction. The potential was changed in increments of 20 mV for the first 100 mV on each side of the rest or open circuit potential and then increased to 100 mV and

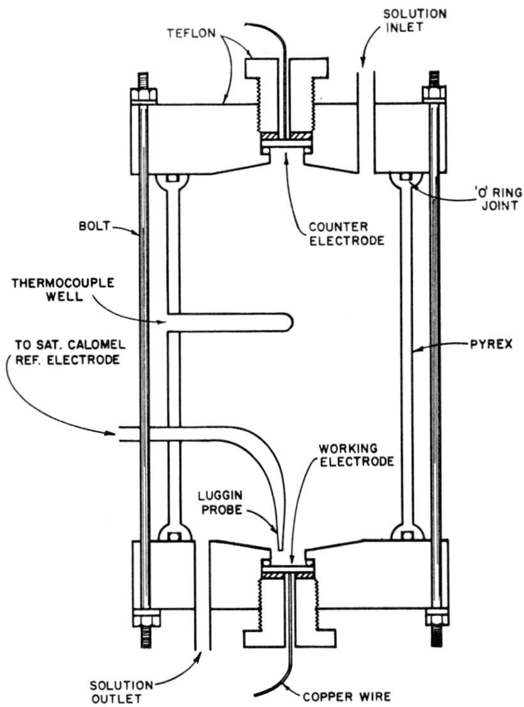

FIG. 1—*Polarization cell.*

finally 250 mV as the distance from the rest potential increased. Three minutes were allowed at each potential setting before the current was read. Different specimens were used for the anodic and cathodic portions of the curve. This was done to avoid any effect from hydride or oxide films formed in the preceding portion of the curve.

HUE data were obtained by cathodically charging specimens of titanium which had been freshly pickled in nitric acid-hydrofluoric acid (HNO_3-HF) solution at a current density of 9 mA/cm^2 for 1 h in each of the caustic solutions investigated. These specimens were then analyzed for hydrogen content by vacuum extraction at 1400 C.

The alloys used in this investigation were as follows:

Alloy	Composition	Heat Treatment
Ti-50A	0.08% max C; 0.05% max N; 0.0019% H_2; 0.20% O_2; 0.09% Fe	mill annealed
Ti-0.2Pd	0.08% max C; 0.05% max N; 0.0019% H_2; 0.20% max Fe	mill annealed
Ti-2Ni	0.08% max C; 0.05% max N; 0.0036% H_2	mill annealed

Results and Discussion

A typical polarization diagram is shown in Fig. 2. Because of passivation effects the anodic portion of the curve was not used for determining I_c, the corrosion current. However, the cathodic portion of the curve has a clearly discernible linear region which can be extrapolated to the open circuit or corrosion potential (E_c). The intercept then gives the corrosion current (I_c) [8].

The HUE of an alloy is simply the percentage of the total hydrogen available which was absorbed by the specimen. From our cathodic charging experiments we know the total amount of hydrogen that was generated

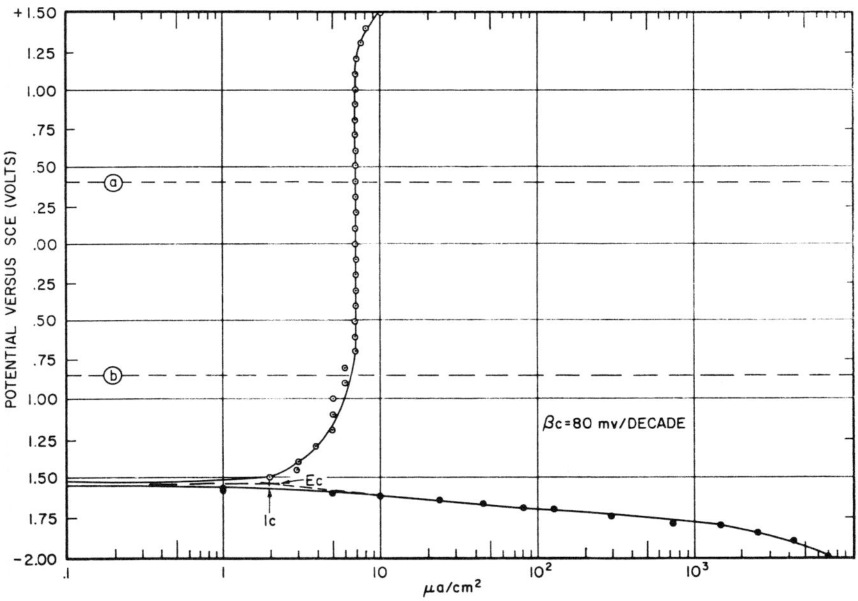

FIG. 2—*Polarization curve for Ti-50A in 59 percent NaOH at 90 F.* $I_c = 2$ $\mu A/cm^2$. $E_c = -1.556$ *V.*

on the surface of the specimen from the ampere minutes of current that was passed. If we assume that the cathodic reaction is entirely the discharge of hydrogen ions, which should be true in a deaerated pure solution of NaOH, then we know from Faraday's law that 1 g of hydrogen will be generated for each 96,500 c or 1608.3 A min of current passed. A current density of 9 mA/cm^2 maintained for 1 h, therefore, would generate 0.000355 g of hydrogen per square centimeter of specimen surface. A 1 cm^2 specimen totally immersed however, would, have 2 cm^2 of surface area (two sides) and, therefore, would be exposed to 0.00071 g of hydrogen. Analysis of the cathodically charged specimens will give the amount of hydrogen (ΔH) expressed as weight percent which was picked up. Knowing the weight (W) of 1 cm^2 of the specimen, the HUE can be calculated from the following relationship:

$$\% \text{ HUE} = \frac{(\Delta H)\ (W)}{0.00071}$$

Table 1 lists the HUE, I_c and corrosion rate values obtained for each condition of NaOH concentration and temperature used in our investigation. From Faraday's law we have calculated that a corrosion current of 1 μA/cm^2 will generate 0.000327 g of H_2/year/cm^2 of specimen

TABLE 1—*HUE, I_c and corrosion rate values for Ti-50A in NaOH solutions.*

NaOH, %	Temperature, deg F	I_c μA/cm^2	HUE, %	Corrosion Rate,[a] mpy	ppm H_2/Year 1 × HUE for 3/16 in. Sheet	ppm H_2/Year 1 × HUE for 1/8 in. Sheet
10.9	70	0.8	0.5	0.28	1	1
10.9	90	1.0	0.6	0.34	1	1.4
10.9	130	1.7	1.3	0.59	3	5
10.9	180	4.0	3.7	1.39	23	34
10.9	220	6.0	6.1	2.08	58	84
16.8	70	5.2	0.7	1.89	6	8
16.8	90	7.5	1.0	2.60	12	17
16.8	130	13.0	2.7	4.51	56	80
16.8	180	30.0	5.8	10.4	276	398
16.8	220	60.0	8.4	20.8	801	1154
24.0	70	2.2	0.3	0.76	1	1.5
24.0	90	20.0	0.5	6.93	16	23
24.0	130	50.0	1.3	17.3	103	149
24.0	180	90.0	13.1	31.2	1874	2700
24.0	220	220.0	16.1	76.3	5629	8100
36.8	70	5.0	0.3	1.73	2	3.4
36.8	90	10.0	0.4	3.45	6	9
36.8	130	23.0	2.0	7.97	73	105
36.8	180	85.0	9.0	29.5	1216	1750
36.8	220	250.0	16.3	86.6	6470	9327
59.0	70	0.15	7.2	0.05	2	2.5
59.0	90	0.4	6.8	0.14	4	6
59.0	130	0.9	6.1	0.31	9	12
59.0	180	9.0	6.0	3.13	86	124
59.0	220	40.0	7.9	13.9	502	723

NOTE—

$$\frac{0.000327 \times \text{HUE} \times I_c \times 10{,}000}{2.058} = \text{ppm } H_2\text{/year for 3/16-in. sheet Ti-50A.}$$

$$\frac{0.000327 \times \text{HUE} \times I_c \times 10{,}000}{1.429} = \text{ppm } H_2\text{/year for 1/8-in. sheet Ti-50A.}$$

[a] Calculated from I_c values using the relationship $R_{(mpy)} = 0.13\ I_c\ e/p$.

where

$R_{(mpy)}$ = corrosion rate in mils per year,
I_c = corrosion current density in μA/cm^2,
e = equivalent weight of titanum in grams = 12, and
p = density of titanium in g/cm^3 = 4.5.

surface. The amount of hydrogen pickup that could be expected from a year's exposure in each environment was calculated as follows:

$$\text{ppm } H_2/\text{year} = \frac{(0.000327)\ (I_c)\ (\text{HUE})\ (10{,}000)}{W}$$

These values in Table 1 were calculated on the basis of 3/16-in. Ti-50A sheet weighing 2.058 g/cm^2 and 1/8-in. Ti-50A sheet weighing 1.429 g/cm^2 since most titanium process equipment uses material in this thickness range.

A plot of log ppm H_2/year versus the temperature at which the value was determined gives an approximate linear relationship for each solution, as shown in Fig. 3. By plotting the data for each solution it is possible to construct a family of curves that permit one to estimate the hydrogen pickup per year at any temperature and NaOH concentration within the ranges covered by our data.

Figure 4 displays these data in the form of hydrogen isobars with

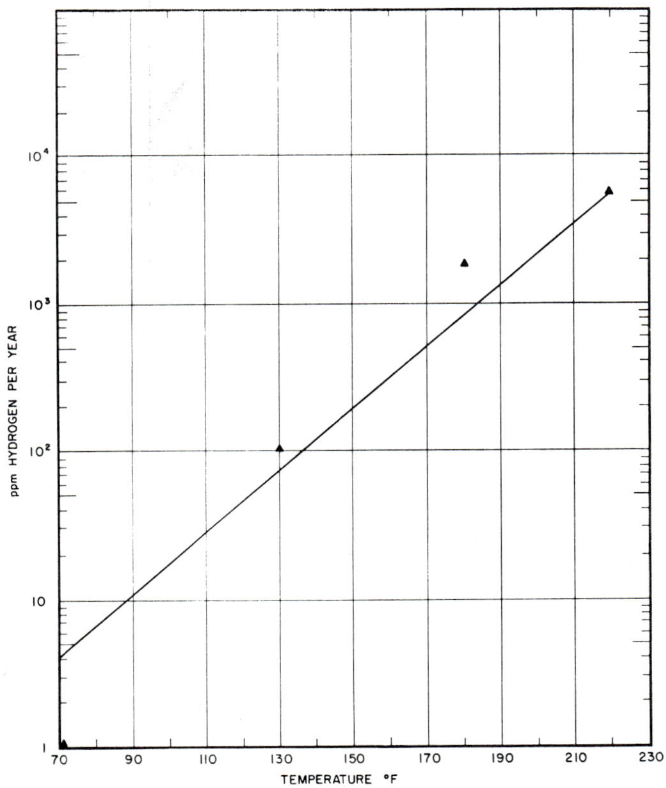

FIG. 3—*Predicted hydrogen pickup for Ti-50A 3/16-in. sheet in 24 percent NaOH solution.*

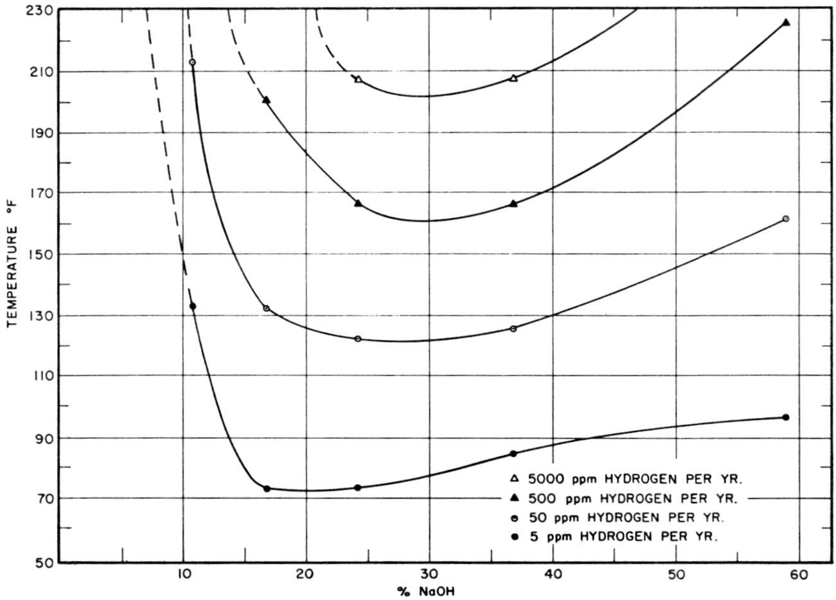

FIG. 4—*Hydrogen isobars for 3/16-in. Ti-50A sheet.*

temperature as the ordinate and percent NaOH as the abscissa. It is interesting to note that the point of maximum hydrogen pickup occurs in solutions in the 20 to 30 percent NaOH range. At higher concentrations there seems to be a passivation effect that reduces hydrogen absorption. In this connection, we have visually observed that a film builds up on the titanium in these solutions, and it appears to grow thicker with increasing temperature and solution concentration. This probably accounts for the passivating effect indicated by the data.

The question arises as to the relationship between the HUE measured at a current density of 9 mA/cm^2 and that existing in a corrosion situation where the current density is two to three orders of magnitude smaller. It would be most accurate if the HUE in each case could be measured at current densities corresponding to the measured corrosion current densities. However, this is impractical due to the long elapsed time that would be required to accumulate sufficient hydrogen in the specimens to be measured accurately.

An experiment was performed in which the percent HUE of Ti-50A was determined at current densities of 9, 6, 3, and 1.5 mA/cm^2 in 59 percent NaOH at 180 F. The results are plotted in Fig. 5. This gives an approximately linear relationship over the range covered. If this straight line is extrapolated to zero current density, it indicates that the value for HUE in the I_c current density region would be increased by a factor of about three over that measured at 9 mA/cm^2. However, it seems unlikely

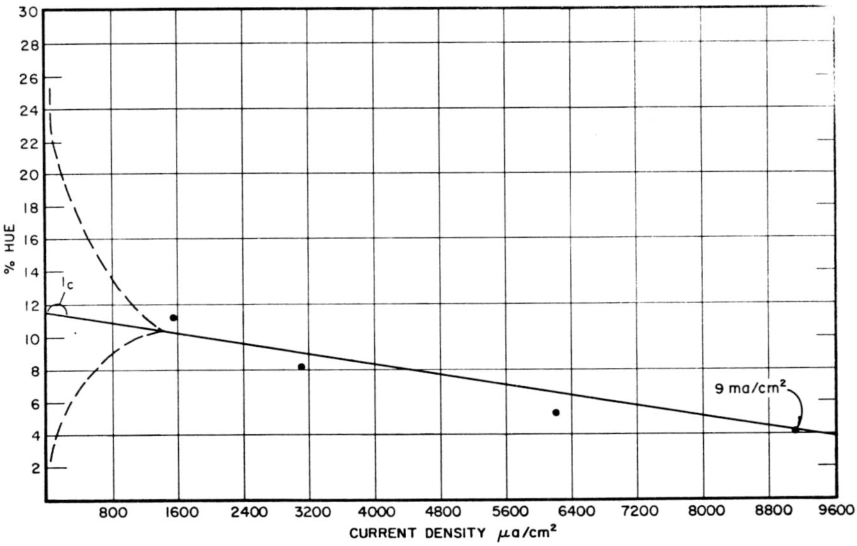

FIG. 5—*Percent HUE versus current density in 59 percent NaOH at 180 F.*

that the relation would continue to be a straight line function as the current density approaches zero. The line may approach zero current density asymptotically so that HUE values approach 100 percent or zero, depending on the mechanism involved as indicated by the dotted lines in Fig. 5.

Several investigators [*9,10*] agree that hydrogen is absorbed in metals in the atomic form. A hydrogen ion (H^+) is nothing more than a hydrogen atom stripped of its electron. The hydrogen molecule has a diameter of 2.12 A. The diameter of the hydrogen atom is 1.058 A [*9*]. The diameter of the hydrogen ion or proton is about 10^{-5} A or one hundred thousand times smaller than the size of the atom.

The titanium surface is covered with an oxide film which is an insulator and will not readily conduct electrons. Therefore, the hydrogen ions (H^+) must diffuse through the film and contact the metal surface in order to pick up an electron and be converted into a hydrogen atom, in which form it diffuses into the metal lattice. Because of its extremely small size the hydrogen ion should pass through the oxide film with ease. However, when it picks up an electron its size is suddenly increased one hundred thousand times, which may make it much more difficult for the hydrogen atom to pass back through the oxide film again. This would suggest that at higher current densities the hydrogen may build up considerable pressures under the film. This is supported by the observation that at high-current densities, scales on the surface of a titanium cathode will frequently spall off due to the action of the evolving hydrogen.

This concept may also explain the rectifying behavior of titanium oxide (TiO_2) films. As an anode, titanium offers considerable resistance to the flow of current in an electrolyte because an oxide film forms on the surface which acts as an insulator. It will not readily conduct electrons, and anions diffuse only slowly through the film to the metal surface. As a cathode, however, TiO_2 films do not block current flow because the hydrogen ions (H^+) diffuse through the film with ease.

The diffusion rate of hydrogen into titanium is proportional to the square root of the hydrogen pressure [*9*]. At low-current densities, diffusion rates through the oxide film might nearly equal the hydrogen generation rate so the hydrogen pressure would be very low and little or no absorption would take place. However, at higher current densities the hydrogen generation rate would exceed the diffusion rate through the film with the result that the excess hydrogen would build up a pressure that would drive it into the metal in accordance with the square root rule $S = K \sqrt{P}$, where: S = solubility, K = constant, and P = pressure.

Hydrogen analyses performed on specimens of Ti-50A immersed in 50 percent NaOH at 200 F for three months indicated a HUE of about 1 percent based on corrosion currents calculated from weight loss measurements. The HUE measured from cathodic charging experiments at 9 mA/cm^2 in similar environments gave values near 13 percent. This would indicate that for Ti-50A, the HUE tends to drop off towards zero at low-current densities.

Since an accurate measurement of HUE seemed to be impractical for a corrosion situation and since it appeared that HUE values measured at 9 mA/cm^2 would tend to be higher than those encountered under corrosion conditions, it was decided to use these values in estimating hydrogen pickup in a corrosion environment. Any error involved would tend to be on the high side, thus providing a safety factor, and it is believed these values would tend to vary in the same manner from the effects of temperature and solution concentration as would HUE values obtained at corrosion current densities.

It is generally agreed [*5,6,9*] that for most applications hydrogen contents of less than 300 ppm do not degrade the physical properties of titanium sufficiently to cause any problems. Figure 4 indicates that Ti-50A will endure several years exposure in solutions containing up to 59 percent NaOH and at temperatures up to 130 F before hydrogen contents begin to approach this level. Therefore, it could be expected to have a useful life in NaOH solutions. Our experiment did not determine the effect of time on the passivating film buildup. It seems reasonable to assume that the film thickness would continue to increase with time to some equilibrium value and that the corrosion current and hence the hydrogen pickup would decrease in a corresponding manner. It would seem logical, therefore, that the hydrogen pickup estimated from Fig.

4 would be a maximum value and in actual practice might be found to be considerably lower. The low-corrosion rates reported by Gegner and Wilson [2] in industrial exposures would tend to support this idea.

In order to establish some correlation between industrial exposures and the predictions based on these electrochemical measurements, arrangements were made with a major chemical company to expose specimens of Ti-50A and Ti-2Ni in the caustic effluent from their chlorine cells. This solution was 50 percent NaOH at a temperature of 200 F. Specimens were weighed at one month intervals, and two specimens were withdrawn and analyzed for hydrogen at the end of the second month.

Weight loss measurements indicated a corrosion rate of 10 mils per year (mpy) for Ti-50A and 9.8 mpy for Ti-2Ni for the first month which decreased to 5.5 mpy for Ti-50A and 4.7 mpy for Ti-2Ni the second month. The hydrogen analyses at the end of the second month indicated the Ti-50A had picked up 60-ppm hydrogen in two months, while the Ti-2Ni had increased by about 5500 ppm and was severely embrittled. The Ti-50A showed no localized corrosion and was still fully ductile. The hydrogen pickup of 60 ppm on the Ti-50A, 0.040-in.-thick specimen would have amounted to only 20 ppm on $\frac{1}{8}$-in.-thick sheet or 14 on $\frac{3}{16}$-in. sheet. Also, these specimens were totally immersed so hydrogen penetrated from both sides. In service, probably only one side would be exposed so the hydrogen pickup could be expected to be less. The exposures are continuing, and, if the corrosion rate continues to drop, the Ti-50A should show sufficient passivation by the end of the third or fourth month to have a very useful life in this environment.

The embrittlement of the Ti-2Ni alloy in the plant exposure could have been predicted from HUE measurements made in this environment. Table 2 compares values of HUE for Ti-50A and Ti-2Ni in 10.9 percent NaOH at various temperatures. The Ti-2Ni alloy was found to have very high HUE values.

TABLE 2—*Comparison of HUE for various titanium alloys.*

NaOH, %	Temperature, deg F	HUE				
		Ti-50A	Ti-2Ni	Ti-0.2Pd	Ti-50A Smeared with Fe	Anodized Ti-50A
10.9	70	0.5	12.1			
10.9	130	0.9	26.7			
10.9	180	3.7	57.8			
10.9	220	6.1	70.1			
16.8	70	0.0		2.5		
16.8	130	4.8		8.3		
16.8	180	6.3		5.5		
16.8	220	11.9		3.1		
24.0	70	0.3		4.2	9.5	0.5
24.0	130	1.3		20.5	11.5	1.5
24.0	180	13.1		2.8	19.3	3.5
24.0	220	16.1		1.6	34.6	4.5

Nickel alloy additions to titanium form the Ti_2Ni compound which is beneficial in resisting anodic pitting attack in the neutral pH brine environment. Under cathodic control or general corrosion of NaOH, the Ti_2Ni compound provides a local site for hydrogen absorption. The significant increase in HUE of the Ti_2Ni compound was demonstrated in our experiments. Field experiments confirm its unsuitability for service in the caustic environment. However, in neutral brine environments where hydrogen pickup is not a problem, the ability of Ti-2Ni to resist pitting and crevice corrosion can be used to advantage.

Localized Corrosion Due to Iron Contamination

Previous investigators [*11*] have demonstrated that the presence of metallic iron at the titanium surface facilitates hydrogen ingress into the titanium. The mechanism of this attack is based upon the premise that the surface contaminant is initially anodically attacked, and, either at an adjacent site or within the same site, hydrogen is charged into the metallic lattice. If sufficient hydrogen is absorbed to exceed the solubility limit for the system, titanium hydride (TiH_2) is precipitated, providing a phase with very low anodic pitting resistance. At this point the process becomes autocatalytic and can continue to proceed after the initial local area has been consumed.

Iron additions as either smeared iron or alloy addition react the same as nickel under cathodic control. The HUE values in Table 2 demonstrate this point. These values were obtained on specimens of Ti-50A which had been smeared with iron by abrading the surface with a blunt mild steel rod mounted in a drill press. In every case, iron contamination greatly increased the HUE.

The upper solubility limit of iron, in practice, to prevent the Ti_2Fe compound formation and the possibility of localized attack is 0.12 percent maximum. Present U.S. titanium industry practice is to hold iron levels below 0.12 percent in both ASTM Grades 1 and 2 formulations. Of more serious concern in commercial practice is smeared surface iron resulting from material handling and fabrication. Carbon steel wire brushing of welds is a classic mode of failure in these marginal environments where general attack is occurring to the titanium. Steel wire brushing should be avoided. Acid pickling or stone grinding has been found to be effective ways of removing surface iron contamination. Also, anodizing can remove surface iron contamination if it is carried out for a sufficient length of time to permit complete dissolution of the iron.

In addition to removing surface iron, the anodizing process builds up a passive oxide film which is quite effective in suppressing hydrogen absorption. This is also evident from the data in Table 2 which lists HUE values obtained on specimens of Ti-50A that were anodized for 20 min at 90 V in a saturated solution of $NaNH_4HPO_4$.

The demonstrated ability of anodized films to suppress hydrogen pickup of titanium in solution, on first thought, is not consistent with the concept that hydrogen ions because of their small size diffuse through surface films with ease. However, if we consider the subject more carefully we can see why the film retards hydrogen pickup. The cathodic reaction $H^+ + e^- \rightarrow H$ is balanced by an equal and opposite anodic reaction involving anions which must diffuse through the anodized film to the metal surface to give up an electron which then travels through the metal to the cathodic site where it is picked up by a H^+. Thus, in the absence of an external electromotive force (emf), the cathodic reaction can proceed no faster than the anodic reaction supplies it with electrons. Thus, by blocking the access of the relatively large anions to the metal surface, the anodized film effectively suppresses the cathodic reaction also and thus the absorption of hydrogen.

The data in Table 2 are graphically shown in Fig. 6 where we have HUE values plotted versus temperature for the various titanium alloys. From this it is apparent the titanium-palladium alloy has a reduced HUE under general corrosion attack. The alloy's significant commercial application over the years has demonstrated its usefulness when the system has an occasional excursion into the highly acidic or basic range. It also appears to be promising for applications where hydrogen embrittlement

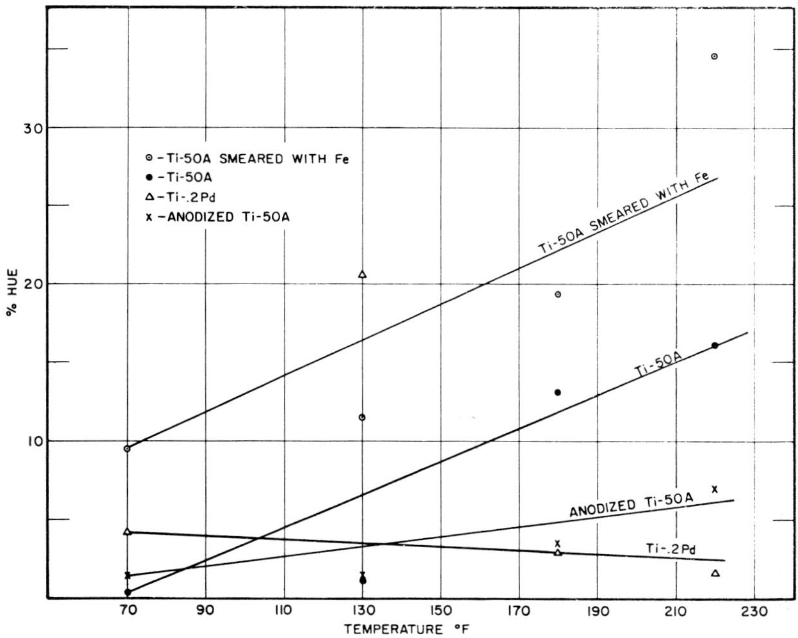

FIG. 6—*Comparison of HUE values for various titanium specimens in 24 percent NaOH.*

may present a problem. The variability in our testing program is most probably related to the high-current density imposed and the surface being unable to cope with the solid hydride layer. At lower current densities, the majority of the surface film is probably a passive oxide with the palladium-rich areas being the site of hydrogen ion discharge.

Summary

The mechanism of titanium's corrosion in caustic solutions is slow, general anodic dissolution of the surface film with the rate of attack accelerated by an increase in temperature. This reaction, although not detrimental in metal wastage from the corroded surface, is the cause of hydrogen ion discharge to atomic hydrogen at the oxide-metal interface. The basis of this experimental work was to evaluate the anodic dissolution rate with temperature and caustic concentration and predict the hydrogen discharge to atomic hydrogen at the metal surface.

The test program excludes the passivating effect of oxide film which will significantly reduce the rate of attack and operates at current densities considerably higher than the predicted corrosion current I_c. The test procedure allows the investigator the opportunity to segregate one step; the hydrogen uptake at the metal-oxide interface, and compare the differences between alloys and surface finish preparations under the worst possible test conditions.

The experimental technique described measures over a 1-h period the formation of a hydride film and the bulk diffusion of hydrogen with time at temperature through the interface. HUE in the simplistic terms of the test program, is the amount of hydrogen absorbed on the surface versus the amount evolved on the surface.

This work has demonstrated that electrochemical methods can be used to rapidly estimate the suitability of a particular titanium alloy and surface film condition for service in caustic environments. The data point up the wide variability of HUE dependent upon surface film conditions and indicate what direction may be taken to mitigate the rate of attack. The short-term laboratory data reflect the worst conditions possible in the NaOH environment.

Acknowledgment

The experimental work reported in this paper was carried out by J. J. Henderson of TIMET's Henderson Technical Laboratories.

References

[*1*] Sheppard, R. S., Hise, D. R., Gegner, P. J., and Wilson, W. L., "Performance of Titanium vs. Other Materials in Chemical Plant Exposures," *Corrosion*, SBIIA, Vol. 18, June 1962, pp. 211t–217t.

[*2*] Gegner, P. J. and Wilson, W. L., "Corrosion Resistance of Titanium and

Zirconium in Chemical Plant Exposures," *Corrosion,* SBIIA, Vol. 15, July 1959, pp. 341t–350t.

[*3*] Stearn, M. and Geary, A. L., 'Electrochemical Polarization," *Journal,* Electrochemical Society, JESOA, Vol. 104, No. 1, Jan. 1957, pp. 56–63.

[*4*] Greene, N. D., "Predicting Behavior of Corrosion Resistant Alloys by Potentiostatic Polarization Methods," *Corrosion,* SBIIA, Vol. 18, April 1962, pp. 136t–142t.

[*5*] Williams, D. N., Schwartzberg, F. R., and Jaffee, R. I., "Hydrogen Contamination in Titanium and Titanium Alloys, Part V: Hydrogen Embrittlement," WADC Technical Report 54–616, Wright Air Development Center, Feb. 1959.

[*6*] Lenning, G. A., Berger, L. W., and Jaffee, R. I., "The Effect of Hydrogen on the Mechanical Properties of Titanium and Titanium Alloys," Battelle Memorial Institute, Fourth Summary Report to Watertown Arsenal, Ordnance Department, Department of the Army, Contract No. DAI-33-019-505-ORD-(P)-1.

[*7*] France, W. D., Jr., *Journal,* Electrochemical Society, Vol. 114, No. 8, Aug. 1967, pp. 818–819.

[*8*] Henthorne, M., "Polarization Data Yield Corrosion Rates," *Chemical Engineering,* CHEEA, 26 July 1971.

[*9*] Livanov, V. A. et al, *Hydrogen in Titanium,* published by the Israel Program for Scientific Translations, 1965, Daniel Davey & Co., New York.

[*10*] Evans, U. R., "The Corrosion and Oxidation of Metals: Scientific Principles and Practical Applications," Edward Arnold Publishers, Ltd., 1960, p. 397.

[*11*] Cotton, J. B., "Using Titanium in the Chemical Plant," *Chemical Engineering Progress,* CEPRA, Vol. 66, No. 10, Oct. 1970, pp. 57–62.

C. C. Nathan,[1] *C. L. Dulaney,*[1] *and M. J. Leary*[1]

Prevention of Hydrogen Blistering and Corrosion by Organic Inhibitors in Hydrocarbon Systems of Varying Composition

REFERENCE: Nathan, C. C., Dulaney, C. L., and Leary, M. J., "**Prevention of Hydrogen Blistering and Corrosion by Organic Inhibitors in Hydrocarbon Systems of Varying Composition,**" *Localized Corrosion—Cause of Metal Failure, ASTM STP 516,* American Society for Testing and Materials, 1972, pp. 236–249.

ABSTRACT: The inhibition of the corrosion of steel, caused by hydrogen sulfide, was studied at various pH ranges in the presence of an oil phase. Oil-soluble filming amine-type inhibitors are effective in stopping acid corrosion. At 7.7 pH, only two of the oil-soluble inhibitors tested had any effect. At 10 pH, in the presence of cyanide ion, only the water-soluble inhibitors are effective.

KEY WORDS: corrosion, corrosion tests, corrosion prevention, cyanides, inhibitors, pH, evaluation, hydrogen sulfide, tests

Numerous papers have been written on the effects of hydrogen sulfide on steel [*1–23, 25–33*].[2] Effects are almost uniformly bad, and generally due to the handling or processing of sour (sulfur containing) petroleum or natural gas.

Approximately 40 percent of the oil and gas wells in the United States produce a sour product [*12*]. Sulfide caused corrosion occurs in well equipment, crude storage tanks, and pipelines [*1, 12*], and also in crude distillation units, catalytic cracking units, and other units of refineries and chemical plants [*2, 3, 15, 22, 23*].

Previous laboratory studies have been made on acidic systems [*5–7, 9, 10, 14, 16, 19–22, 25, 26, 29, 30*], or on basic systems in the absence of oil [*3, 4, 11, 17, 27, 31*].

This paper presents a study of the effects of corrosion inhibitors on

[1] Senior research associate, senior research chemist, and research chemist, respectively, Betz Laboratories, Inc., Trevose, Pa. 19047.

[2] The italic numbers in brackets refer to the list of references appended to this paper.

the hydrogen sulfide induced corrosion of steel (Fig. 1). Tests were made in the presence of oil at various pH ranges up to 10.

Procedures

Corrosion Test

The apparatus used, and the procedures followed, for the corrosion tests were the same as those of Gardner [*16*] and Gardner et al [*5*]. Briefly, a glass reaction bottle was charged with test coupon, aqueous solution, oil, and hydrogen sulfide. The reaction vessel was then rotated at 15 rpm in a cabinet heated to 80 C for a 20-h period.

Distribution Coefficients

No coupons were used. The oil/water distribution for cyanide and for hydrogen sulfide was determined.

Analytical Procedures

Cyanide was analyzed by ASTM Tests for Cyanides in Water (D 2036–68), employing the modified Liebig titration procedure. Hydrogen sulfide was analyzed by the method of Shannon and Boggs [*14*].

Experimental Work: Theoretical Basis

Hydrogen Sulfide Corrosion: Redox Potential

In addition to pH, another method of arriving at a corrosive tendency is to measure the oxidation-reduction, or redox, potential of an iron electrode immersed in the corrodent [*17*]. In the absence of oxygen, as is the usual case in a refinery, the redox potential is essentially a linear function of the pH. Therefore, throughout this paper, the pH is used as a measure of the corrosive tendency.

Corrosion caused by hydrogen sulfide is pH dependent. In general, four pH ranges may be distinguished [*3, 27, 31*].

1. At a pH level below 4.5, acid corrosion and hydrogen blistering occur. Both increase in severity with decreasing pH. Hydrogen penetration occurs because of poisoning of the surface by sulfide ions (see next section for discussion of hydrogen blistering).

2. In the pH range of 4.5 to 6.0, because of low concentration of both hydrogen and bisulfide ions, low rates of corrosion and hydrogen penetration are observed.

3. The least protective sulfide films are formed [*27*] at pH 6.0 to 8.0. These films consist of Mackinawite (ferrous sulfide, previously called Kansite) [*28*]. If the film is mechanically broken, corrosion may occur.

4. At a pH level above 8.0, in the absence of cyanide ion, very low corrosion rates occur because of the protective sulfide film. In the presence of cyanide ion, hydrogen blistering and corrosion can occur. Cyanide

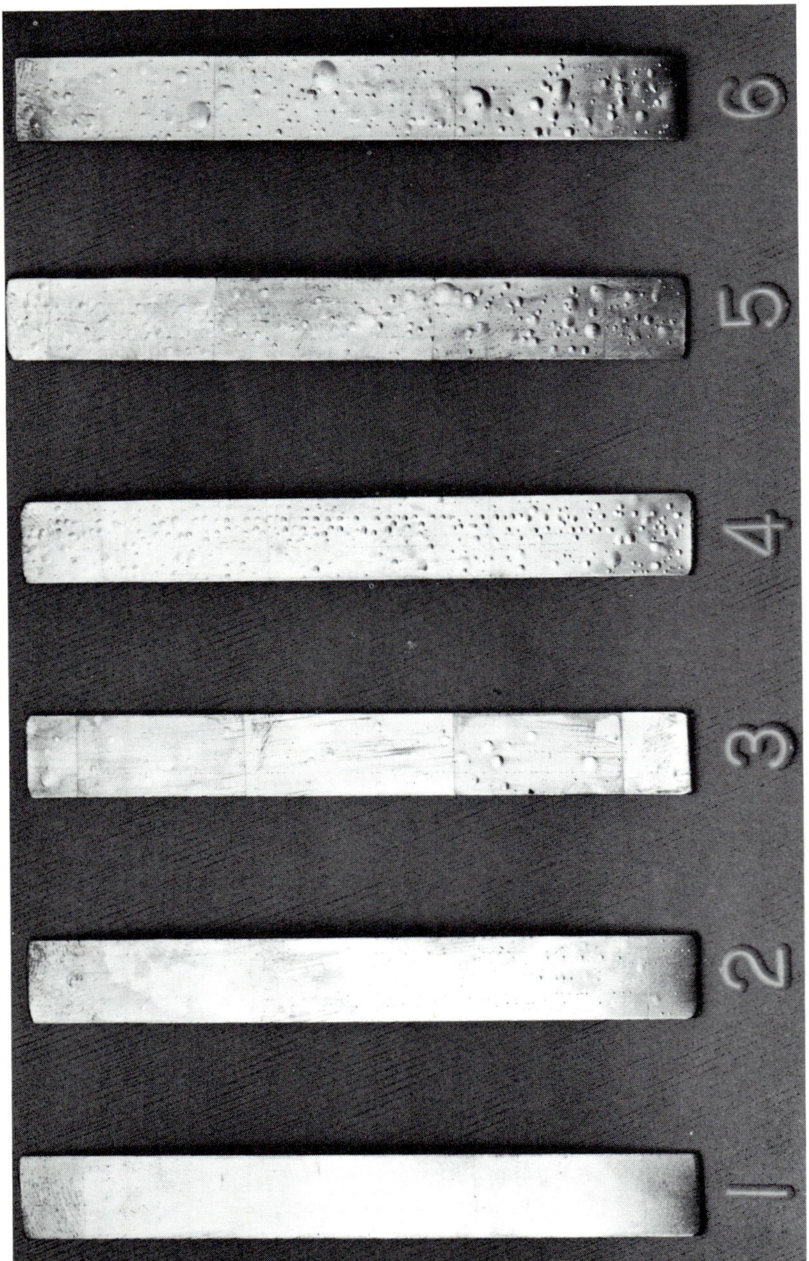

FIG. 1—*Effect of corrosion inhibitors on the hydrogen sulfide induced corrosion of steel.*

ion is very strongly absorbed on steel; the heat of absorption is approximately 20 kg·cal/mole [*24*]. Cyanide ion also dissolves sulfide films to give iron-cyanide complex ions.

Hydrogen Blistering

Hydrogen atoms are formed on a steel surface by corrosion processes, even in basic systems. If the surface is poisoned by strong adsorption of an ion, such as sulfide, cyanide, or arsenic, the hydrogen atoms cannot combine to form molecules. The atoms penetrate the surface and may diffuse completely through fairly thick metal walls.

If imperfections are present, hydrogen atoms may combine to form molecules which cannot diffuse farther. Concentrations of molecular hydrogen can occur which are sufficient to exceed the yield strength of the metal.

Corrosion Inhibitors: Effect of Oil

Acid Corrosion—The majority of the organic corrosion inhibitors are of the filming amine type. These are molecules that contain amine groups or nitrogen ring compounds, with a long chain hydrocarbon. The nitrogen is adsorbed on a metal surface, and the hydrocarbon tail helps make the surface hydrophobic [*33*].

The presence of oil has a marked effect on inhibitor performance. In the complete absence of oil, filming amines have been found to be ineffective [*9*]. Nathan [*18*] found the inhibition by filming amines to be proportional to the concentration of inhibitor in the oil phase. Boies [*6*] found that inhibition increased as the aromaticity of the oil phase increased.

An oil phase is always present in the case of refinery or chemical plant corrosion. It thus appears that studies of filming amine inhibitors should be made in the presence of an oil phase.

Almost all filming amines are formulated in oil. In laboratory tests in aqueous systems, the amount of oil present in the inhibitor may be sufficient to provide the necessary oil phase [*10*].

Basic Corrosion Inhibition

Most filming amines work best at a pH of 4.5 to 6.5, and little or no work has been reported on the effect of inhibitors with an oil phase in basic systems.

Experimental Results

Chemicals Used

All inorganic chemicals used were chemically pure and were used as received. Water was demineralized. The oil used was "Odorless Insecticide

Base" from American Mineral Spirits Corp. This oil is a highly purified, dearomatized kerosine.

Aqueous Solution for Corrosion Tests

The aqueous solution for the pH 7.7 test contained 0.77 percent ammonium sulfide ($(NH_4)_2S$); 5 percent ammonium chloride (NH_4Cl). Any pH adjusted to 7.7 by addition of 6 *N* hydrochloric acid (HCl). Any sulfur that precipitated was filtered out before use of the solution. The $(NH_4)_2S$ was Fisher light ammonium sulfide, 20 to 24 percent by weight. The pH 10 aqueous solution consisted of 4 percent ammonia (NH_3) (from ammonium hydroxide (NH_4OH)) and 1000 parts CN^- (from sodium cyanide). All tests were run with 60 ml of aqueous phase and 30 ml of oil, unless stated otherwise.

Inhibitors Used

With the exception of two, the corrosion inhibitors were oil soluble. All were of the filming amine type. Both of the water soluble amines have some oil solubility. Table 1 shows the inhibitors used and their composition as determined by infrared analysis.

TABLE 1—*Inhibitors used and their composition as determined by infrared analysis.*

Inhibitor	Solvent	Composition
A	oil	cyclic amine + linear amine
B	oil	cyclic amine + linear amine
C	oil	cyclic amine + linear amine
D	oil	linear amide amine
E	oil	linear amide amine
F	oil	cyclic amine + linear amide
G	water	linear amine
H	water	linear amine

Test Program

Distribution Experiment

The water-oil distribution coefficients for cyanide and hydrogen sulfide were determined, as shown in Table 2.

TABLE 2—*Water-oil distribution coefficients for cyanide and hydrogen sulfide.* Conditions: 30-ml AMSCO odorless insecticide base and 60-ml water rotated for 20 h at 15 rpm at a temperature of 80 C.

Added	Found in Water	Oil	CN in Oil / CN in Water
CYANIDE[a]			
526	478	0	0
1072	930	0	0
HYDROGEN SULFIDE			
Added	Found in Water	Oil	H_2S in Oil / H_2S in Water
352	91	261	2.87
680	158	522	3.3
760	170	590	3.46
1112	249	863	3.46
1787	407	1380	3.4
2508	588	1920	3.3
4070	1040	3030	2.91

[a] NOTE:—No cyanide was found in any oil layer.

Effect of pH on Water-Oil Hydrogen Sulfide System

The pH of the system was varied by adding on HCl or NH_4OH to adjust the pH. Data are shown in Table 3 and Fig. 2.

TABLE 3—*Effect of pH on water-oil hydrogen sulfide system.* Conditions: 30-ml oil and 60-ml aqueous phase rotated 20 h at 80 C. Hydrogen sulfide maintained constant at 1100 ppm.

Aqueous Phase	pH as Prepared	pH After Test	Average Weight Loss, mg	Appearance of Specimen
38 000-ppm HCl	1.2	1.0	690	...[a]
19 000-ppm HCl	1.6	1.2	367	...[a]
10 000-ppm HCl	1.5	1.4	242	...[a]
5 000-ppm HCl	2.1	1.6	149	2[b]
1 000-ppm HCl	2.5	2.4	54	3
Demineralized water	6.6	4.4	26	3
500-ppm NH_4OH	9.6	5.1	10	2
1 000-ppm NH_4OH	9.4	5.9	8	2
5 000-ppm NH_4OH	9.9	6.6	3.5	1
10 000-ppm NH_4OH	10.2	7.7	3	1
50 000-ppm NH_4OH	10.7	8.3	2.5	1
100 000-ppm NH_4OH	10.9	8.7	4	1

[a] General acid attack.
[b] See Fig. 1 for comparison of degree of hydrogen blistering of coupons.

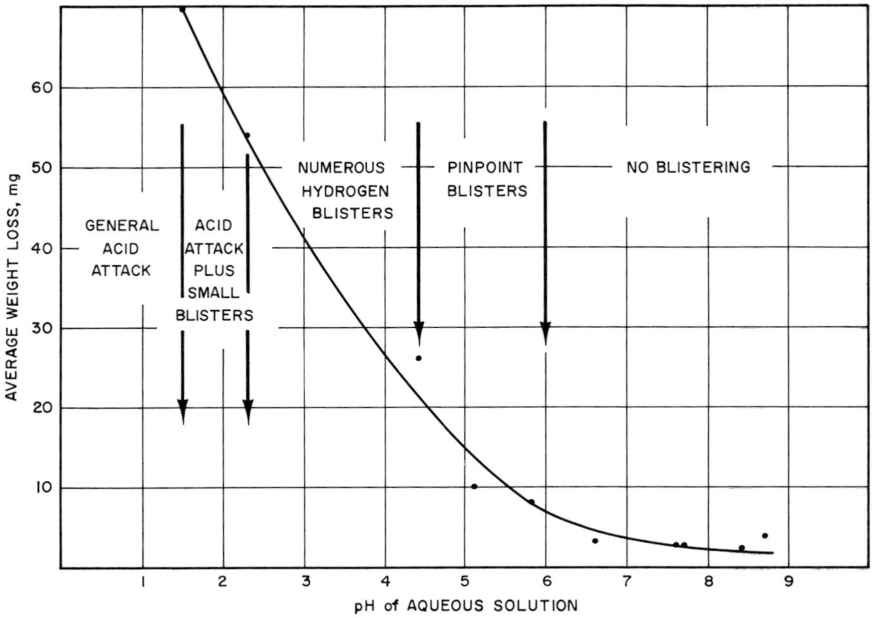

FIG. 2—*Effect of pH on corrosion and hydrogen blistering in presence of H_2S.*

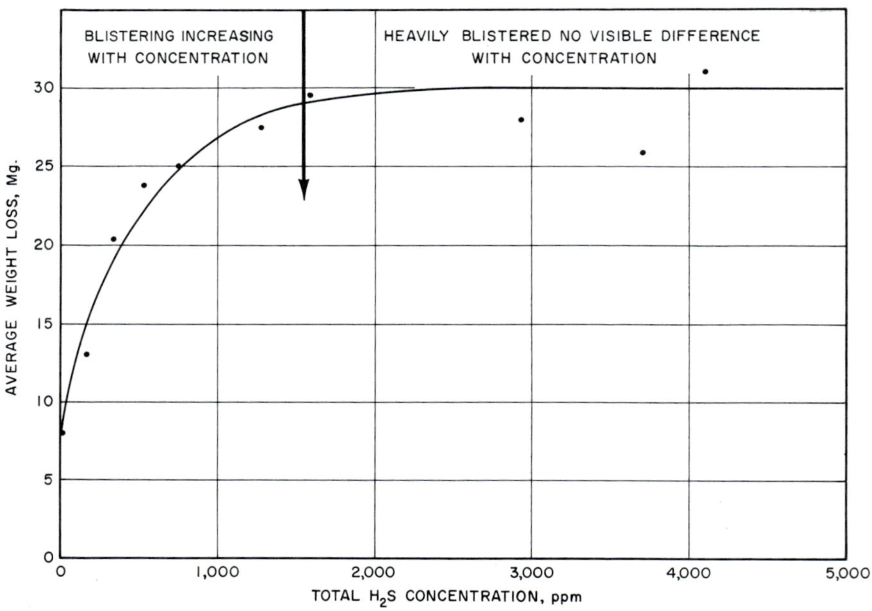

FIG. 3—*Effect of H_2S concentration on corrosion in oil-water system.*

Effect of Hydrogen Sulfide Concentration on the Water-Oil System

Only hydrogen sulfide was added to the water and oil. Effects of the inhibitors were studied, as shown in Table 4 and Fig. 3.

TABLE 4—*Effect of hydrogen sulfide concentration on water-oil system.*

Conditions: 30-ml oil and 60-ml water, rotated 20 h at 80 C. Inhibitors added at 100 ppm, based on total fluid. Inhibitors 25 percent active (weight basis).

	Hydrogen Sulfide, ppm										
Inhibitor	0	160	338	508	723	1284	1525	2271	2949	3705	4102
None:											
Weight loss, mg	4	19	17	29	29	32	40	29	30	26	30
Appearance of specimen	1[a]	2	2	4	4	5	5	6	6	6	6
Inhibitor A:											
Weight loss, mg	4	7	9	8	7	11	7	7	7	5	4
Appearance of specimen	1	1	1	1	1	1	1	1	1	1	1
Inhibitor B:											
Weight loss, mg	5	6	9	3	21	6	8	10	8	11	4
Appearance of specimen	1	1	1	1	E[a]	1	1	1	1	1	1
Inhibitor C:											
Weight loss, mg	6	8	7	7	7	7	7	8	10	7	9
Appearance of specimen	E[b]	1	1	1	1	1	1	1	1	1	1
Inhibitor D:											
Weight loss, mg	12	6	8	9	5	9	10	7	4	7	6
Appearance of specimen	1	1	1	1	1	1	1	1	1	1	1
Inhibitor E:											
Weight loss, mg	7	9	11	6	27	17	11	11	12	9	9
Appearance of specimen	1	1	1	1	E	1	1	1	1	1	1
Inhibitor F:											
Weight loss, mg	6	14	11	6	7	10	9	9	7	8	6
Appearance of specimen	1	1	1	1	1	1	1	1	1	1	1
Inhibitor G:											
Weight loss, mg	14	14	11	16	13	33	25	16	11	13	9
Appearance of specimen	E	E	E	E	E	E	E	E	E	E	E
Inhibitor H:											
Weight loss, mg	7	6	7	6	9	6	17	8	17	6	6
Appearance of specimen	E	E	E	E	E	E	E	E	E	E	E

[a] Compare with specimens pictured in Fig. 1.
[b] Indicates specimen etched.

Intermediate 7.7 pH Test

The effect of inhibitors on 7.7 pH is shown in Table 5.

TABLE 5—*Effect of cyanide concentration in intermediate pH 7.7 test.* Conditions: 30-ml oil and 60-ml aqueous phase, rotated 20 h at 80 C. Inhibitors added at 100 ppm, based on total fluid.

Inhibitors	Weight Loss, mg	Appearance of Specimen
None	43	etched
A	1	smooth
B	39	etched
C	random (13,21,49,51)	etched
D	60	etched
E	42	etched
F	51	etched
G	39	etched
H	47	etched

Effect of Hydrogen Sulfide Concentration in pH 10.0 System with Cyanide

Results with the various inhibitors are shown in Table 6 and Fig. 4.

TABLE 6—*Effect of hydrogen sulfide concentration in pH 10.0 system with cyanide.* Conditions: 30-ml oil and 60-ml aqueous phase. Cyanide constant at 1000 ppm. Inhibitors added at 100 ppm, based on total fluid. Bottles rotated 20 h at 80 C.

	Hydrogen Sulfide Added, ppm									
Inhibitors	160	338	508	723	1284	1525	2270	2950	3705	4100
None:										
Weight loss, mg	8	10.5	9.5	9	9.5	15	14	17.5	17	16.5
Specimen[a] appearance	1	2	2	3	3	3	4	5	5	6
Inhibitor A:										
Weight loss, mg	15	10	12	15	14	19	18	22	21	21
Specimen appearance	1	2	2	3	3	4	5	5	6	6
Inhibitor B:										
Weight loss, mg	11	13	16	13	15	11	12	22	24	21
Specimen appearance	1	2	2	3	3	3	5	5	6	6
Inhibitor C:										
Weight loss, mg	17	15	15	16	16	12	17	24	23	22
Specimen appearance	1	2	2	3	3	3	4	5	6	6
Inhibitor D:										
Weight loss, mg	14	17	15	14	17	15	14	22	23	22
Specimen appearance	1	3	2	2	3	4	4	5	6	6
Inhibitor E:										
Weight loss, mg	7	12	11	13	13.5	19	18	21.5	21	20
Specimen appearance	1	2	2	3	3	3	4	5	5	6
Inhibitor F:										
Weight loss, mg	9	14	13	13	15	16	15	20	22	21
Specimen appearance	1	2	2	3	3	4	4	5	6	6
Inhibitor G:										
Weight loss, mg	5	7	8	10	9	9	12	13	16	15
Specimen appearance	1	1	2	2	3	3	3	3	3	3
Inhibitor H:										
Weight loss, mg	4	4	4	4	5	3	4	5	5	3
Specimen appearance	1	1	1	1	1	1	1	1	1	1

[a] Figure 1 shows typical specimen appearance.

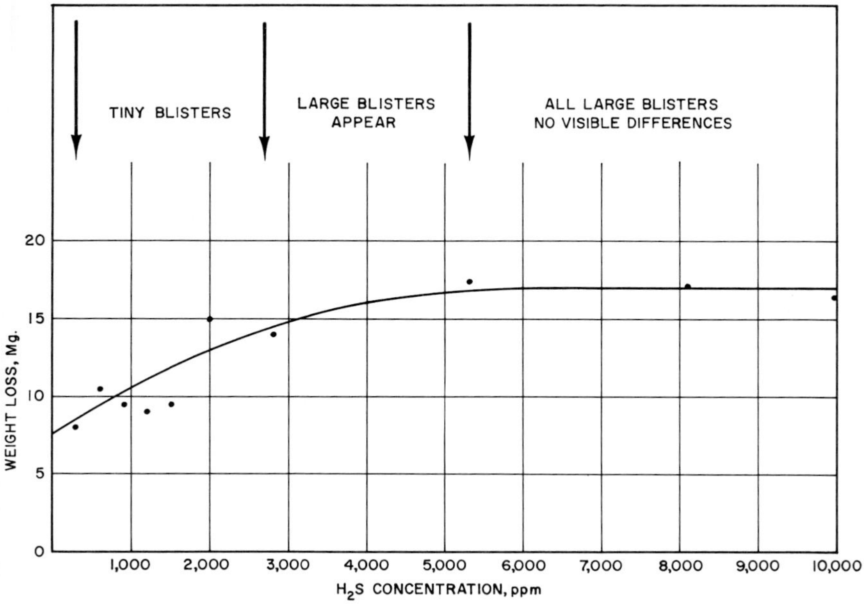

FIG. 4—*Effect of H_2S concentration on corrosion in pH 10 aqueous-oil system.*

Effect of Cyanide Concentration in pH 10.0 Test

A constant hydrogen sulfide content of 4100 ppm was used. Results are shown in Table 7 and Fig. 5.

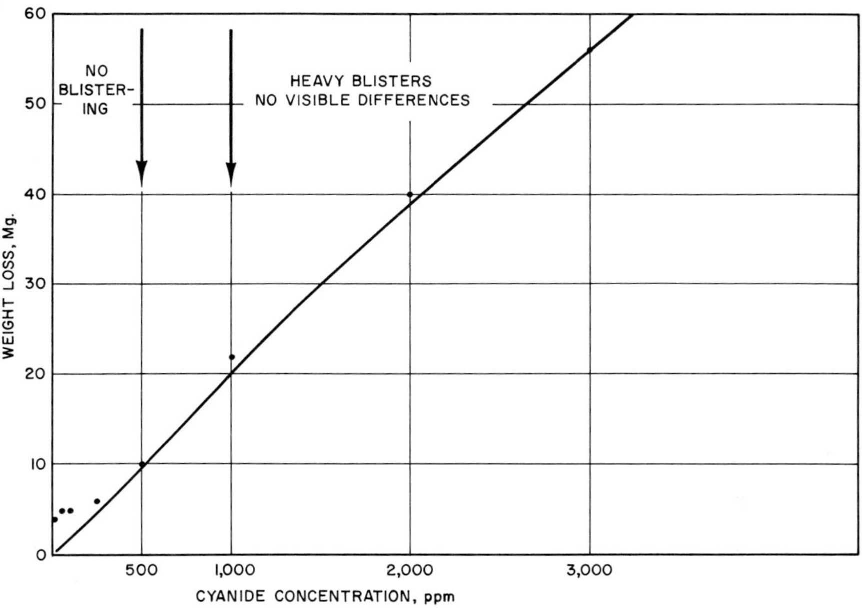

FIG. 5—*Effect of cyanide concentration in pH 10 system, 4100-ppm H_2S.*

TABLE 7—*Effect of cyanide concentration in pH 10.0 test.*
Conditions: 30-ml oil and 60-ml aqueous phase, with hydrogen sulfide held constant at 4100 ppm. Inhibitors added at 100 ppm, based on total fluid. Test bottles rotated 20 h at 80 C.

	Cyanide Added, ppm							
Inhibitors	0	50	100	250	500	1000	2000	3000
None:								
Weight loss, mg	4	5	5	6	10	22	40	56
Specimen appearance	1[a]	1	1	1	2	5	6	6
Inhibitor A:								
Weight loss, mg	...	...	...	...	15	22	37	48
Specimen appearance	...	...	...	...	2	5	6	6
Inhibitor B:								
Weight loss, mg	...	...	...	...	16	22	32	49
Specimen appearance	...	...	...	...	2	5	6	6
Inhibitor C:								
Weight loss, mg	...	...	...	...	16	23	30	46
Specimen appearance	...	...	...	...	3	6	6	6
Inhibitor D:								
Weight loss, mg	...	...	...	...	18	26	39	47
Specimen appearance	...	...	...	...	3	6	6	6
Inhibitor E:								
Weight loss, mg	...	...	...	...	13	19	39	55
Specimen appearance	...	...	...	...	2	5	6	6
Inhibitor F:								
Weight loss, mg	...	...	...	...	15	26	41	41
Specimen appearance	...	...	...	...	3	6	6	6
Inhibitor G:								
Weight loss, mg	...	...	...	...	10	17	25	11
Specimen appearance	...	...	...	...	2	2	3	2
Inhibitor H:								
Weight loss, mg	...	...	...	...	3	5	5	8
Specimen appearance	...	...	...	...	1	1	1	1

[a] Compare with Fig. 1 for degree of blistering.

Discussion

Distribution Experiments

Data in Table 2 show that cyanide is not extracted into the oil phase, although a slight amount may be adsorbed on the glass vessel. The constancy of the coefficient for hydrogen sulfide indicates that only the total concentration need be considered.

Because of sulfide interferences with the cyanide titration, it was impossible to determine joint coefficients.

Effect of pH on Hydrogen Sulfide Corrosion

These experiments show that the same sort of results are obtained in the presence of oil as in water alone [*31*]. Some hydrogen blistering is shown down to a pH of 6. There is still some corrosion at a pH of 8.

Effect of Hydrogen Sulfide Concentration

Here, only hydrogen sulfide was added to an oil water system. Hydrogen blistering increased with the hydrogen sulfide concentration. However, corrosion rates levelled off at approximately 1500-ppm hydrogen sulfide and seemed to decrease slightly at still higher concentrations.

The water-soluble inhibitors were ineffective. Oil-soluble inhibitors were, on the whole, very satisfactory in stopping this acid sulfide attack.

Intermediate pH (7.7) Experiments

These solutions were buffered heavily to maintain a pH of 7.7. The corrosion rates are not exceptionally high, and no hydrogen blistering occurred. Only two of the oil-soluble inhibitors, Inhibitors A and C, seemed to have any appreciable effect. At the present, no firm explanation can be given for the better performance of Inhibitors A and C.

Filming amine type inhibitors are generally reaction products of mixtures of long-chain amines and long-chain acids. The resulting products are complex mixtures which are very difficult to analyze.

Effect of Hydrogen Sulfide in Presence of Cyanide at pH 10.

In this high pH range, corrosion rates are not high. Hydrogen blistering increases with the increase in hydrogen sulfide concentration.

The oil-soluble inhibitors have essentially no effect at this pH range. The water-soluble, linear amine, type inhibitors seem to perform best. It was noted that noninhibited coupons were water wet, while inhibited coupons had hydrophobic surfaces.

Variation of Cyanide in the pH 10 Test

Hydrogen sulfide was constant at 4100 ppm. Corrosion and blistering increased linearly with the cyanide concentration. This indicates that, even in the presence of oil, cyanide is able to dissolve the sulfide from the steel.

As in the case of variation of hydrogen sulfide, water-soluble Inhibitor H was by far the most effective.

Conclusions

The general conclusions are:

1. There does not seem to be any major differences in uninhibited hydrogen sulfide corrosion in the presence or absence of oil, at least for the nonaromatic oil used.

2. Oil-soluble inhibitors in acid systems generally inhibit both hydrogen blistering and corrosion.

3. At the intermediate pH level, some inhibitors appear to work, although the corrosion load is not high.

4. At pH above 8 in the presence of cyanide, corrosion rates can be fairly high. Oil-soluble corrosion inhibitors seem to be completely ineffective under these conditions, but the water-soluble materials tested gave good results in decreasing both overall weight loss and hydrogen blistering. The difference in effectiveness between oil-soluble and water-soluble inhibitors may be a function of solubility, or of chemical properties of the inhibitor molecules.

References

[*1*] Bartz, M. H. and Rawlins, C. E., *Corrosion,* SBIIA, Vol. 4, 1948, pp. 187–206.
[*2*] Schofer, N., *Corrosion,* SBIIA, Vol. 5, June 1949, pp. 182–188.
[*3*] Skei, T., Wachter, A., Bonner, W. A., and Burnham, H. D., *Corrosion,* SBIIA, Vol. 9, 1953, pp. 163–172.
[*4*] Ewing, S. P., *Corrosion,* SBIIA, Vol. 11, 1955, pp. 497t–501t.
[*5*] Gardner, G. S., Faigen, H. L., Gibson, G. L., and Hall, W. S., *Journal of the Franklin Institute,* JFINA, Vol. 262, 1956, pp. 369–384.
[*6*] Boies, D. B., *Corrosion,* SBIIA, Vol. 12, 1956, pp. 371t–375t.
[*7*] Biehl, J. A. and Schnake, E. A., *American Petroleum Institute, Division of Refining,* PDRAA, Vol. 37 (III), 1957, pp. 129–134.
[*8*] Neumaier, B. W. and Schillmoller, C. M., *Petroleum Refiner,* PEREA, Sept. 1957, pp. 319–326.
[*9*] Roebuck, A. H., Gant, P. L., Riggs, O. L., and Sudbury, J. D., *Corrosion,* SBIIA, Vol. 13, 1957, pp. 733t–738t.
[*10*] Bradley, B. W. and Dunne, N. R., *Corrosion,* SBIIA, Vol. 13, 1957, pp. 238t–242t.
[*11*] Meyer, F. H., Riggs, O. L., McGlasson, R. L., and Sudbury, J. D., *Corrosion,* SBIIA, Vol. 14, 1958, pp. 109t–115t.
[*12*] "Corrosion of Oil and Gas-Well Equipment," *American Petroleum Institute, Division of Production,* APIPB, Dallas, 1958, pp. 26–39.
[*13*] Comeaux, R. V., *Corrosion,* SBIIA, Vol. 15, 1959, pp. 189t–193t.
[*14*] Shannon, D. W. and Boggs, J. E., *Corrosion,* SBIIA, Vol. 15, 1959, pp. 299t–302t.
[*15*] Ehmke, E. F., *Corrosion,* SBIIA, Vol. 16, 1960, pp. 246t–252t.
[*16*] Gardner, G. S., *Corrosion,* SBIIA, Vol. 16, 1960, pp. 312t–318t.
[*17*] Hutchison, C. B. and Hughes, W. B., *Corrosion,* SBIIA, Vol. 17, 1961, 514t–518t.
[*18*] Nathan, C. C., *Corrosion,* SBIIA, Vol. 18, 1962, pp. 282t–286t.
[*19*] Hughes, W. B. and Stromberg, V. L., *Corrosion,* SBIIA, Vol. 19, 1963, pp. 9t–11t.
[*20*] Riggs, O. L. and Radd, F. J., *Corrosion,* SBIIA, Vol. 19, 1963, pp. 1t–8t.
[*21*] Sardisco, J. B., Wright, W. B., and Greco, E. C., *Corrosion,* SBIIA, Vol. 19, 1963, pp. 354t–359t.
[*22*] Storey, W. D., *Oilweek,* OLWKA, 20 May 1963, pp. 48–53.
[*23*] Auer, L. and Hewes, F. W., *Materials Protection,* MAPRA, Aug. 1964, pp. 10–14.
[*24*] Beck, W., Glass, A. L., and Taylor, E., *Journal,* Electrochemical Society, JESOA, Vol. 112, 1965, pp. 53–59.
[*25*] Brickell, W. F., Greco, E. C., and Sardisco, J. B., *Corrosion,* SBIIA, Vol. 21, 1965, pp. 48–52.
[*26*] Sardisco, J. B. and Pitts, R. E., *Corrosion,* SBIIA, Vol. 21, 1965, pp. 245–253.
[*27*] Sardisco, J. B. and Pitts, R. E., *Corrosion,* SBIIA, Vol. 21, 1965, pp. 350–354.
[*28*] Milton, C., *Corrosion,* SBIIA, Vol. 22, 1966, pp. 191–194.
[*29*] Greco, E. C. and Brickell, W. F., *Materials Protection,* MAPRA, Oct. 1966, pp. 29–33.
[*30*] Chance, M. R., *Materials Protection,* MAPRA, April 1967, pp. 33–37.

[*31*] Gutzeit, J., *Materials Protection.* Dec. 1968, pp. 17–23.
[*32*] Hudgins, C. M., *Materials Protection,* Jan. 1969, pp. 41–47.
[*33*] Hausler, R. H., Goeller, L. A., Zimmerman, R. P., and Rosenwald, R. H., "Contribution to the Filming Amine Theory: Interpretation of Experimental Results," paper presented at the National Meeting, National Association of Corrosion Engineers Philadelphia, 2–6 March 1970.
[*34*] Barnett, L. M., Betz, Trevose, Pa., 24 March 1971, personal communication.

A. P. Bond[1]

Pitting Corrosion—A Review of Recent Advances in Testing Methods and Interpretation

REFERENCE: Bond, A. P., **"Pitting Corrosion—A Review of Recent Advances in Testing Methods and Interpretation,"** *Localized Corrosion—Cause of Metal Failure, ASTM STP 516,* American Society for Testing and Materials, 1972, pp. 250–261.

ABSTRACT: There is widespread disagreement as to the applicability of various test procedures in the study of pitting corrosion in alloy systems. A literature survey has been made to identify areas of agreement and to rationalize areas of disagreement, especially in the area of test methods, such as potentiodynamic scanning to determine the breakthrough pitting potential, potentiostatic tests, and other electrochemical tests. The variables causing disagreement and the lack of reproducibility of test results are discussed. The current status of attempts to formulate a detailed mechanism for pitting corrosion in a given system is reviewed.

KEY WORDS: corrosion, stainless steel, tests, evaluation, pitting, aluminum, aluminum alloys, polarization, electrochemistry

Pitting corrosion has long been a serious problem for metals that owe their corrosion resistance to passivity. Localized breakdown in passivity can lead to corrosion of very small anodes coupled to relatively large cathodes, a situation that creates very high rates of penetration even when the overall corrosion rate is not high. Since the phenomenon of passivity is still not fully understood, it is not surprising that the localized breakdown of passivity resulting in pitting is even less well understood.

A great deal of experimental work has been devoted to the study of pitting corrosion. Much data have been accumulated using immersion tests, but a completely satisfactory test never has been developed. A given immersion test cannot differentiate between the pitting resistance of metals that do not pit in it and usually can give only a crude ranking among those that do pit. In order to rank accurately the pitting resistance of a group of alloys, a series of immersion tests of varying severity is required in order that a pitting threshold can be established for each alloy. In order to change the severity of the environment, the concentration of the aggressive anion, the type or concentration of the oxidizing agent, or the

[1] Research supervisor, Research Laboratory, Climax Molybdenum Company of Michigan (a subsidiary of American Metal Climax, Inc.), Ann Arbor, Mich. 48106.

temperature can be changed. An example of this approach is shown in Table 1, in which the results give a relative ranking of the pitting resistance of some rather resistant alloys [*1*].[2] Tests of this type could be rather time consuming and open to the objection that numerous changes in the test environment are made. Immersion tests are also greatly influenced by edge effects. Pitting is much more severe at the edges for most alloys, and, in many cases when edge pitting occurs and no pits are formed away from the edges, it is impossible to know whether the edge pitting protected the rest of the specimen or whether the major surface really was resistant to pitting in the given environment.

TABLE 1—*Results of exposure of 25Cr ferritic stainless steels to* $10FeCl_3 \cdot 6H_2O$ (*Ref* 1).

Test Temperature, deg C	Corrosion Rate (mdd)[a] of Indicated Alloys			
	0Mo	2Mo	3.5Mo	5Mo
25	40	nil	nil	nil
50		300	nil	nil
70			400	nil

[a] mdd = milligrams per square decimeter per day.

In attempts to avoid these problems, several investigators [*2–4*] introduced electrolytic methods. In this way, the potential difference between the metal and the electrolyte can be varied at will without changing the electrolyte or the temperature. Breakthrough potentials were determined for various stainless steels by applying an external electromotive force to a cell made up of the test alloy as anode and a nonpolarizable cathode [*2*] or use of an arrangement such that the polarization that did occur at the auxiliary electrode did not interfere with the measurements [*3,4*]. These measurements indicated that increased chromium or molybdenum content improved resistance of stainless steels to pitting in agreement with practical experience. However, electrolytic determination of breakthrough potentials was not generally used by other workers. Streicher [*5*] used a somewhat different approach in that he applied a potential sufficient to cause pitting to initiate on all the alloys tested and counted the number of pits formed under standard conditions. In the general case, interpretation of the results obtained by his technique requires knowing how overall pitting resistance would be expected to correlate with the number of pits found in this type of test.

None of the electrolytic tests for pitting corrosion received wide acceptance until the electronic potentiostat came into wide use. Since that time, measurements of pitting or breakthrough potentials have been made

[2] The italic numbers in brackets refer to the list of references appended to this paper.

on a wide variety of metals and under a wide variety of conditions. A typical determination of pitting potential is shown in Fig. 1. Unfortunately, the great variety of materials and conditions has led to many apparent contradictions. As a result, pitting corrosion has become one of the very controversial areas of corrosion research. The purpose of the present paper is to review recent work on some representative alloy systems in an effort to identify areas of agreement and to rationalize some of the areas of disagreement, especially in test methods.

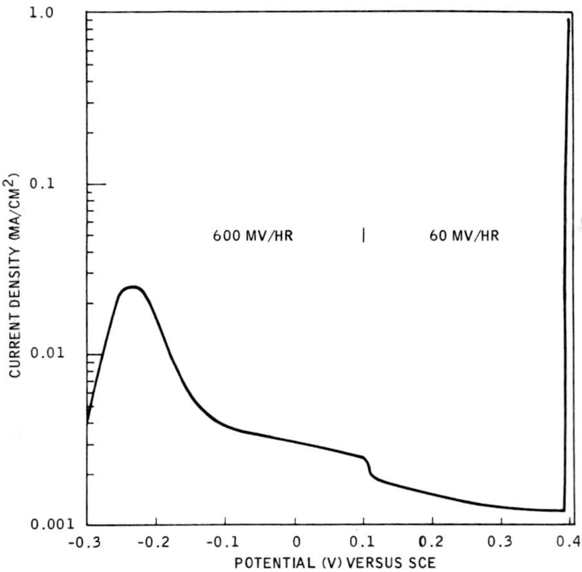

FIG. 1—*Polarization curve obtained for 17.6Cr-13.6Ni-1.99Mo austenitic stainless steel in argon-saturated 0.1* N *HCl. The scanning rate was 600 mV/h from —0.3 to 0.1 V; it was decreased to 60 mV/h from 0.1 V to the pitting potential* (*Ref* 31).

Aluminum and Aluminum Alloys

Several investigators have used controlled potential techniques to determine pitting potentials of high-purity aluminum in various halide containing environments [*6–9*]. The results obtained are shown in Table 2. The various determinations agree remarkably well in spite of many differences in experimental technique. In some cases, the test specimens were electropolished; in others, abraded and then etched in sodium hydroxide. Various electrode mounting techniques were used, and no investigator reported any necessity to take precautions against crevice corrosion. The details of the polarization technique varied considerably, and there was no evidence that variations in the rate of change of potential had any

TABLE 2—*Pitting potentials determined on aluminum at 25 C.*

Electrolyte	Reference	Pitting Potential, V versus NHE[a]
	HIGH-PURITY AL	
0.1 *M* NaCl	8	−0.40
0.1 *M* NaCl	9	−0.46
0.5 *M* NaCl	7	−0.50
1 *M* NaCl	6	−0.48
1 *M* NaCl	9	−0.52
0.1 *M* NaBr	8	−0.29
1.0 *M* KBr	9	−0.42
1.0 *M* KBr	6	−0.35
1.0 *M* KI	9	−0.26
1.0 *M* KI	6	−0.20
	99.4AL	
0.1 *M* NaCl	8	−0.41
	AL-2.4MG	
0.1 *M* NaCl	8	−0.44
	AL-0.2CU	
0.1 *M* NaCl	9	−0.46
	AL-4CU (SOLUTION TREATED)	
0.1 *M* NaCl	9	−0.36

[a] NHE = normal hydrogen electrode.

large effect on the results obtained. It was, in fact, found that pit growth could be stopped and started at will by changing the potential from 20 mV active to the pitting potential to 20 mV noble to the pitting potential [*9*]. This process could be repeated many times on the same specimen. Oxygen content of the test solutions was not reported to be an important factor.

From these results, it is clear that there is a definite pitting potential for aluminum which is readily measured reproducibly by different investigators (Table 2). The results obtained from immersion tests are in excellent agreement with predictions made on the basis of pitting potentials. When aluminum is undergoing pitting corrosion, it exhibits a corrosion potential equal to the pitting potential in that environment, and when no pitting corrosion is occurring, the corrosion potential is more active than the pitting potential [*7,8*].

Apparently the pitting potential observed for aluminum is relatively insensitive to purity or deliberate alloy additions (Table 2.) An exception is copper in solid solution, since it shifts the pitting potential in the noble direction [*9*]. This is not necessarily beneficial to the corrosion resistance of the alloy; in fact, according to Gavele and DeMicheli [*9*], it leads to intergranular corrosion in aged aluminum-copper alloys. Intergranular corrosion results because the alloy becomes depleted in copper as precipitation occurs at the grain boundaries and thus has a more active pitting potential in the areas immediately adjacent to grain boundaries than is the case for the areas of the grain faces.

The results on aluminum and its alloys indicate that other factors besides pitting potential are extremely important in determining whether or not pitting will occur. In many cases it is the cathodic polarization behavior of aluminum that determines whether or not the pitting potential will be reached in an exposure test. For example, it has been found that in relatively pure aluminum (99.99 and 99.999 percent) it is the distribution of impurities, mainly iron and copper, that determines whether or not pitting will occur in 0.5 *M* sodium chloride (NaCl) containing 0.3 percent hydrogen peroxide (H_2O_2) even though distribution of impurities has no effect on the pitting potential [7]. It was found that microsegregation of iron and copper reduced the cathodic overvoltage greatly, allowing the pitting potential to be reached and pitting to initiate. That copper and iron in aluminum reduce cathodic polarization has been well established [*9–11*].

In summary, several investigators have demonstrated that a well-defined pitting or breakdown potential can be determined experimentally for aluminum and some of its alloys. This information, coupled with knowledge of cathodic polarization behavior, has proved useful in interpreting the corrosion behavior of aluminum.

Stainless Steels

In contrast to the rather well-defined situation for aluminum, at least from an empirical standpoint, electrochemical studies of pitting of stainless steels are fraught with many contradictions because of the pronounced effect of many experimental factors on the results obtained. Unfortunately, the importance of some experimental variables is only now being appreciated.

As might be expected, surface preparation exerts a decided effect on the results obtained on stainless steel. It has been reported often that resistance to pitting increases with the degree of mechanical polishing [*5,12,13*], while electropolishing or etching leads to less resistance to pitting [*5*] (Fig. 2). In a study of the effects of surface preparation on pitting potentials of a series of iron-chromium alloys, Pessall et al [*14*] showed that differences in surface preparation explain an apparent contradiction in the results of several workers [*15–17*] on this system. The reasons for the pronounced effect of surface preparation are rather obscure. One suggestion is that mechanical polishing covers grain boundaries and inclusions with flowed metal. Pitting which was observed to initiate preferentially at grain boundaries and inclusions may explain the effect of mechanical polishing [*5*]. On the other hand, Standifer [*18*], as quoted by Greene and Fontana [*19*], reported that pit sites were distributed randomly on stainless steel except on application of "higher" anodic current densities. In a study of surface films on stainless steel, Vernon et al [*20*] found that the chromium content of the oxide film increased with the degree of mechanical polishing. This effect could be important in

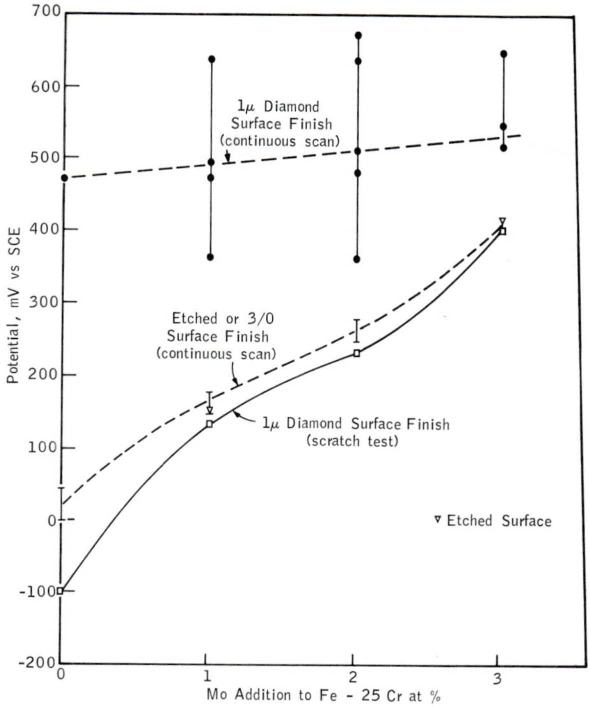

FIG. 2—*Critical pitting potentials of Fe-Cr-Mo alloys in deaerated synthetic sea-water at 90 C, comparing the effects of differing surface preparation and also comparing results from a continuous scan and from the scratch test* (*Ref* 37).

establishing the initial conditions in a pitting test. It has been noted also that the influence of surface treatment is much less pronounced in tests carried out at 80 C than it is at 20 C [*13*].

A rather surprising result is the rather large effect of gases dissolved in the test solution on pitting potentials as shown in Table 3 [*21*]. These results have been used [*22*] to explain the lack of agreement between pitting potential measurements and exposure tests for pitting of zirconium and Type 430 stainless steel reported by France and Greene [*23*]. Thus,

TABLE 3—*Effects of dissolved gases on the pitting potentials of some stainless steels in 1* M *NaCl at 25 C* (*Ref* 18).

Gas	Pitting Potential, V versus NHE,[a] of the Indicated Steels	
	Type 430	Type 304
H_2	0.06	0.19
N_2	0.11	0.22
A	0.14	0.30
O_2	0.21	0.31

[a] NHE = normal hydrogen electrode.

in carrying out exposure tests, it must be kept in mind that oxidizing agents added to change the redox potential of the solution may themselves have an effect on the pitting potential.

The rate of change of potential and the initial potential are of considerable importance in the determination of pitting potentials of stainless steels. Many investigators have reported that there is an induction time that varies with chloride concentration and potential [*24*] for pit initiation and that for this reason a continuous potentiodynamic scan overshoots the pitting potential for 18 Cr–8 to 10Ni austenitic stainless steels by an amount that increases with increasing scanning rate [*25–29*]. On the other hand, other investigators working with ferritic stainless steels [*30*] or molybdenum containing austenitic stainless steels [*31*] have reported that the pitting potential is shifted in the active direction as the scanning rate is increased. The latter effect could be explained by considering that, as the potential difference between the bulk metal and the bulk solution is increased, the passive film grows until the field across it has the same gradient as it did before the change in potential difference. Thus, the potential difference across the film/solution interface, which controls the reaction between the solution and film, remains relatively constant. On the other hand, if the potential difference is increased more rapidly than the film can thicken, then the film/solution potential difference is increased. Therefore, the faster the scan, the greater is the potential difference favoring film breakdown [*30*].

Depending on the relative importance of effects of induction time for pitting breakdown and of film thickening, it appears that pitting potentials determined by potentiodynamic scanning may either increase or decrease as scanning rate is increased. The observed effect probably depends on surface preparation, starting potential, and the nature of the test alloy. For these reasons, it has been suggested that potentiostatic techniques should be used for measuring pitting potentials [*27,28*]. However, this technique is very time consuming, since a separately prepared and activated electrode must be used to determine each point on the polarization curve. Apparently, for this reason, the technique has been little used. A novel experimental approach to this problem has been suggested by deWaard et al [*32*] who used a wire specimen as the test electrode and passed a constant current the length of the wire by means of a separate auxiliary circuit so that IR drop creates a potential gradient within the wire from one end to the other. In most cases 1000-mV gradients were used. The potential at one point on the wire was controlled, with respect to a reference electrode, with the aid of a potentiostat and counter electrode so that the metal-solution potential difference varied over the length of the test electrode in a known way, as long as the current flowing in the electrolytic circuit controlled by the potentiostat was much smaller than the current in the auxiliary circuit involving only the wire. By noting the position of any pits that form, the pitting potential under potentiostatic conditions can be determined in a single experiment.

Another approach has been recommended by Pourbaix and co-workers [*33–35*] who introduce the concept of the protection potential, that is, the potential at which actively growing pits are stopped completely from further growth. Thus, what actually is determined is a sort of hysteresis loop for pitting with the most important value being the potential at which the pits that were initiated at the breakdown potential repassivate as the potential is scanned back in the active direction.

This hysteresis has been used by Wilde and Williams [*22*] to explain the observation by Steigerwald [*16*] that the corrosion potential exhibited by iron-chromium alloys was more active than the pitting breakdown potential in an environment of equivalent chloride concentration. Wilde and Williams [*22*] monitored the corrosion potentials of several alloys in ferric chloride solution and found that the corrosion potential was more noble than the pitting breakdown potential for a fairly short time, then drifted to potentials more active than the breakdown potential, but remained more noble than the repassivation potential as long as pitting continued.

Determination of the protection potential appears to present some experimental difficulties. Once pits are initiated and begin to grow, the solution within the pits becomes more concentrated in the aggressive ion, and the pH shifts in the acid direction. In fact, it has been shown that opening actively growing pits and allowing free circulation of the bulk electrolyte into the pits causes these pits to stop growing in many cases [*36*]. Therefore, if pitting is allowed to develop, the protection potential that is determined must be characteristic of the solution within the pits, not of the bulk electrolyte. Thus, the protection potential would be placed at a more active potential than would be expected from the bulk solution composition.

A technique recently described by Pessall and Liu [*37*] may make it possible to determine what amounts to the protection potential without first initiating actual pits. This technique consists of polarizing the test electrode to the desired potential and then lightly scratching the electrode surface. While the potential is held constant, the current is recorded as a function of time after scratching until the electrode either repassivates or stable pit growth is taking place. This process is repeated until the minimum potential is found at which the electrode does not repassivate after scratching. This value was reported to be reproducible and independent of the surface preparation and prior history of the electrode. By eliminating the initiation step in pitting, most of the factors that have caused the lack of agreement among pitting potential measurements of several investigators are also eliminated.

The results obtained by this method are, in general, qualitatively comparable to those obtained by the scanning technique as far as the effects of alloy composition is concerned [*33*]. In fact, for some alloys, especially those containing molybdenum, only small differences were observed in the pitting potentials determined by the scratch test and by scanning tests (Fig. 2). From a practical standpoint, the pitting potential measured in this way should be a very useful guide to pitting corrosion resistance since it takes into account the possibility of mechanical damage that would be likely to occur to the passive layer under service conditions. The scratch technique for determining pitting potential is very promising, but more experience with results from different investigators is required before the method can be assessed completely.

Several investigators [*17,24,32,34,35*] have reported that anions such as sulfate, nitrate, chromate, and chlorate inhibit pitting in chloride containing solutions. Inhibition takes different forms with the different anions. Sulfate moves the pitting potential to more noble values [*24,34*], nitrate is only effective at more noble potentials [*28,38*], and chromate acts mainly as a cathodic inhibitor [*28*], preventing attainment of the pitting potential in various environments.

In general, pitting potential measurements obtained by all techniques are in qualitative agreement with exposure tests and practical experience as far as the effects of major factors on the pitting resistance of stainless steels are concerned. Thus, increasing chromium and molybdenum content and decreasing exposure temperature and concentration of the activating ion, such as chloride or bromide, all contribute to greater resistance to pitting as has been indicated by pitting potential measurements.

Other Metals

Electrochemical determinations of pitting potential have been made on several other metals as summarized in Table 4. Several investigators report reproducible results for zirconium [*17,23,40,41*], but it has been reported also that the electrochemical tests do not correlate well with exposure tests [*23*]. Too little information is available to indicate how the results on the other metals in Table 4 compare with exposure tests.

Mechanism of Pit Initiation

No mechanism of the pitting of passive metals has been developed to the point that it is capable of making quantitative predictions as to the pitting resistance of any metal. In rough terms, most of the proposed mechanisms involve either an activating ion, such as chloride, penetrating the passive film and destroying it locally or the activating ion being preferentially adsorbed at the metal solution interface displacing the adsorbed passive layer and thus locally destroying passivity.

The mechanism involving breakdown of an oxide film has been further subdivided by Hoar [*30*] into the "ion-migration" and the "mechanical"

TABLE 4—*Pitting potentials of Zr, Mg, and Ni.*

Metal	Electrolyte	Pitting Potential, V versus NHE[a]	Ref.
Zr	1 *M* H_2SO_4 + 1 *M* NaCl	0.4	*20*
Zr	1 *M* NaCl	0.38	*17*
Zr	1 *M* KCl	0.4	*36*
Zr	0.5 *M* NaCl	0.42	*37*
Mg	1 *M* NaCl	−1.4	*17*
Ni	0.1 *M* NaCl	0.28	*15*

[a] NHE = normal hydrogen electrode.

models. In the ion-migration model, the activating anion enters the oxide film lattice without exchange [*5,42*]. This greatly increases the ionic conductivity of the film, leading to locally high anodic dissolution rates and pitting. In the mechanical model, anions adsorb at the oxide/solution interface, lowering the interfacial energy to the point that cracks or splits develop in the protective film under the influence of the electrostatic repulsion of the adsorbed anions. In either model, it is assumed that the initial step involves adsorption of the activating ion on the oxide surface and that it is this step that is affected by the applied potential and the presence of other anions that may inhibit pitting.

The direct adsorption model has been discussed by many investigators [*8,15,17,25,26,38*]. This mechanism is able to explain qualitatively the effects of potential and especially of inhibiting anions.

In both mechanisms, adsorption of anions plays a key role, but not enough information on the energies involved is available for any quantitative treatment to be made. However, based on another approach, Vermilyea [*43*] has developed a theory that allows a semiquantitative prediction of pitting potentials. He assumes that the pitting potential is the potential at which the protective metal oxide and the salt of the metal and the aggressive anion are in equilibrium. To achieve this equilibrium, the activity of the aggressive anion in the pit is increased by the potential difference between the inside of the pit and the bulk solutions. By using these assumptions and some approximations concerning the behavior of concentrated solutions, the pitting potential can be estimated. Reasonable agreement with experimental data was achieved for aluminum, magnesium, iron, and nickel, but not for zirconium, titanium, or tantalum. This theory does not predict which ions should be aggressive. It also appears difficult to apply this treatment to alloys. However, it represents a promising approach that should receive a thorough test.

Conclusion

Great care must be taken in applying electrochemical techniques to a study of pitting corrosion to ensure that the technique selected is appropriate to the metal under study and the results required. If the proper technique is selected, electrochemical results are just as valid as carefully accelerated exposure tests and, in most cases, much more convenient. With the present state of knowledge, no laboratory technique can be used to predict the exact corrosion behavior that will be encountered in a complex practical installation. Although some progress is being made in the study of the theory of pitting corrosion, empirical investigation will remain of prime importance in this area for some time to come.

References

[*1*] Lizlovs, E. A. and Bond, A. P., *Journal,* Electrochemical Society, JESOA, Vol. 118, 1971, p. 22.
[*2*] Uhlig, H. H. and Wulff, J., *Transactions,* American Institute of Mining and Metallurgical Engineers, TAMMA, Vol. 135, 1939, p. 494.
[*3*] Brennert, S. J., *Journal,* Iron and Steel Institute, JISIA, Vol. 135, 1937, p. 379.
[*4*] Mahla, E. M. and Nielsen, N. A., *Transactions,* Electrochemical Society, JESOA, Vol. 89, 1946, p. 167.
[*5*] Streicher, M. A., *Journal,* Electrochemical Society, JESOA, Vol. 103, 1956, p. 375.
[*6*] Kaesche, H., *Zeitschrift für Physikalische Chemie Neve Folge,* ZPYFA, Vol. 26, 1960, p. 138 and Vol. 34, 1962, p. 87.
[*7*] Bond, A. P., Bolling, G. F., Domian, H. A., and Biloni, H., *Journal,* Electrochemical Society, JESOA, Vol. 113, 1966, p. 773.
[*8*] Bohni, H. and Uhlig, H. H., *Journal,* Electrochemical Society, JESOA, Vol. 116, 1970, p. 906.
[*9*] Galvele, J. R. and DeMicheli, S. M., *Corrosion Science,* CRRSA, Vol. 10, 1970, p. 795.
[*10*] Hansen, D. J. and Wetmore, F. E. W., *Canadian Journal of Chemistry,* CJCHA, Vol. 34, 1956, p. 659.
[*11*] Ketcham, S. J. and Haynie, F. H., *Corrosion,* SBIIA, Vol. 19, 1963, p. 242.
[*12*] Smith, H. A., *Metal Progress,* MEPOA, Vol. 33, 1938, p. 596.
[*13*] Steensland, O., *Corrosion Prevention and Control,* CRPCA, Vol. 15, 1968, p. 25.
[*14*] Pessall, N., Hull, F. C., and Liu, C., "Development of a Low-Cost Iron-Base Alloy to Resist Corrosion in Hot Sea Water," Research and Development Report No. 478, Office of Saline Water, Westinghouse Electric Corp., Pittsburgh, 1969.
[*15*] Horvath, J. and Uhlig, H. H., *Journal,* Electrochemical Society, JESOA, Vol. 115, 1968, p. 791.
[*16*] Steigerwald, R. F., *Corrosion,* SBIIA, Vol. 22, 1966, p. 107.
[*17*] Kolotyrkin, Ja. M., *Corrosion,* SBIIA, Vol. 19, 1963, p. 261t.
[*18*] Standifer, L. R., Jr., "A Study of the Relations between Micro and Gross Solution Potentials and Susceptibility to Intergranular Corrosion of a 24S Aluminum Alloy and Some Austenitic Stainless Steels," Ph.D. dissertation, The Ohio State University, Columbus, 1952.
[*19*] Greene, N. D. and Fontana, M. G., *Corrosion,* SBIIA, Vol. 15, 1959, p. 25t.
[*20*] Vernon, W. H. J., Wormwell, F., and Nurse, T. J., *Journal,* Iron and Steel Institute, JISIA, Vol. 150, 1944, p. 81p.
[*21*] Wilde, B. E. and Williams, E., *Journal,* Electrochemical Society, JESOA, Vol. 116, 1969, p. 1539.
[*22*] Wilde, B. E. and Williams, E., *Journal,* Electrochemical Society, JESOA, Vol. 117, 1970, p. 775.
[*23*] France, W. D., Jr., and Green, N. D., *Corrosion,* SBIIA, Vol. 26, 1970, p. 1.
[*24*] Hoar, T. P. and Jacob, W. R., *Nature,* NATUA, Vol. 216, 1967, p. 1299.
[*25*] Leckie, H. P. and Uhlig, H. H., *Journal,* Electrochemical Society, JESOA, Vol. 113, 1966, p. 1262.
[*26*] Leckie, H. P., *Journal,* Electrochemical Society, JESOA, Vol. 117, 1970, p. 1152.

[*27*] Schwenk, W., *Corrosion Science,* CRRSA, Vol. 5, 1965, p. 245.
[*28*] Herbsleb, G., *Werkstoffe und Korrosion,* WSKRA, Vol. 16, 1965, p. 929.
[*29*] Herbsleb, G., *Werkstoffe und Korrosion,* WSKRA, Vol. 17, 1966, p. 649.
[*30*] Hoar, T. P., *Corrosion Science,* CRRSA, Vol. 7, 1967, p. 341.
[*31*] Bond, A. P. and Lizlovs, E. A., *Journal,* Electrochemical Society, JESOA, Vol. 115, 1968, p. 1131.
[*32*] deWaard, C., Nicholson, J. W., and Posch, W., *Werkstoffe und Korrosion,* WSKRA, Vol. 19, 1968, p. 782.
[*33*] Pourbaix, M., Klimzack-Mathieiu, L., Mertens, Ch., Meunier, J., Vanleugen-Haghe, Cl., DeMunck, L., Laureys, J., Neelemans, L., and Warzee, M., *Corrosion Science,* CRRSA, Vol. 3, 1963, p. 239.
[*34*] Pourbaix, M., *Corrosion Science,* CRRSA, Vol. 5, 1965, p. 677.
[*35*] Pourbaix, M., *Corrosion Science,* CRRSA, Vol. 26, 1970, p. 431.
[*36*] Rosenfeld, I. L. and Danilov, I. S., *Corrosion Science,* CRRSA, Vol. 7, 1967, p. 129.
[*37*] Pessall, N. and Liu, C., *Electrochimica Acta,* ELCAA, Vol. 16, 1971, p. 1987.
[*38*] Schwenk, W., *Corrosion,* SBIIA, Vol. 20, 1964, p. 192t.
[*39*] Uhlig, H. H. and Gilman, J. R., *Zeitschrift für Physikalische Chemie,* ZPCFA, Vol. 226, 1964, p. 127.
[*40*] Maraghini, M., Adams, G. B., Jr., and VanRysselberghe, P., *Journal,* Electrochemical Society, JESOA, Vol. 101, 1954, p. 400.
[*41*] Hackerman, N. and Cecil, O. B., *Journal,* Electrochemical Society, JESOA, Vol. 101, 1954, p. 419.
[*42*] Hoar, T. P., Mears, D. C., and Rothwell, G. P., *Corrosion Science,* CRRSA, Vol. 5, 1965, p. 279.
[*43*] Vermilyea, D., *Journal,* Electrochemical Society, JESOA, Vol. 118, 1971, p. 529.

M. J. Johnson[1]

Relative Critical Potentials for Pitting Corrosion of Some Stainless Steels

REFERENCE: Johnson, M. J., **"Relative Critical Potentials for Pitting Corrosion of Some Stainless Steels,"** *Localized Corrosion—Cause of Metal Failure, ASTM STP 516,* American Society for Testing and Materials, 1972, pp. 262–272.

ABSTRACT: The breakthrough potentials of stainless steel alloy Types 409, 304, and HWT were determined in solutions containing various concentrations of chloride ion, ranging from 10 to 40,000 ppm. Four different salts were used to introduce the chloride ion: sodium chloride, calcium chloride, cupric chloride, and ferric chloride. The breakthrough potential values obtained decreased with increasing chloride ion concentration. These breakthrough potential values were plotted against the logarithm of the chloride ion concentration and a straight line drawn through the lowest potential points for each alloy. This line described the minimum pitting potentials for each alloy and was used to compare the three alloys studied. Using this method, HWT proved to be a more pit resistant alloy than Type 304 which in turn was more pit resistant than Type 409.

KEY WORDS: corrosion, pitting, stainless steels, austenitic stainless steels, polarization, anodic polarization

Since the design of the potentiostat and its initial use for corrosion studies, technicians and particularly researchers have worked toward establishing the potentiostat as a tool for rapid, reproducible testing of the corrosion resistance of metals. In 1968, after considerable work by all involved, ASTM Committee G-1/VI, Section I, issued a "Recommended Practice for a Standard Reference Method for Making Potentiostatic and Potentiodynamic Anodic Polarization Measurements."[2] This method is well accepted and widely used for general corrosion resistance determinations. However, investigating and establishing the pitting susceptibility of a metal by potentiostatic or potentiodynamic methods remain undefined.

Many investigators have used the breakthrough potential as a measure

[1] Senior research chemical engineer, Allegheny Ludlum Industries, Inc., Research Center, Brackenridge, Pa. 15014.

[2] Designation: G 5–71, *1971 Annual Book of ASTM Standards,* Part 31, American Society for Testing and Materials, pp. 1047–1056.

of susceptibility to pitting [*1–6*][3]. Others have used current density differences taken at the same potential from anodic polarization curves, obtained in acid solutions with and without chlorides [7]. Still others have investigated and proposed the use of the ratios of current density values as a method of evaluating the pitting susceptibility of metals and alloys [*8*].

For this investigation the breakthrough potential was used as a measure of pitting susceptibility. Although no attempt is made to establish or refute any of the methods mentioned above as valid or invalid, it is felt that this investigation, by use of the breakthrough potential, establishes the relative pitting susceptibility of three stainless steel alloys.

Procedure

Materials

The alloys investigated were Types 409, 304, and HWT. All specimens were taken from commercial heats. The chemical compositions of the heats used in this investigation appear in Table 1.

TABLE 1—*Chemical compositions of alloys studied.*

Alloy	Weight Percent			
	C	Cr	Ni	Ti
Type 409	0.049	11.50	0.20	0.53
HWT	0.041	18.55	0.30	0.70
Type 304	0.062	18.60	9.25	0.012

The specimens were sheared from 0.060-in.-thick strip which was in the asannealed and pickled condition produced by the mill. No additional preparation was used on the specimens except to degrease them in a hot alkaline solution, rinse them thoroughly in water, and dry them.

Polarization Tests

The potentiodynamic polarization tests were performed using standard equipment, except for the electrochemical cell, which was designed to accept and expose 1 cm^2 of a flat specimen.

The cell consists of a 2.5 in. diameter by 5-in.-high glass cylinder which has been ground flat along its length about 1 in. wide. (Fig. 1). A hole was blown in the center of the ground surface and the insides tapered so that gases or heavy corrosion products would not be trapped. A Teflon gasket with a 1-cm^2 hole fit flat on the ground surface of the cell. The Teflon gasket was machined flat except for a ring around the 1-cm^2 hole, which was left raised $\frac{1}{16}$ in. The specimen was exposed by placing it flat on the Teflon gasket and pressing it with a clamp. This arrangement is

[3] The italic numbers in brackets refer to the list of references appended to this paper.

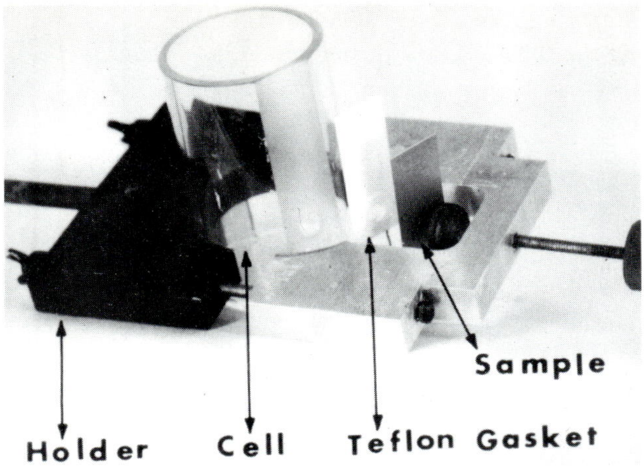

FIG. 1—*Cell assembly and clamping system used for potentiodynamic polarization tests.*

very similar to the Stern-Makrides specimen holder specified in the ASTM Recommended Practice, the only difference being the much larger Teflon-to-specimen contact area [*9*]. Electrical contact to the specimen was made externally.

The platinum counter electrode, salt bridge, and saturated calomel electrode were of the usual design. A Wenking potentiostat was used in conjunction with a two-way recorder.

All anodic polarization curves were conducted automatically at a scanning speed of about 0.7 V/h. The solutions were used "as mixed," without any aeration or deaeration, and at a temperature of 72 F $\pm$ 1. The pH of the solutions was not adjusted, except where the effect of pH was being investigated. The specimens were allowed to remain in contact with the solution for ½ h before the curve was started from the corrosion potential.

For the potentiostatic holding experiments, the desired potential was set before the specimen was mounted and the electrolyte introduced. After the experiment was prepared, the potential was introduced and held constant for 72 h. Subsequently, the specimens were examined for evidence of pitting by a low power microscope.

Results and Discussion

Typical potentiodynamic polarization curves obtained in this investigation appear in Fig. 2. The lines and arrows point at the potentials where a rapid increase of current occurred with very little change in voltage. This is considered an indication of a breakdown of the passive, corrosion resistant, stainless steel film, and initiation of corrosion, which in this case is in the form of pitting.

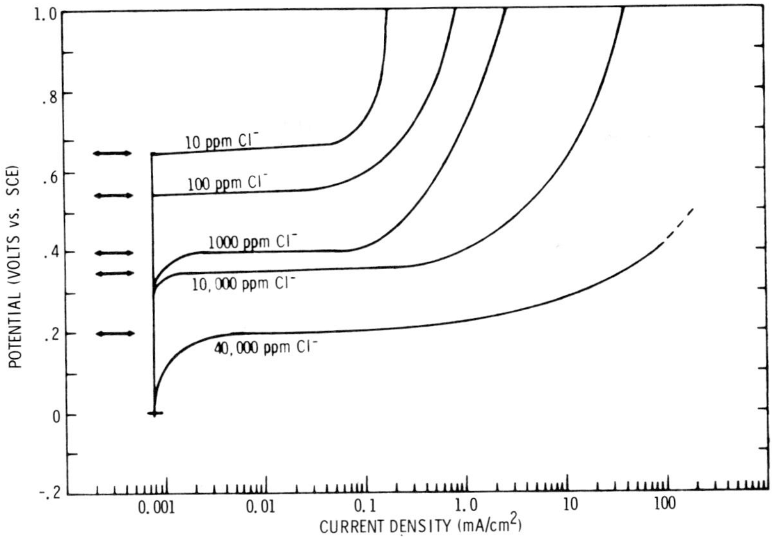

FIG. 2—*Potentiodynamic polarization curves of Type 304 in various chloride ion concentrations using NaCl as the salt.*

Occasionally crevice as well as pitting corrosion occurred on the specimen. The breakthrough potential in this case occurs at a lower potential, and the current increase is slower, until the pitting breakthrough potential is reached (Fig. 3). Then both curves essentially merge and form similar

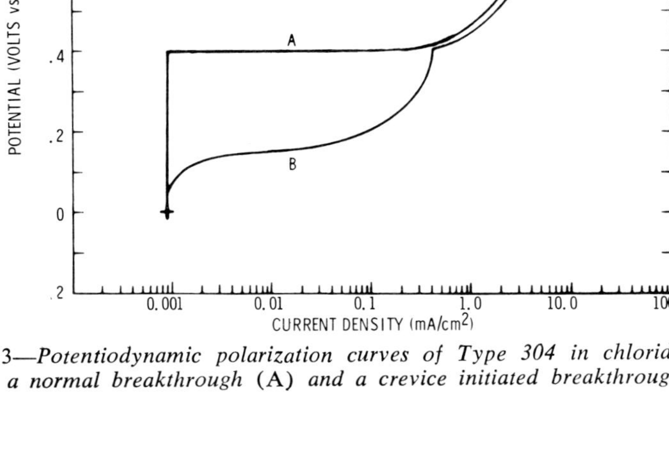

FIG. 3—*Potentiodynamic polarization curves of Type 304 in chloride solution showing a normal breakthrough* (A) *and a crevice initiated breakthrough* (B).

slopes. It has been shown that crevice corrosion does indeed occur at much lower potentials than pitting. Lizlovs indicated that Type 304 would crevice corrode at 0.00 V saturated calomel electrode (SCE) in 1 *N* sodium chloride (NaCl) [*10*]. When curves indicating crevice corrosion were obtained, the test was repeated until duplicate pitting breakthrough potentials were obtained on each alloy in each solution.

The results obtained are plotted against the chloride ion concentration in the graphs shown in Figs. 4–6. As would be expected, the breakthrough potential values are decreased as the chloride ion is increased. This was true regardless of the type of salt used to introduce the chloride ion. There was no apparent effect of the cation on the breakthrough potential, except possibly on the Type 409, where the breakthrough potentials in the NaCl are generally about 0.1 V (SCE) above the lowest breakthrough potential obtained in any other salt solution. Breakthrough potentials in NaCl on Type 304 resulted in a similar effect, but only in the high-chloride ion concentration.

There is considerable scatter in each concentration. This is attributed somewhat to the nature of the test and mostly to the nature of pitting corrosion. The tests were conducted potentiodynamically, and, although many potentiostatic tests were conducted before the potential change rate of 0.7 V/h was considered adequate, the test method remains dynamic, and the breakthrough potential could be "overshot." However, such overshooting and scatter can occur even in potentiostatic tests, where 5, 10, and even 60-min holds are used. The nature of the initiation of

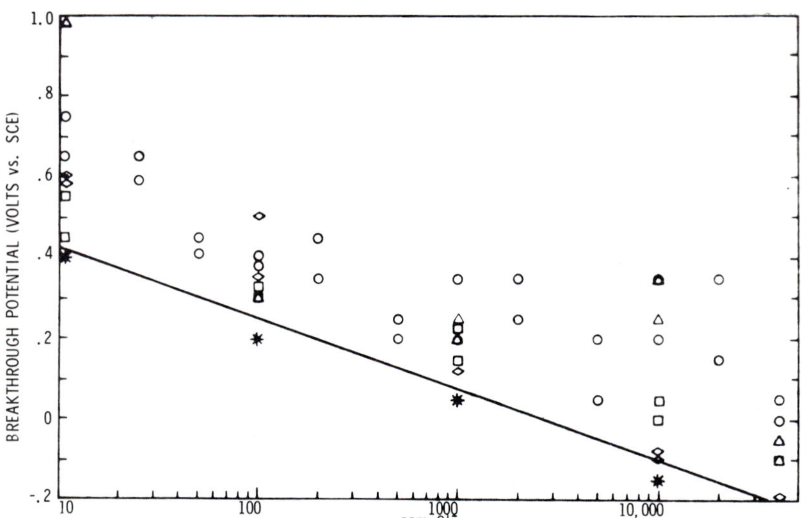

FIG. 4—*Effect of chloride concentration on the breakthrough potential of Type 409 in various chloride solutions:* ○ *NaCl,* △ *$CaCl_2$,* □ *$CuCl_2$,* ◇ *$FeCl_3$.* (**72-h hold test—no pitting.*)

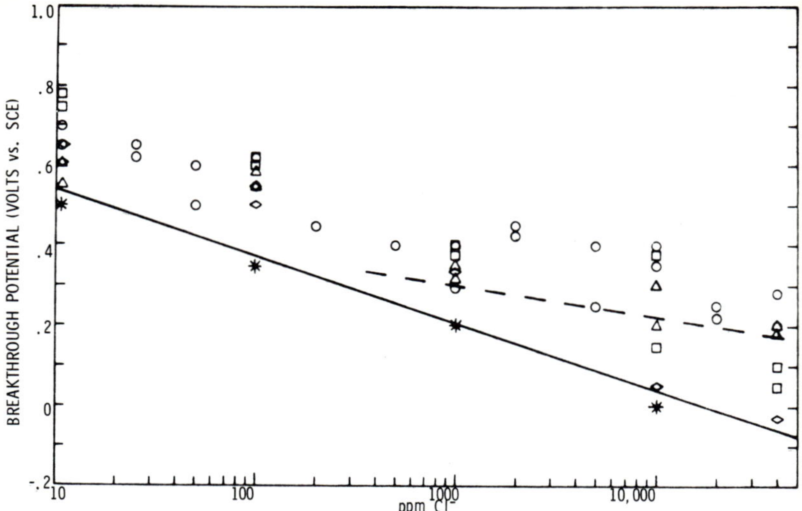

FIG. 5—*Effect of chloride concentration on the breakthrough potential of Type 304 in various chloride solutions:* ○ *NaCl,* △ *$CaCl_2$,* □ *$CuCl_2$,* ◇ *$FeCl_3$. (*72-h hold test—no pitting.)*

pitting on stainless steels is even more of a problem, particularly at the low-chloride ion concentrations studied. It is not unusual to observe only one pit on a 4 by 6-in. stainless steel panel exposed to a mild chloride containing environment. This one spot would then be more susceptible to

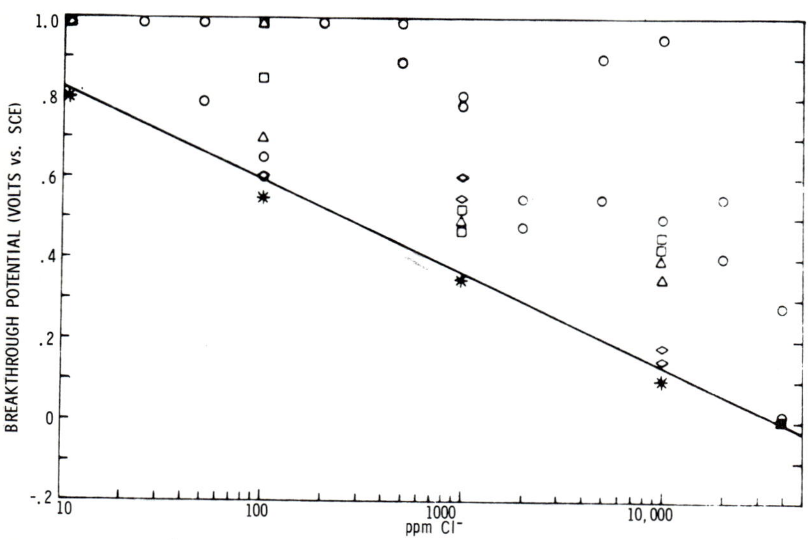

FIG. 6—*Effect of chloride concentration on the breakthrough potential of HWT in various chloride solutions:* ○ *NaCl,* △ *$CaCl_2$,* □ *$CuCl_2$,* ◇ *$FeCl_3$. (*72-h hold test—no pitting.)*

chloride attack, and, if tested as in this investigation, it would result in a lower breakthrough potential than the 1-cm^2 area adjacent to it.

Thus, by testing over 40 cm^2 in various chloride ion concentrations and salts, a trend which is not affected by "scatter" is established. The trend is shown in Figs. 4–6 where a line is drawn through the minimum breakthrough potentials.

The "minimum pitting potential line" describes the potentials below which pitting should not occur in the particular alloy. This was tested by holding at the potentials shown on the graphs, at the chloride ion concentrations indicated, for 72 h without observing any pitting. The potentials and results obtained are listed in Table 2.

The slopes of the minimum pitting potential lines were different for each alloy. There is a shift in the active direction by about 0.17 V (SCE) for a tenfold change in concentration for Type 409, 0.16 V for Type 304, and 0.23 V for HWT. The 0.16 V for Type 304 is considerably higher than the 0.09 V reported by Leckie and Uhlig [*5*] for Type 304 in NaCl solutions. However, if only the NaCl breakthrough potentials are considered (Fig. 5), and a minimum pitting potential line is drawn through the higher chloride ion concentrations (the concentrations investigated by Leckie and Uhlig), the slope obtained, 0.08 V, is in excellent agreement with that observed by these investigators.

The different slopes of the minimum pitting potential lines for the different stainless steel alloys is not unusual. Hospadaruk and Petrocelli [*4*] reported the pitting potential of the austenitic steels Types (201 and 301) to be practically independent of chloride concentration in the range of 0.1 to 5.0 *N* NaCl, while that of Type 434 decreased linearly with the increase in log $[Cl^-]$.

It is evident by comparing Figs. 4–6, that the minimum pitting potential line is characteristic of the alloy and its resistance to pitting in chloride environments. Using this criterion for evaluating alloys, this investigation indicates that Type 304 is more pit resistant than Type 409, and alloy HWT is more pit resistant than both Types 409 and 304 (Fig. 7).

The effect of pH on the breakthrough potentials of these alloys was also investigated. Sodium chloride was used to mix solutions containing 10, 100, 1000, and 10,000-ppm Cl^-. The results obtained are shown in Figs. 8–10.

As reported by other investigators [*4,5*], the breakthrough potential was not greatly affected by pH in the range of 3 to 8 for alloys Types 409 and 304. At pH values below three, the effect was different for the low-chloride solutions than that with high-chloride concentrations. In the 10 and 100-ppm Cl^- the breakthrough potentials increased, while in the high concentrations the breakthrough potentials decreased. In general, increasing the pH above eight tended to increase the breakthrough potentials. No breakthroughs and no pitting was observed on any of the alloys

TABLE 2—*Testing of minimum pitting potential curve 72-h hold test (V versus SCE).*

Chloride Concentration NaCl	Alloy					
	Type 409		Type 304		HWT	
	Hold Potential	Results	Hold Potential	Results	Hold Potential	Results
10-ppm Cl^-	0.40	no breakthrough no pitting	0.50	no breakthrough no pitting	0.80	no breakthrough no pitting
100-ppm Cl^-	0.20	no breakthrough no pitting	0.35	no breakthrough no pitting	0.55	no breakthrough no pitting
1000-ppm Cl^-	0.05	no breakthrough no pitting	0.20	no breakthrough no pitting	0.35	no breakthrough no pitting
10,000-ppm Cl^-	−0.15	no breakthrough no pitting	0.00	no breakthrough no pitting	0.10	no breakthrough no pitting

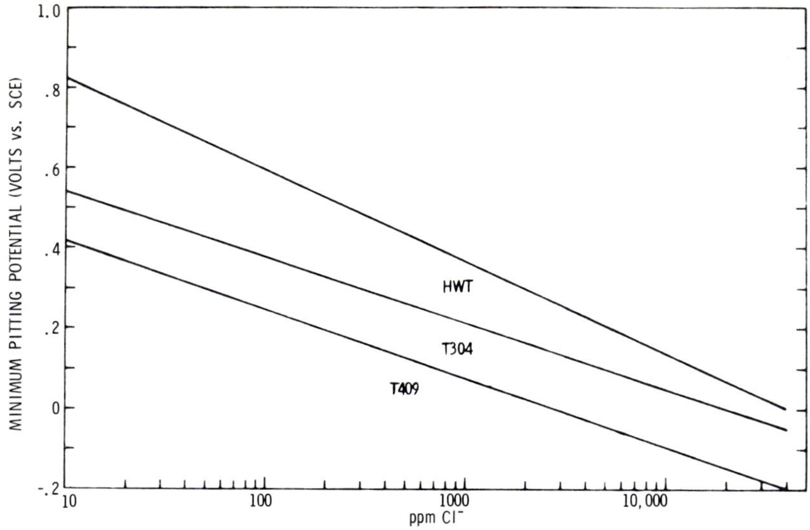

FIG. 7—*Minimum pitting potential lines of Type 409, Type 304, and HWT.*

in solutions adjusted to pH 12. This is in agreement with known information that stainless steels do not pit in strong alkaline solutions.

The breakthrough potential values obtained on each alloy in calcium, cupric, and ferric chloride solutions for establishing the minimum pitting potential lines were also plotted in Figs. 8–10 (filled symbols). The pH's of these solutions were not adjusted. Note, however, how well these

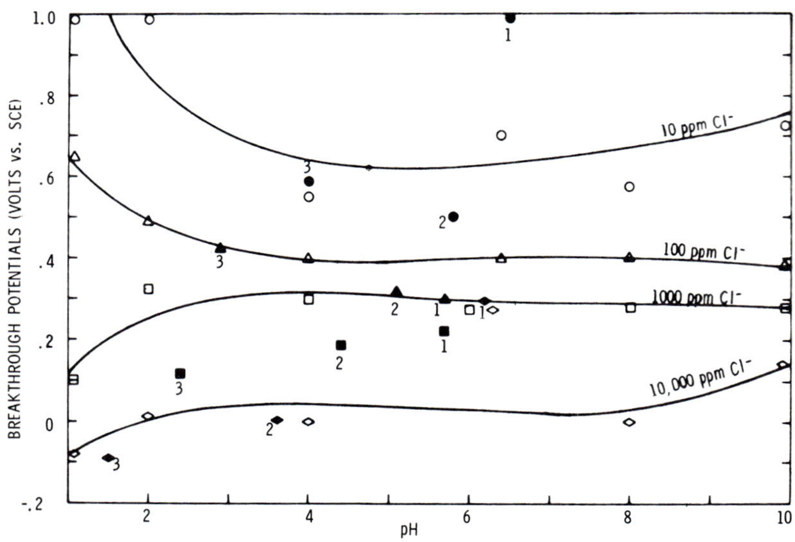

FIG. 8—*Effect of pH on the breakthrough potential of Type 409: open symbols, NaCl; filled symbols, (1) $CaCl_2$, (2) $CuCl_2$, and (3) $FeCl_3$.*

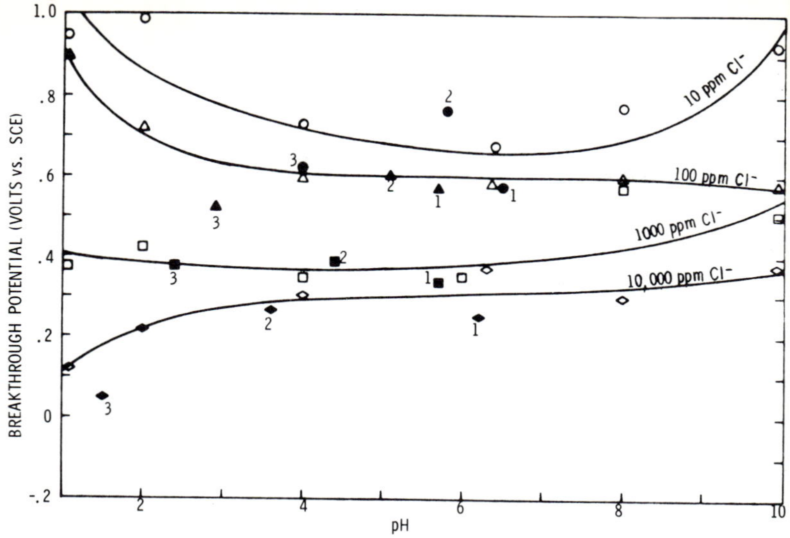

FIG. 9—*Effect of pH on the breakthrough potential of Type 304: open symbols, NaCl; filled symbols, (1) $CaCl_2$, (2) $CuCl_2$, and (3) $FeCl_3$.*

breakthrough potential values fit the curves obtained with the adjusted pH, sodium chloride solutions.

The effect of pH on the breakthrough potentials of HWT is somewhat anomalous (Fig. 10). Although the curves for 10, 100, and 10,000-ppm Cl^- solutions follow the general trend described for Types 409 and 304,

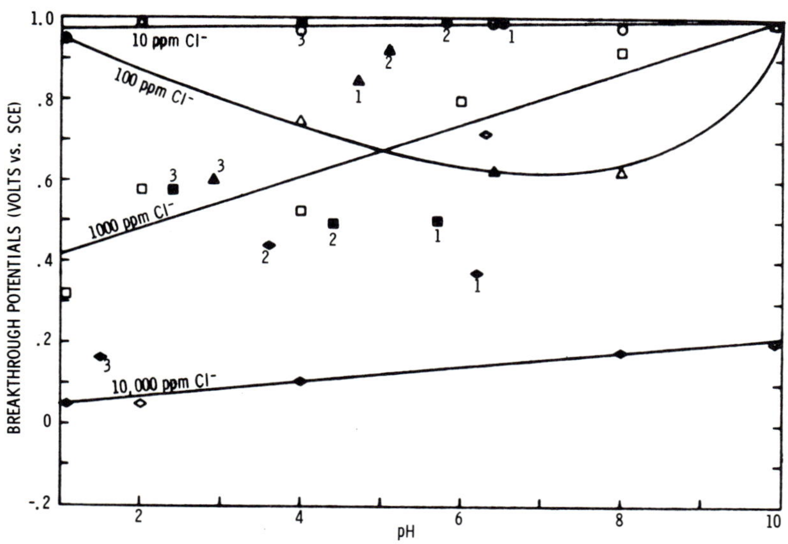

FIG. 10—*Effect of pH on the breakthrough potential of HWT: open symbols, NaCl; filled symbols, (1) $CaCl_2$, (2) $CuCl_2$, and (3) $FeCl_3$.*

the values obtained in 1000-ppm Cl^- solutions are "erratic." No attempt will be made to explain this behavior at this time.

Conclusions

Anodic polarization tests in chloride containing solutions have been described in the literature as methods for measuring pitting susceptibility. Such tests, however, result in substantial scatter due to the nature of the test and the nature of pitting corrosion. Thus, in comparing the pitting resistance of two alloys, it is conceivable that an erroneous choice could be made.

Conducting anodic polarization tests in solutions containing various chloride ion concentrations, a minimum pitting potential line may be drawn for each alloy. The minimum pitting potential line is characteristic of the pitting resistance of the alloy and may be used to compare alloys in general or in some particular chloride concentration. Using this method of evaluation, HWT proved to be a more pit resistant alloy than Type 304 which, in turn, was more pit resistant than Type 409.

There is very little effect on the breakthrough potentials of Types 409 and 304 in the 3 to 8 pH range. The effect of lower pH solutions depends on chloride ion concentration and at higher pH solutions the breakthroughs increase. More data are needed to fully establish the effect of pH on HWT.

Acknowledgment

The author wishes to express his appreciation to James Fenoglietto for his assistance in conducting all experimental laboratory work.

References

[*1*] Nielsen, N. A. and Rhodin, T. N., Jr., *Zeitschrift Electrochemie,* ZEELA, Vol. 6 No. 2, 1958.
[*2*] Kolotyrkin, J. M., *Corrosion,* SBIIA, Vol. 19, Aug. 1963, p. 261.
[*3*] Schwenk, W., *Corrosion Science,* CRRSA, Vol. 5, 1965, p. 245.
[*4*] Hospadaruk, V. and Petrocelli, J. V., *Journal,* Electrochemical Society, JESOA, Vol. 113, 1966, p. 878.
[*5*] Leckie, H. P. and Uhlig, H. H., *Journal,* Electrochemical Society, JESOA, Vol. 113, 1966, p. 1262.
[*6*] Horvath, J. and Uhlig, H. H., *Journal,* Electrochemical Society, JESOA, Vol. 115, Aug. 1968, p. 791.
[*7*] Walker, M. S. and Rowe, L. C., *Corrosion,* SBIIA, Vol. 25, Feb. 1969, p. 47.
[*8*] Greene, N. D. and Judd, G., *Corrosion,* SBIIA, Vol. 21, Jan. 1965, p. 15.
[*9*] Stern, M. and Makrides, A. C., *Journal,* Electrochemical Society, JESOA, Vol. 107, 1960, p. 782.
[*10*] Lizlovs, E. A., *Journal,* Electrochemical Society, JESOA, Vol. 117, Oct. 1970, p. 1335.

S. J. Ketcham[1] *and P. W. Jeffrey*[2]

Exfoliation Corrosion Testing of 7178 and 7075 Aluminum Alloys*

REFERENCE: Ketcham, S. J. and Jeffrey, P. W., **"Exfoliation Corrosion Testing of 7178 and 7075 Aluminum Alloys,"** *Localized Corrosion—Cause of Metal Failure, ASTM STP 516,* American Society for Testing and Materials, 1972, pp. 273–302.

ABSTRACT: As special heat treatments to improve resistance of aluminum alloys to exfoliation were developed, the need for accelerated testing methods to control production heat treatment arose. A joint task group formed by ASTM Committee G01.05 and the Aluminum Association has conducted an interlaboratory testing program of various salt spray and immersion tests for 7000 series alloys. A constant immersion test produced the best results.

KEY WORDS: corrosion, exfoliation corrosion, corrosion tests, salt spray tests, pitting, immersion tests (corrosion), aluminum alloys, etching

The development of heat treatments for 7075 and 7178 aluminum alloys to produce resistance to exfoliation has created the need for a test method sensitive enough to differentiate between susceptible and resistant materials and also capable of detecting borderline cases.

Descriptions of three types of salt spray tests and one total immersion test developed for this purpose have been published [*1–3*].[3] Lack of agreement among producers and users of the exfoliation resistant tempers as to the best test method prompted the formation of a task group in ASTM G01/05 which in cooperation with the Aluminum Association conducted an interlaboratory testing program of several test methods. The object of the program was to prepare an ASTM recommended practice for the method which in the opinion of the participants was the most satisfactory for the purpose.

Initial effort was concentrated on two salt spray tests. It was later expanded to include three total immersion tests and corrosion potential measurements in an organic electrolyte [*4*].

* Report of an ASTM G01/05 Interlaboratory Testing Program in cooperation with the Aluminum Association.

[1] Naval Air Development Center, Warminster, Pa. 18974.

[2] Alcan Research and Development Ltd, Kingston, Ontario, Canada.

[3] The italic numbers in brackets refer to the list of references appended to this paper.

Description of Materials

7178 Alloy Plate

Specimens 3 by 6 by 0.375 in. were supplied by the Alcoa Research Laboratories for the salt spray tests in the as-machined condition. The specimens had been step-machined along the rolling direction to expose both the T/10 and T/2 surfaces on a single specimen. The T/10 surface refers to the plane exposed when one tenth of the thickness is removed from the plate; T/2 when one half of the thickness is removed.

Triplicate ¼ by ¼ by 2-in. long transverse specimens were supplied from each of the above lots for corrosion potential measurements in an organic electrolyte.

For the immersion tests, 3 by 3-in. stepped panels were supplied, but, due to the possibility that in an immersion test one step might cathodically protect another, separate panels 1½ by 3 in. were also supplied. These were taken from the T/10 and the T/2 planes. All were in the as-machined condition.

7075 Alloy Extrusion

Three by three-in. stepped panels were machined from three ½ by 6-in. extrusions supplied by Harvey Aluminum Co.; also 1½ by 3-in. panels with the T/10 and T/2 planes exposed; and ¼ by ¼ by 2-in. specimens for potential measurements.

2024 Alloy Plate

Specimens of 2024–T42 alloy 2 by 5 by 0.500 in. were also supplied to be used as controls for the corrosion tests. These were machined to expose the T/2 plane and had been heat treated to have a low resistance to exfoliation corrosion. Heat treatment conditions for these materials are given in the Appendix.

Experimental Procedures

Cyclic Salt Spray Tests (first phase)

An acidified 5 percent sodium chloride (NaCl) solution (MASTMAA SIS[4]) [*1*] and an acidified (approximately) 5 percent synthetic sea salt solution (SWAAT[4]) [*2*] were evaluated for their efficacy in assessing susceptibility to exfoliation. The operating conditions prescribed for the two salt spray tests are given in the Appendix. For the SWAAT test all participants maintained wet bottom conditions in the cabinet, for the MASTMAASIS, three operated with dry bottom conditions.

[4] MASTMAASIS is an acronym for Modified ASTM Acetic Acid Salt Intermittent Spray; SWAAT for Seawater Acetic Acid Test.

Immersion Tests (second phase)

Solutions used for the immersion tests were:

1. $NaCl-KNO_3-HNO_3$ (EXCO).[5]
2. $NH_4Cl-NH_4NO_3-H_2O_2$.[6]
3. $NaCl-SO_2$[*3*].

Operating conditions for these tests are given in the Appendix.

Specimen Surface Preparation

Both degreased and etched specimens were used in the 7178 program. In an earlier preliminary testing program there had been considerable variation in results which was attributed to surface effects. Etching the surface was proposed as a way to provide a more uniform surface. It seemed worthwhile to the task group members to include the two surface preparations at this point in the interlaboratory program to determine whether etching was necessary to obtain better reproducibility of results.

Acetone or benzene was used for degreasing. Etching was accomplished by a 1-min immersion in 5 percent sodium hydroxide (NaOH) at 82 C, a dip in concentrated nitric acid (HNO_3), and rinsing.

For the 7075 program etching was used throughout. Stepped panels were not used except in two cases. At that stage, however, it was decided to determine whether there were any galvanic effects between the back of the specimens which had the as-extruded surface and the machined surface. Therefore, specimens with the sides and backs masked with electroplaters tape were included as well as unmasked.

System for Rating Degree of Exfoliation Attack

In order to compare results of the exfoliation tests among laboratories it was necessary to have a precise rating system. The system finally adopted is given in Table 1 and illustrated in Fig. 1. The questionable category refers to a condition where raised blisters or pit-blisters are visible which could be due to either mild exfoliation attack or severe pitting. Metallographic examination would be required to determine which type of attack had occurred. In evaluation of heat treatments or materials metallographic

TABLE 1—*Visual examination.*

Code	Explanation
N	no appreciable attack; surface may be etched or discolored
P	pitting
Q	questionable exfoliation (pit blisters)
EM	exfoliation, mild
EI	exfoliation, intermediate
ES	exfoliation, severe

[5] EXCO is an acronym for Exfoliation Corrosion Test. The test was developed by Alcan. The solution was used originally for stress corrosion testing [*5*].

[6] See p. 38.

examination would be desirable in questionable cases. The purpose of this task group however was to compare test methods, and an important criterion for final selection of a test method was that the test produce the fewest visually questionable ratings. Therefore, metallographic examination was not conducted on specimens with questionable ratings.

Corrosion Potential Measurements

A test method involving measurement of corrosion potentials in methanol-carbon tetrachloride ($CH_3OH–CCl_4$) electrolyte to predict exfoliation resistance of 7000 series alloys has been described [*4*]. The method is very quick and simple and would make an ideal production control test. This method was included in the testing program in order to compare results with those from the exfoliation tests.
The method for conducting the test is given in the Appendix.

Results

Salt Spray Tests

Results of the two salt spray tests are given for 7178 plate in Table 2. On the first lot of low-resistance plate material, both tests gave results ranging from pitting only through all degrees of exfoliation. However, the SWAAT gave more reproducible results from laboratory to laboratory. In the majority of cases, the degree of exfoliation produced by the SWAAT test was more severe than that of the MASTMAASIS, particularly on the T/2 plane.

On the low resistance lot 1 plate specimen, six out of eight laboratories reported exfoliation on the T/10 plane with the SWAAT test; only two with the MASTMAASIS. On the T/2 plane both tests produced exfoliation at all eight laboratories, but all eight specimens showed severe exfoliation in the SWAAT test, whereas all degrees of exfoliation occurred in the MASTMAASIS.

On the low-resistance lot 2 plate material seven out of eight laboratories reported exfoliation on the T/10 plane with the SWAAT test, only three with the MASTMAASIS; on the T/2 plane all eight obtained exfoliation with the SWAAT test, five out of eight with the other test. Again the degree of exfoliation was greater in the SWAAT test.

On the high-resistance material, both tests produced either pitting or a questionable rating with the exception of one case of exfoliation reported on the T/2 plane.

Figures 2, 3, and 4 show representative examples of these results. Figures 2 and 3 illustrate the variability obtained with the MASTMAASIS test by presenting results from two different laboratories. Little or no exfoliation is evident in Fig. 2 (Laboratory A) while a greater degree of exfoliation is shown in Fig. 3 (Laboratory B).

TABLE 2—*Results of salt spray tests on 7178 plate (stepped specimens).*

Laboratory	Low Resistance—Lot 1				Low Resistance—Lot 2				High Resistance			
	MASTMAASIS		SWAAT		MASTMAASIS		SWAAT		MASTMAASIS		SWAAT	
	Degreased	Etched	Degreased	Etched	Degreased	Etched	Degreased	Etched	Degreased	Etched	Degreased	Etched
					Part A—T/10 Plane							
A	Q	...	EI	...	Q	...	EI	...	P	...	Q	...
B	EM	...	...	EM	EM	...	...	EM	P	...	...	P
C	...	N	...	EI	...	N	...	EI	...	N	...	P
D	Q	...	ES	...	EM	...	ES	...	P	...	Q	...
E	P	...	...	Q	P	...	...	EI	P	...	...	P
F	P	...	ES	...	P	...	EI	...	P	...	P	...
G	EM	...	P	...	Q	...	EM	...	P	...	P	...
H	Q	...	EI	...	EM	...	Q	...	P	...	P	...
Summary	2 exfoliation 6 other		6 exfoliation 2 other		3 exfoliation 5 other		7 exfoliation 1 other		7 pitting 1 other		6 pitting 2 other	
					Part B—T/2 Plane							
A	EI	...	...	ES	Q	...	...	ES	P	...	...	Q
B	ES	...	...	ES	EI	...	...	EM	Q	...	...	P
C	...	N	...	ES	...	N	...	ES	...	N	...	Q
D	EI	...	ES	...	EI	...	ES	...	Q	...	EM	...
E	EM	...	...	ES	Q	...	...	ES	P	...	...	Q
F	ES	...	ES	...	EI	...	ES	...	P	...	P	...
G	EI	...	ES	...	EI	...	ES	...	Q	...	Q	...
H	ES	...	ES	...	EI	...	EI	...	P	...	P	...
Summary	7 exfoliation 1 other		8 exfoliation		5 exfoliation 3 other		8 exfoliation		4 pitting 4 other		3 pitting 5 other	

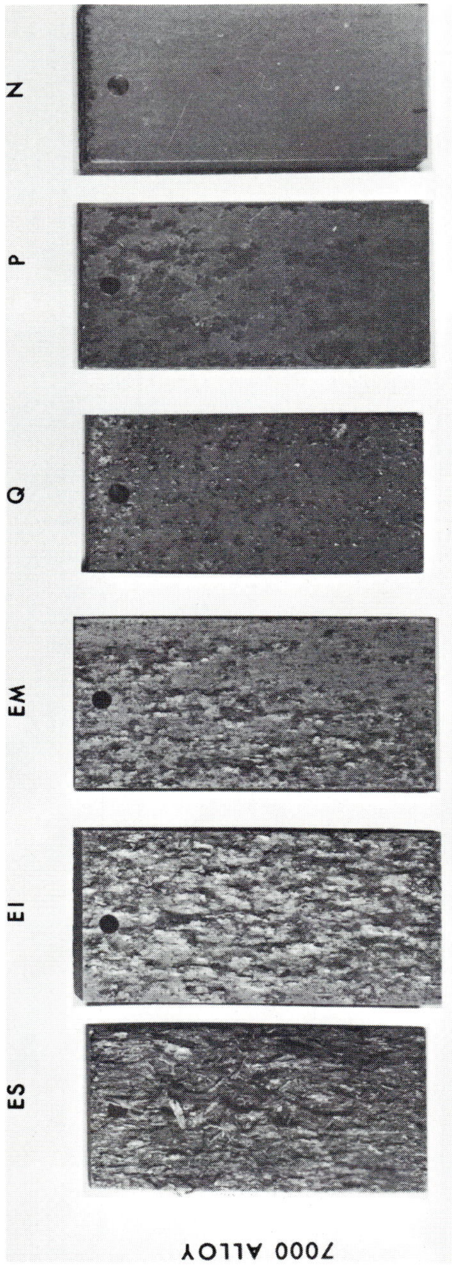

FIG. 1—*Visual rating of corrosion for exfoliation test specimens.*

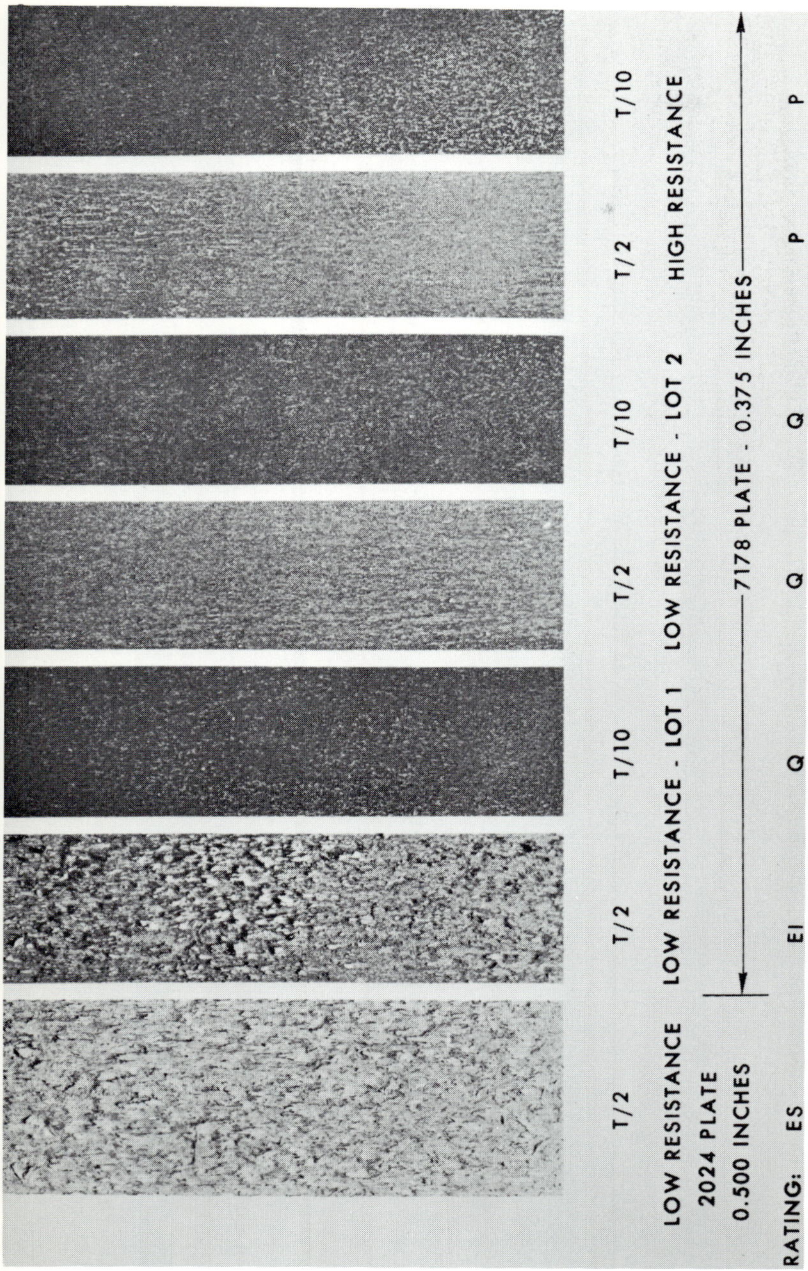

FIG. 2—*2024 and 7178 alloy specimens subjected to MASTMAASIS test at Lab A.*

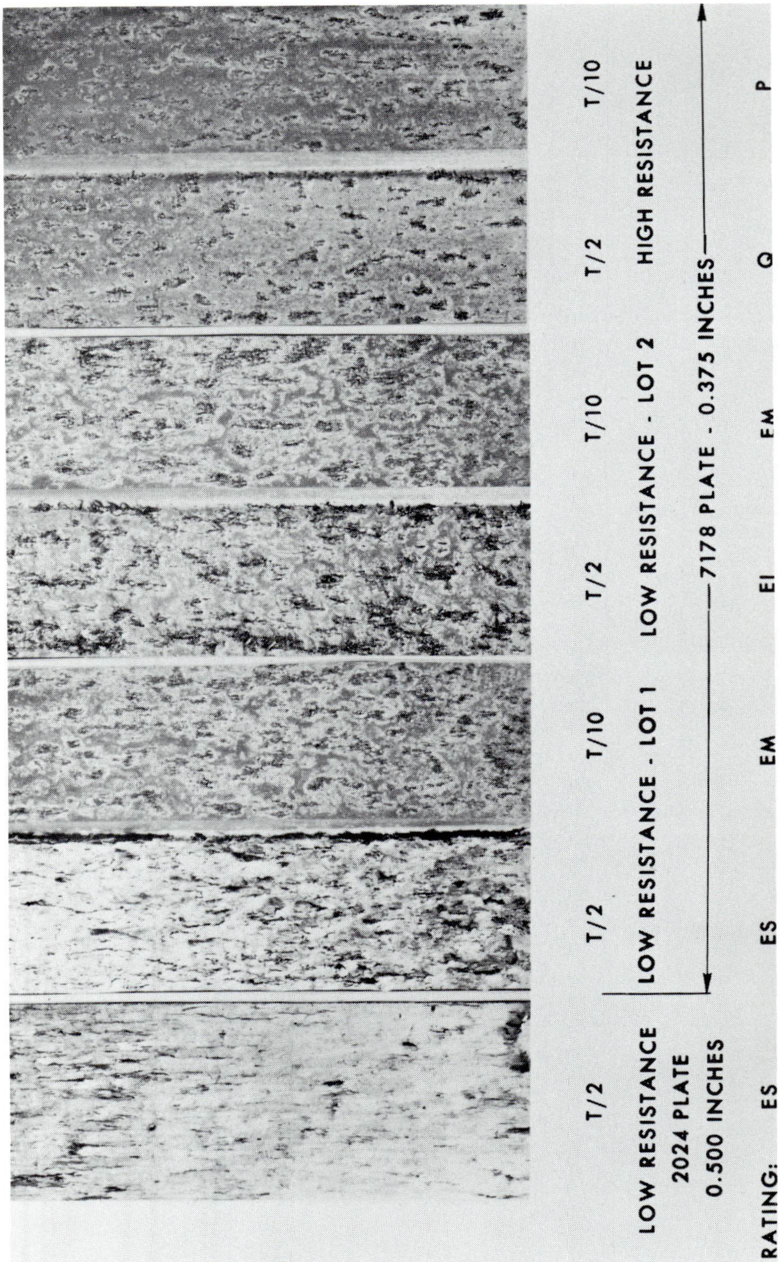

FIG. 3—*2024 and 7178 alloy specimens subjected to MASTMAASIS test at Lab B.*

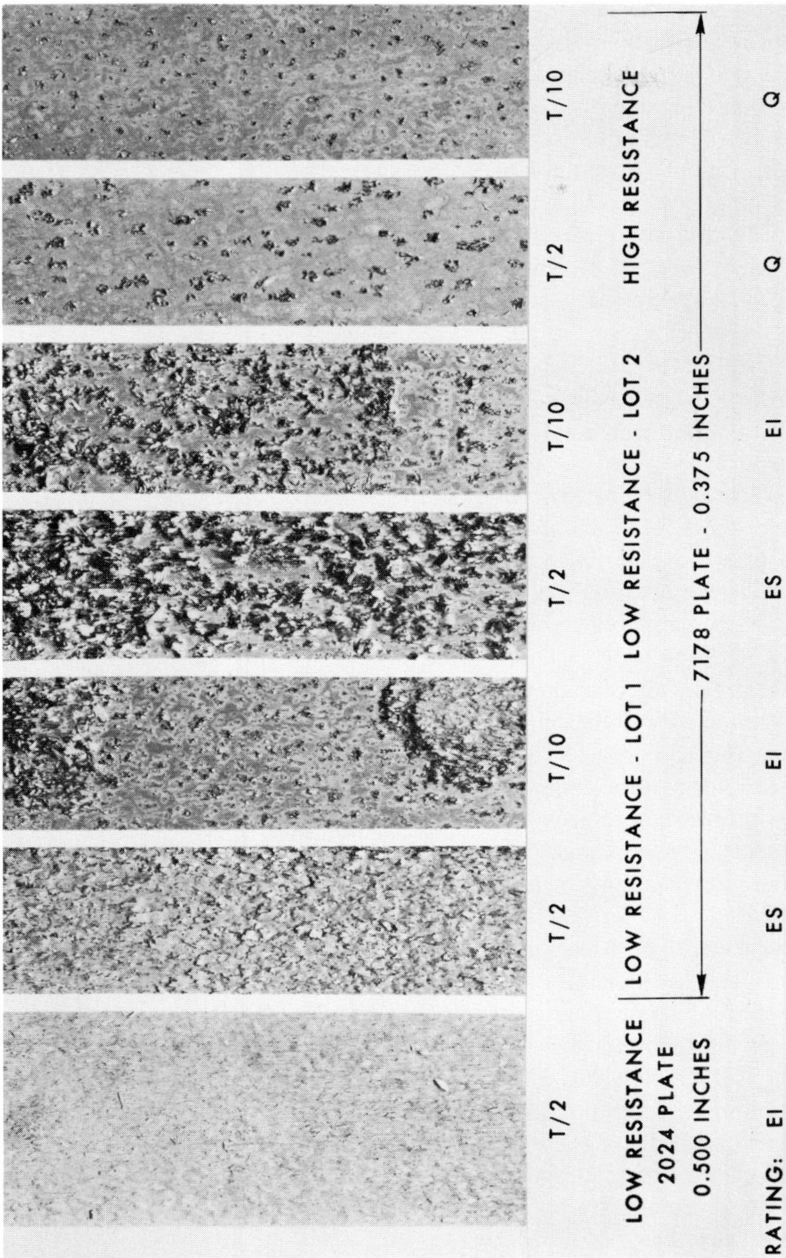

FIG. 4—*2024 and 7178 alloy specimens subjected to SWAAT test.*

The salt spray results for 7075 are given in Table 3. The data for 7075 are less complete than that for 7178 because by the time the 7075 extrusion was tested primary interest of the task group had shifted from salt spray tests to immersion tests.

The two surface preparations, degreasing or etching, did not appear to have any significant effect on the results.

Results of the salt spray tests on the 2024 specimens are not presented because this alloy was not of prime interest in this program. It is worth mentioning however that the MASTMAASIS test produced better reproducibility and more severe exfoliation on 2024 than did the SWAAT test, the reverse of the results on the 7000 alloys. The appearance of some 2024 specimens is shown in Figs. 2, 3, and 4.

Immersion Tests

7178 Plate Material—Results obtained by the various laboratories in the three immersion tests are presented in Tables 4, 5, and 6 and Figs. 5, 6, and 7.

Low-Resistance Materials (*Lot 1*) (*Table 4, Fig. 5*)—On this lot of material the EXCO test produced some degree of exfoliation on the T/10 plane in six out of seven laboratories; on the T/2 plane all laboratories reported severe exfoliation.

With the ammonium chloride-ammonium nitrate-hydrogen peroxide (NH_4Cl-NH_4NO_3-H_2O_2) test seven out of eight laboratories obtained some degree of exfoliation on the T/10 plane; on the T/2 plane all laboratories reported exfoliation, but the degree varied from mild to severe.

With the sodium chloride-sulfur dioxide ($NaCl$-SO_2) test five out of six laboratories reported exfoliation on the T/10 plane; all six reported either intermediate or severe on the T/2 plane.

On this low-resistance material the effects of pretreatment and of whether the planes were tested separately or as a stepped panel seemed negligible.

Low-Resistance Material (*Lot 2*) (*Table 5, Fig. 6*)—On the T/10 plane of this lot four out of seven laboratories which conducted the EXCO test reported some degree of exfoliation. On the T/2 plane all seven laboratories reported some degree of exfoliation.

The NH_4Cl-NH_4NO_3-H_2O_2 test on the T/10 produced a variety of results ranging from no appreciable attack to intermediate exfoliation. On the T/2 plane all six reported exfoliation, either intermediate or severe.

With this lot of material it was again difficult to ascribe any significant effect to pretreatment or to testing the two planes separately or as a stepped specimen.

High-Resistance Material (*Table 6, Fig. 7*)—On the high-resistance 7178 all laboratories reported either pitting or no attack in all three solutions on the T/10 plane (with one exception, from the $NaCl$-SO_2 test).

TABLE 3—*Results of salt spray tests on 7075 extrusions (separate panels).*

Laboratory	Low Resistance—Lot 1				Low Resistance—Lot 2				High Resistance			
	MASTMAASIS		SWAAT		MASTMAASIS		SWAAT		MASTMAASIS		SWAAT	
	Degreased	Etched	Degreased	Etched	Degreased	Etched	Degreased	Etched	Degreased	Etched	Degreased	Etched
					PART A—T/10 PLANE							
A	...	...	...	...	...	...	...	...	...	...	...	...
B	...	...	...	...	...	...	...	...	...	...	...	...
C	EM	EI	...	...	EI	EI	...	...	P	P	...	...
D	...	...	...	...	...	...	...	...	...	...	...	...
E	...	...	...	ES[a]	...	...	...	ES[a]	...	...	...	Q[a]
F	...	...	...	...	...	...	...	...	...	...	...	...
G	...	...	...	...	...	...	...	...	...	...	...	...
H	...	...	...	...	...	...	...	...	...	...	...	...
Summary	2 exfoliation		1 exfoliation		2 exfoliation		1 exfoliation		2 pitting		1 questionable	
					PART B—T/2 PLANE							
A	...	...	...	...	...	...	...	...	...	...	...	...
B	ES	...	...	...	EM	...	...	...	Q	...	...	...
C	EM	EI	...	...	EM	EI	...	...	Q	Q	...	...
D	...	...	...	...	...	...	...	...	...	...	...	...
E	...	...	...	ES[a]	...	...	...	ES[a]	...	...	...	Q[a]
F	...	EI	...	...	...	P	...	...	...	P	...	...
G	...	...	...	...	...	...	...	...	...	...	...	...
H	...	...	...	...	...	...	...	...	...	...	...	...
Summary	4 exfoliation		1 exfoliation		3 exfoliation 1 pitting		1 exfoliation		3 questionable 1 pitting		1 questionable	

[a] Specimens masked with tape on backside and edges.

TABLE 4—*Results of immersion tests on 7178 plate, low-resistance material—LOT 1.*

Laboratory	EXCO Test (48 h)				NH_4Cl-NH_4NO_3-H_2O_2 (16 to 24 h)				$NaCl$-SO_2 Test (72 h)	
	Degreased		Etched		Degreased		Etched		Degreased	Etched
	Separate	Stepped	Separate	Stepped	Separate	Stepped	Separate	Stepped	Separate	Separate
					PART A—T/10					
A	EI	EI	...	...	EM	EM	...	...	EI	...
B	EM[a]	EM[a]	EI	EI	EM	EM	EM	EM	...	EM
C	ES	EI	ES	ES	N	Q	N	N	ES	...
D	...	...	EI	ES	...	...	EM	EM	...	...
E	...	...	ES	Q	...	...	N	EM	...	...
									(stepped panels)	
F	...	...	...	...	...	ES	...	...	EM	...
G	...	...	EM	EI	...	...	N	EM	...	P
H	Q	...	...	...	EM	P	EM	...	EI	...
Summary	15 exfoliation, 2 questionable				13 exfoliation, 7 other				5 exfoliation, 1 pitting	
					PART B—T/2					
A	ES	ES	...	...	EI	EI	...	...	ES	...
B	ES[a]	ES[a]	ES	ES	ES	ES	EI	EI	...	EI
C	ES	ES	ES	ES	EM	EM	EM	EM	ES	...
D	...	...	ES	ES	...	...	EI	EI	...	...
E	...	...	ES	ES	...	...	EI	EI	...	...
									(stepped panels)	
F	...	...	...	...	...	ES	...	...	EI	...
G	...	...	ES	ES	EI	...	Q	EI	EI	EI
H	ES	...	...	...	EI	EM	ES	...	ES	...
Summary	17 exfoliation				20 exfoliation, 1 questionable				7 exfoliation	

[a] 24-h test.

TABLE 5—*Results of immersion tests on 7178 plate, low-resistance material—LOT 2.*

Laboratory	EXCO Test (48 h)				NH_4Cl-NH_4NO_3-H_2O_2 (16 to 24 h)				$NaCl$-SO_2 Test (72 h)	
	Degreased		Etched		Degreased		Etched		Degreased	Etched
	Separate	Stepped	Separate	Stepped	Separate	Stepped	Separate	Stepped	Separate	Separate
					Paet A—T/10					
A	EM	EM	...	...	EM	P	...	...	Q	...
B	EM (24 h)	EM (24 h)	EM	EM	EM	N	EM	EM	...	EI
C	EI	EI	Q	ES	N	N	N	N	ES	...
D	...	...	EI	ES	...	...	Q	EM	...	...
E	...	...	Q	P	...	...	P	EM	...	...
F	...	...	...	...	...	EI	...	...	(stepped Q panels)	...
G	...	...	P	EI	...	...	N	EM	...	EM
H	P	...	...	...	Q	N	Q	...	EM	...
Summary	12 exfoliation 5 other				8 exfoliation 12 other				4 exfoliation 2 questionable	
					Part B—T/2					
A	ES	EI	...	...	EM	EM	...	...	EI	...
B	EI (24 h)	EI (24 h)	ES	EI	EI	EM	EI	EI	...	ES
C	ES	ES	ES	ES	EM	EM	EM	EM	ES	...
D	...	...	ES	ES	...	...	EM	EI	...	...
E	...	...	EM	EI	...	...	EI	EI	...	...
F	...	...	...	...	...	EI	...	...	(stepped EI panels)	...
G	...	...	ES	ES	EM	...	EM	EI	EI	EI
H	EI	...	...	...	EI	P	EI	...	EI	...
Summary	17 exfoliation				20 exfoliation 1 pitting				7 exfoliation	

TABLE 6—*Results of immersion tests on 7178 plate, high-resistance material.*

Laboratory	EXCO Test (48 h)				NH_4Cl-NH_4NO_3-H_2O_2 (16 to 24 h)				$NaCl$-SO_2 Test (72 h)	
	Degreased		Etched		Degreased		Etched		Degreased	Etched
	Separate	Stepped	Separate	Stepped	Separate	Stepped	Separate	Stepped	Separate	Separate
					Part A—T/10					
A	P	P	...	...	P	P	...	...	P	...
B	P (24 h)	P (24 h)	P	P	N	N	P	P	...	P
C	N	N	N	N	N	N	N	N	Q	...
D	...	...	P	P	...	...	P	P	...	...
E	...	...	P	P	...	...	P	P	...	...
F	...	...	...	...	...	P	...	...	(stepped P panels)	...
G	...	...	P	P	...	...	P	P	...	P
H	P	...	...	...	P	N	P	...	P	...
Summary	13 pitting 4 no attack				13 pitting 7 no attack				5 pitting 1 questionable	
					Part B—T/2					
A	P	Q	...	...	P	Q	...	...	Q	...
B	P (24 h)	P (24 h)	P	P	P	P	P	P	...	Q
C	P	P	P	P	EM	EM	EM	EM	Q	...
D	...	...	Q	Q	...	...	P	P	...	...
E	...	...	P	Q	...	...	Q	P	...	...
F	...	...	...	...	...	P	...	...	(stepped P panels)	...
G	...	...	P	P	P	...	P	P	P	P
H	P	...	...	...	P	P	Q	...	P	...
Summary	13 pitting 4 questionable				14 pitting 7 other				4 pitting 3 questionable	

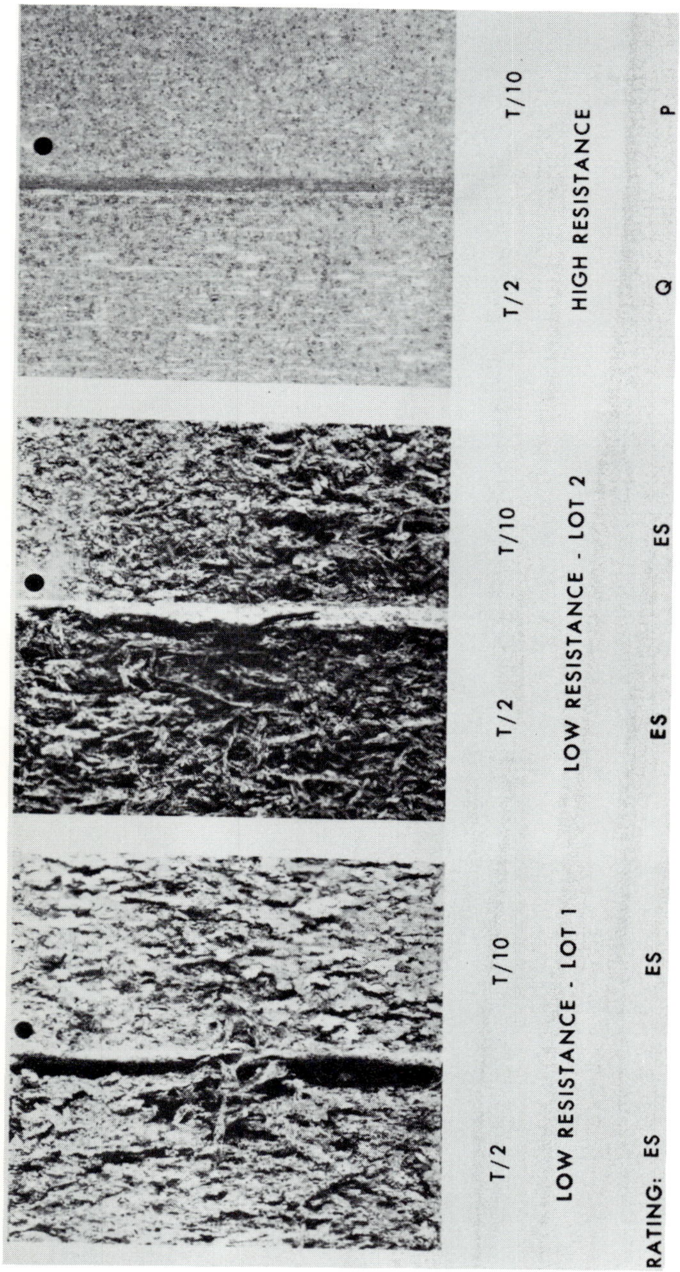

FIG. 5—7178 alloy plate specimens subjected to $NaCl$-KNO_3-HNO_3 immersion test.

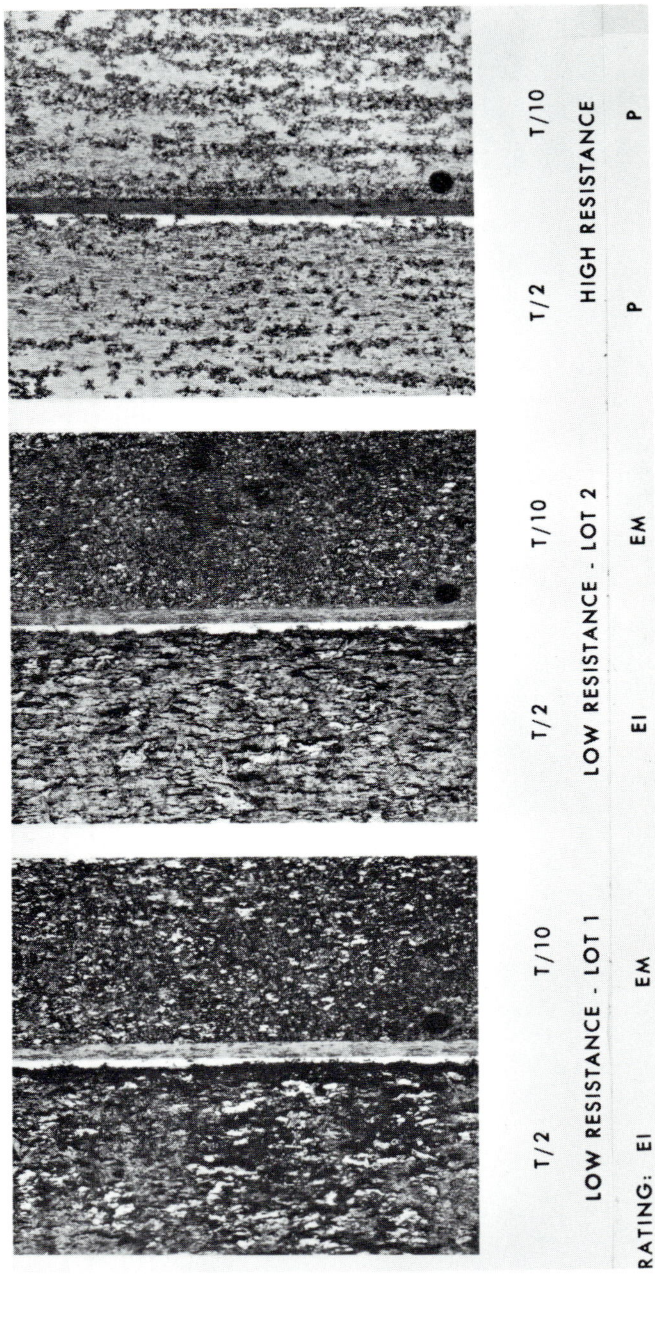

FIG. 6—*7178 alloy plate specimens subjected to* NH_4Cl-NH_4NO_3-H_2O_2 *immersion test.*

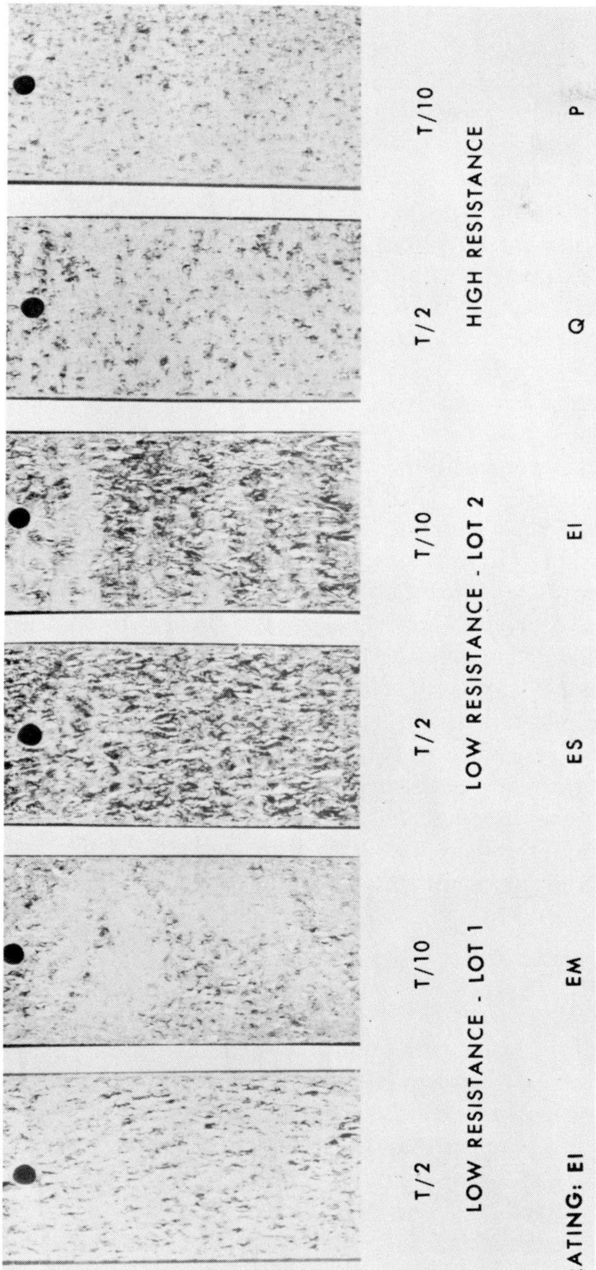

FIG. 7—*7178 plate specimens exposed to NaCl-SO_2 immersion test. Specimens were caustic etched prior to exposure.*

On the T/2 plane either pitting or questionable ratings were reported from all three tests with the exception of mild exfoliation from the NH_4Cl-NH_4NO_3-H_2O_2 test reported by one laboratory.

7075 Extrusion—Results of the three immersion tests on 7075 are presented in Tables 7, 8, and 9. Since the results on masked and unmasked specimens were similar, they will be treated together in the following discussion.

Low-Resistance Material (Lot 1) (Table 7, Fig. 8)—The EXCO test provided the most severe attack and the best reproducibility in results of the three solutions. On both the T/10 and T/2 planes all laboratories reported severe exfoliation with one exception and that was intermediate. Results with the NH_4Cl-NH_4NO_3-H_2O_2 test ranged from questionable to severe. With the NaCl-SO_2 test mild and intermediate exfoliation were reported.

It is interesting to note that very little difference in exfoliation susceptibility was evident between the T/10 and T/2 plane of the low-resistance 7075 extrusion as opposed to the low-resistance 7178 plate (Table 4).

Low-Resistance (Lot 2) (Table 8, Fig. 9)—Again, most severe exfoliation was evidenced with the EXCO test, and reproducibility in results from laboratory to laboratory was very good; all reported either severe or intermediate exfoliation. The NH_4Cl-NH_4NO_3-H_2O_2 solution again gave results ranging from questionable to severe exfoliation; the NaCl-SO_2 solution either mild or intermediate.

High-Resistance Material (Table 9, Fig. 10)—With the exception of two laboratories, all participants reported either pitting or a questionable rating on the high-resistance material, pitting being predominant.

Corrosion Potential Measurements—Potentials measured in the CH_3OH-CCl_4 electrolyte are presented in Table 10. Reproducibility of values among the laboratories was excellent. The method clearly distinguishes between material of low and high susceptibility to exfoliation.

Discussion

Salt Spray Tests

The SWAAT test was more effective than the MASTMAASIS test in detecting exfoliation susceptibility of 7178 alloy with different levels of resistance. Although both tests showed variability among the eight laboratories testing common stocks of material, the variability was less with the SWAAT test.

The greater variability of the MASTMAASIS results could be due to any number of reasons. Since four laboratories operated the cabinet under wet bottom conditions and four under dry bottom conditions, variations in humidity during the purge and soak cycles could result. The concentration of acetic acid might have an effect. Increasing the concentration of

TABLE 7—*Results of immersion tests on 7075 extrusions, low-resistance material*—LOT 1.

	EXCO Test (48 h)				NH_4Cl-NH_4NO_3-H_2O_2 (16 to 24 h)				$NaCl$-SO_2 Test (72 h)			
	Unmasked		Masked		Unmasked		Masked		Unmasked		Masked	
Laboratory	t/10	t/2	t/10	t/2	t/10	t/2	t/10	t/2	t/10	t/2	t/10	t/2
A	ES	ES	ES	ES	EI	EI	EI	EI	EM	EM	EM	EM
B	ES	ES	ES	ES	EI	ES	EI	ES	EI	EI	EI	EI
C	ES[a]	ES[a]	ES[a]	ES[a]	Q	Q	Q	Q	...	...	...	...
D	ES	ES	ES	ES	EI	EI	EM	EI	...	...	...	...
E	...	...	ES	ES	...	...	EI	EI	...	...	...	...
F	...	...	ES	ES	...	...	ES	ES	...	...	...	...
G	...	...	ES	ES	...	...	EM	EM	...	...	EI	EI
H	EI	ES	ES	ES	EM	EM	EM	EM	EI	EI	EI	EI
I	ES	ES	ES	ES	...		...		...		...	
Summary	6 exf	6 exf	9 exf	9 exf	4 exf 1 q	4 exf 1 q	7 exf 1 q	7 exf 1 q	3 exf	3 exf	4 exf	4 exf

[a] 24-h test.

TABLE 8—*Results of immersion tests on 7075 extrusions, low-resistance material*—LOT 2.

Laboratory	EXCO Test (48 h)				NH_4Cl-NH_4NO_3-H_2O_2 (16 to 24 h)				$NaCl$-SO_2 Test (72 h)			
	Unmasked		Masked		Unmasked		Masked		Unmasked		Masked	
	t/10	t/2	t/10	t/2	t/10	t/2	t/10	t/2	t/10	t/2	t/10	t/2
A	ES	EI	ES	EI	EI	EI	EI	EI	EM	EM	EM	EM
B	ES	EI	ES	EI	EM	EI	EM	EI	EI	EM	EM	EI
C	ES[a]	ES[a]	ES[a]	ES[a]	Q	Q	Q	Q	...	...	...	...
D	ES	ES	EI	ES	EM	EM	EM	EM	...	...	...	...
E	...	...	ES	ES	...	...	EI	EI	...	...	...	...
F	...	...	ES	ES	...	...	ES	ES	...	...	...	...
G	...	...	ES	EI[b]	...	...	EM[c]	Q	...	...	EI	EI
H	EI	EI	ES	EI	EM	Q	EM	Q	EI	EI	EI	EI
I	ES	ES	ES	ES	...		...		...		...	
Summary	6 exf	6 exf	9 exf	9 exf	4 exf 1 q	3 exf 2 q	7 exf 1 q	5 exf 3 q	3 exf	3 exf	4 exf	4 exf

[a] 24-h test.
[b] ES with stepped specimen.
[c] Q with stepped specimen.

TABLE 9—*Results of immersion tests on 7075 extrusions, high-resistance material.*

Laboratory	EXCO Test (48 h)				NH_4Cl-NH_4NO_3-H_2O_2 (16 to 24 h)				$NaCl$-SO_2 Test (72 h)			
	Unmasked		Masked		Unmasked		Masked		Unmasked		Masked	
	t/10	t/2	t/10	t/2	t/10	t/2	t/10	t/2	t/10	t/2	t/10	t/2
A	P	P	P	P	P	P	P	P	P	Q	P	P
B	P	P	P	P	P	P	P	P	P	Q	P	Q
C	Q[a]	Q[a]	Q[a]	Q[a]	EM	EM	EM	EM	...	...	...	...
D	N	Q	Q	Q	P	Q	P	Q	...	...	...	...
E	...	...	P	P	...	...	P	P	...	...	...	...
F	...	...	P	P	...	...	P	P	...	...	...	...
G	...	...	P	P	...	...	P	P	...	...	P	Q
H	P	P	P	P	P	P	P	P	Q	Q	Q	Q
I	EM	EM	EM	EM	...		...		...		...	
Summary	3 pitting 3 other	3 pitting 3 other	6 pitting 3 other	6 pitting 3 other	4 pitting 1 exf	3 pitting 2 other	7 pitting 1 exf	6 pitting 2 other	2 pitting 1 q	3 q	3 pitting 1 q	3 q 1 p

[a] 24-h test.

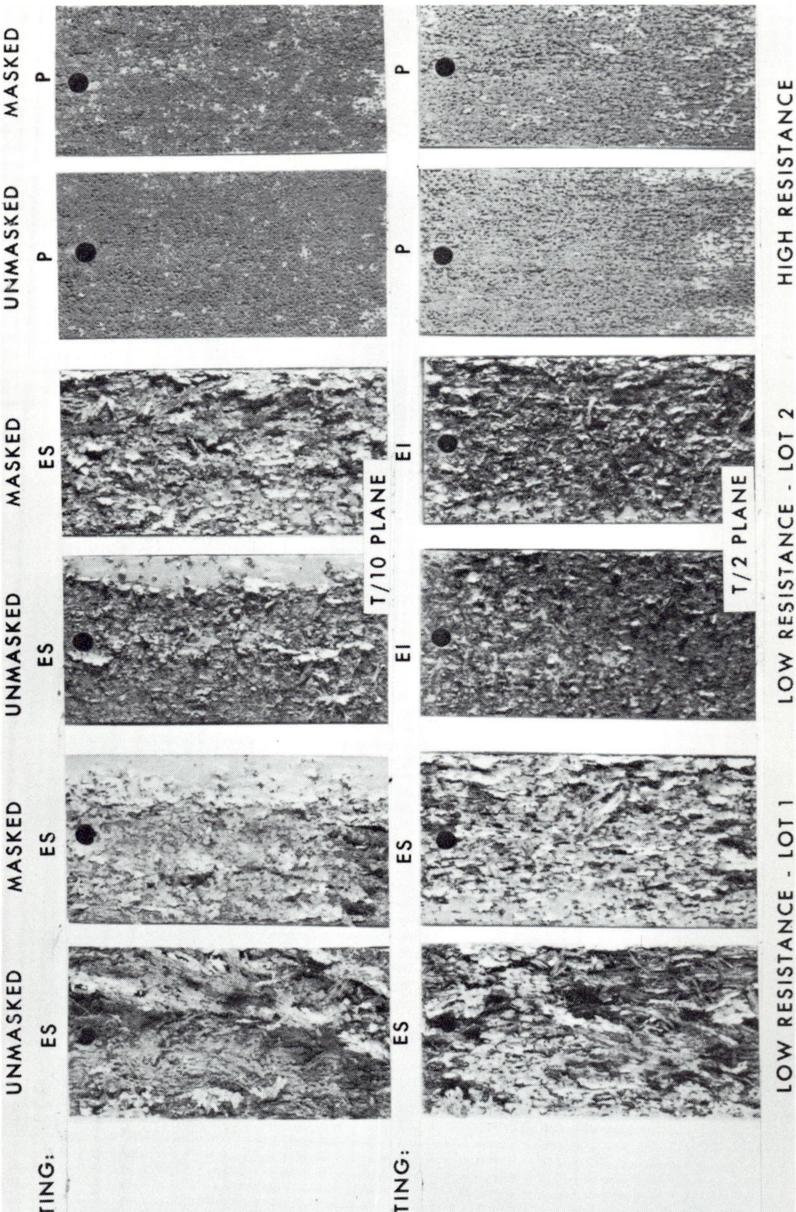

FIG. 8—*7075 extrusion specimens exposed 48 h to NaCl-KNO_3-HNO_3 immersion test. Specimens caustic etched prior to exposure.*

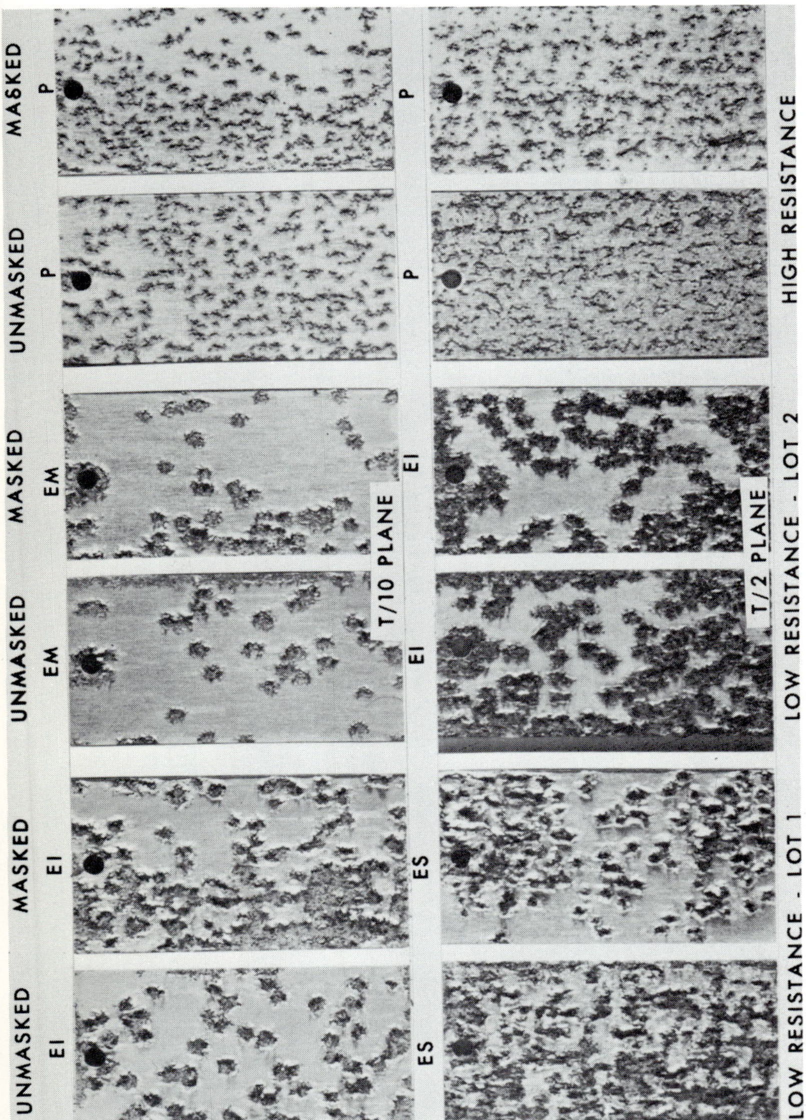

FIG. 9—*7075 extrusion specimens exposed 16 h to* NH_4Cl-NH_4NO_3-H_2O_2 *immersion test. Specimens caustic etched prior to exposure.*

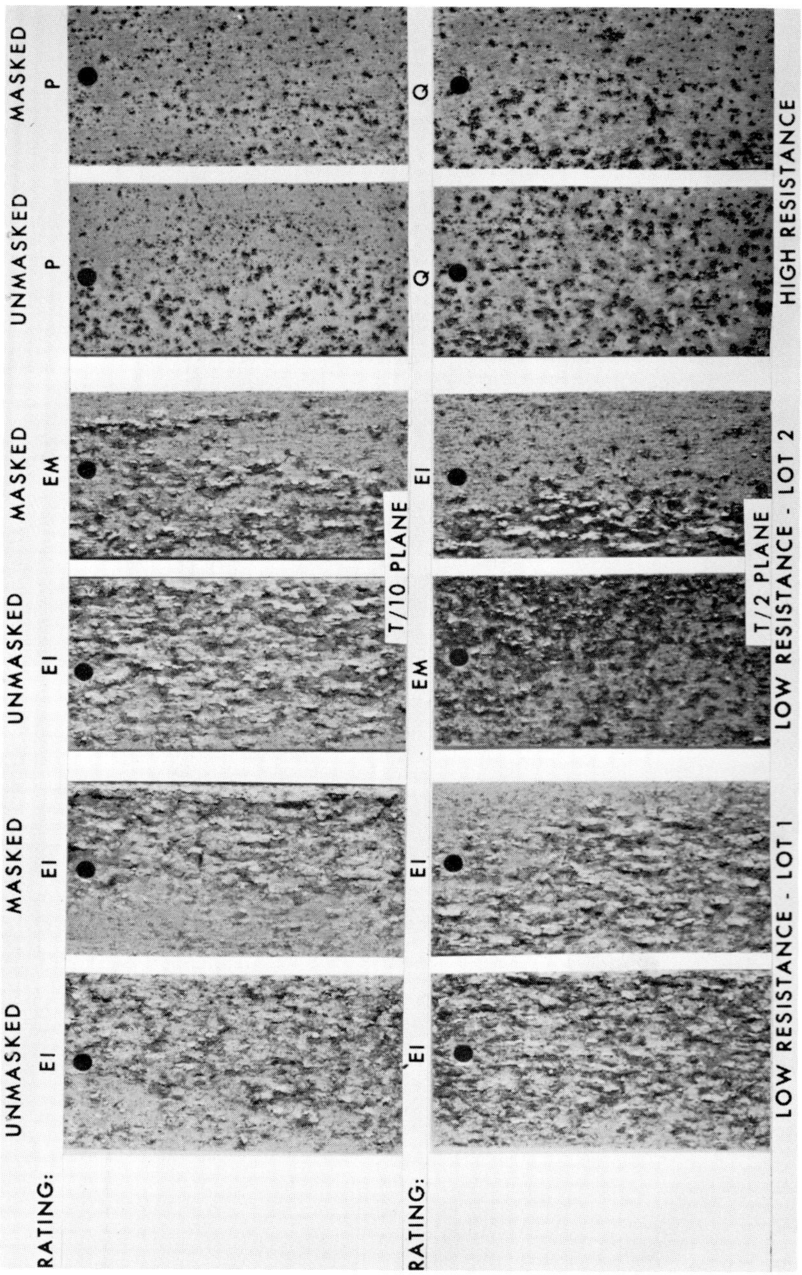

FIG. 10—*7075 extrusion specimens exposed 72 h to $NaCl$-SO_2 immersion test. Specimens caustic etched prior to exposure.*

TABLE 10—*Corrosion potential measurements in CH_3OH/CCl_4 versus saturated calomel electrode.*

	7178 (T/2)			7075 (T/2)		
	Potentials (mV)					
	Exfoliation Resistance			Exfoliation Resistance		
	Low 1	Low 2	High	Low 1	Low 2	High
A	280	280	1030	300	280	1024
B	290	270	1033	293	380	1030
C	258[a]	268[a]	1057[a]	242	270	1047
D	280	277	1050	287	250	1023
				265[b]	275[b]	1120[b]
E	289	275	1045	325[c]	390[c]	1130[c]
F	...	...	...	102	112	1015
G	245	250	1050	270	272	1043
H	282	298	1029	265	270	1011

[a] Values from corrosion test specimens; values from specimens received later were 267, 258, and 890 mV for low- and high-resistance material, respectively.
[b] After 10 min.
[c] After 35 min.

acetic acid to that used for the SWAAT test might improve reproducibility of the MASTMAASIS test.

The number of tests conducted on the 2024 material were insufficient to provide for definite conclusions as to the best salt spray test, but the available evidence suggests the MASTMAASIS test is more effective than the SWAAT for this alloy.

Immersion Tests

The EXCO test proved to be most effective of the immersion tests in distinguishing high and low susceptibility to exfoliation, and good reproducibility was obtained in the various laboratories. The NH_4Cl-NH_4NO_3-H_2O_2 test results had greater variability. The $NaCl$-SO_2 test showed good reproducibility but is the most complicated to perform.

As with the salt spray tests, surface preparation is not critical when conducting immersion tests, as both degreased and etched specimens gave similar results. Masking back and sides of specimen in total immersion tests proved not to be necessary since masked and unmasked specimens gave similar results.

Similar results were obtained whether T/10 and T/2 planes were exposed as a stepped specimen or separately.

Salt Spray Tests Versus Immersion Tests

With regard to reproducibility, the test results show the three immersion tests to be better than the MASTMAASIS but not better than the SWAAT. The switch in emphasis, therefore, toward immersion tests was based on considerations other than reproducibility, such as equipment cost, speed, simplicity, and better discrimination of borderline cases by reducing the number of visual questionable ratings.

Corrosion Potential Measurements

The corrosion potential measurements in CH_3OH-CCl_4 electrolyte can clearly identify material of a high and low resistance to exfoliation. As a quality control test for heat treatment it has the advantages of being a rapid and easily performed test. The chief disadvantage is that it produces no visual evidence of exfoliation.

Conclusions

Based on the results of this round robin testing program it has been concluded by the task group members that the EXCO immersion test is the best laboratory exfoliation test method for the copper bearing 7000 series alloys.

Status

A draft of a recommended practice for the EXCO test has been prepared and is being reviewed by the task group.

Natural environmental tests of the material used in this round robin in marine and industrial atmospheres have been started by three of the laboratories represented in the task group. The Alcoa test site is at Point Judith, R. I., a marine atmosphere. Exposure dates were July 1970 for the 7178 plate and June 1971 for the 7075 extrusion. The Kaiser exposure site is at Daytona Beach, Fla., with the 7178 plate specimens exposed in Aug. 1970, and the 7075 extrusions in Oct. 1970. The Reynolds site is at McCook, Ill., a severe industrial environment, and exposure began there in Oct. 1970.

Acknowledgments

The authors gratefully acknowledge the participation and contributions of the following collaborators in the interlaboratory testing program and their companies: Chairman: D. O. Sprowls, Alcoa Research Laboratory; S. G. Epstein, The Aluminum Association; B. R. Pathak, Alcan Research and Development Ltd.; H. L. Craig, Jr., Reynolds Metals Co.; A. W. Morris, McDonnell-Douglas Corp.; G. W. Kelly, LTV Aero Corp.; J. F. Pashak, Dow Chemical Co.; T. J. Summerson, Kaiser Aluminum and Chemical Corp.; W. C. Rotsell, Harvey Aluminum Inc.; and I. S. Shaffer, Naval Air Development Center.

APPENDIX

MASTMAASIS and SWAAT Salt Spray Tests

Apparatus

1. A commercial fog cabinet designed to comply with ASTM B 117–64 specification for salt spray testing and modified to permit automatic cycling. Both tests operated at 49 C.
2. 6-h cycle for MASTMAASIS test: 4 per day as follows:
   (*a*) 45-min spray.
   (*b*) 2-h purge with air.

(c) 3-h, 15-min soak period at 40 to 95 percent relative humidity.
3. 2-h cycle for SWAAT test: 12 per day as follows:
Alternately spray 30 min and soak 1 h and 30 min at 100 percent humidity.

Solution

1. MASTMAASIS—5 percent NaCl (distilled water) buffered to pH 3 with acetic acid.

2. SWAAT—Salt solution prepared by dissolving 42 g of synthetic sea salt which meets ASTM D 1141–52 specification without heavy metal additions and 10 ml of glacial acetic acid in 1 liter of distilled water.

Test Procedure

1. Degrease specimens.[7]
2. Support or suspend the specimens with the 6 in. dimension 45 deg from the vertical, with the machined surfaces upward.
3. Test duration: 168 h.
4. Clean the corroded specimens gingerly by immersing in concentrated nitric acid only for as long as necessary to remove all corrosion products, taking care not to dislodge anymore flakes of exfoliated metal than can be helped.

Immersion Tests

All the immersion tests had the following procedures in common:
1. Degrease specimens.
2. Use reagent grade chemicals and distilled or deionized water.
3. After completion of tests, clean corroded specimens the same as after salt spray tests.

EXCO Immersion Test[8]

Solution: 4.0 *M* NaCl.
0.5 *M* KNO_3.
0.1 *M* HNO_3.
pH: 0.4.
Test temperature: 25 C ± 2.
Ratio of solution volume to specimen surface area: 50 ml/in.^{2}
Test duration: normally 48 h although 24 h is adequate to demonstrate mode of attack.
Note—In the interlaboratory testing program the specimens were suspended vertically. The ASTM recommended practice will specify the horizontal position.

NH_4Cl-NH_4NO_3-H_2O_2 Immersion Test

Solution: 0.5 *M* NH_4Cl.
0.125 *M* NH_4NO_3.
10 g/liter of a 30 percent stock solution of H_2O_2.
pH: 5.2

[7] In the interlaboratory testing program both etched and unetched specimens were tested.

[8] Different solution concentrations, various chloride/nitrate ratios, time periods and temperatures (20 to 35 C) were explored by Alcan. The conditions recommended were found to be the optimum.

Ratio of solution volume to specimen surface area: 40 to 100 ml/in.^{2} (including edge).

Test duration: 16 h.

Specimens suspended vertically with nonmetallic hangers. Top end of specimen at least 1 in. below surface of solution.

$NaCl$-SO_2 Immersion Test

In this method SO_2 gas is produced, via chemical reaction of Na_2SO_3 + H_2SO_4, above specimens immersed in a 5 percent NaCl solution. The chemical reaction occurring is:

$$Na_2SO_3 + H_2SO_4 -> Na_2SO_4 + H_2SO_3$$
$$-> H_2O + SO_2$$

This test was developed to simulate the environment of an aircraft carrier with sea spray and stack gases containing sulfur.

Apparatus: See Fig. 11.

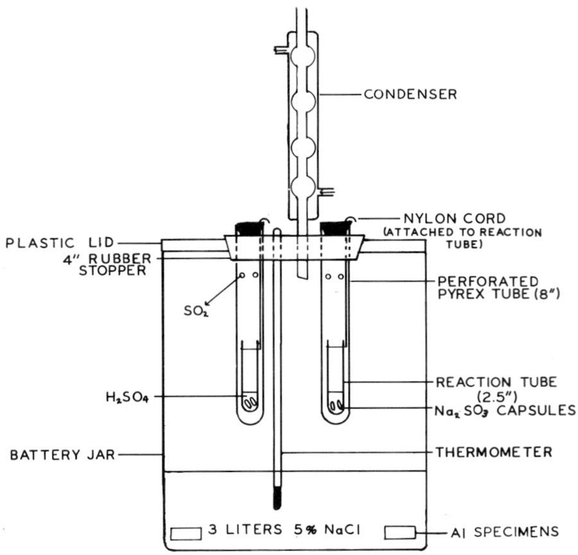

FIG. 11—*$NaCl$-SO_2 immersion test.*

Solution: 5 percent NaCl—in each reaction tube 2 capsules of Na_2SO_3 (1 capsule contains 1.4 ± 0.2 g) are mixed with not less than 8 ml of concentrated H_2SO_4.

Test temperature: 42 C ± 2.

Ratio of solution volume to specimen surface area: 50 to 100 ml/in.^{2}

Test duration: 72 h.

Specimens tested in horizontal position.

Corrosion Potential Measurements

Apparatus: See Fig. 12. The voltmeter can be a pH meter generating on the 0 to 1400 mV scale.

Solution: CH_3OH, absolute—350 ml; CCl_4—150 ml; $CuCl_2$, anhydrous—2.75 g.

Specimen selection: Triplicate specimens from each lot shall be tested. Speci-

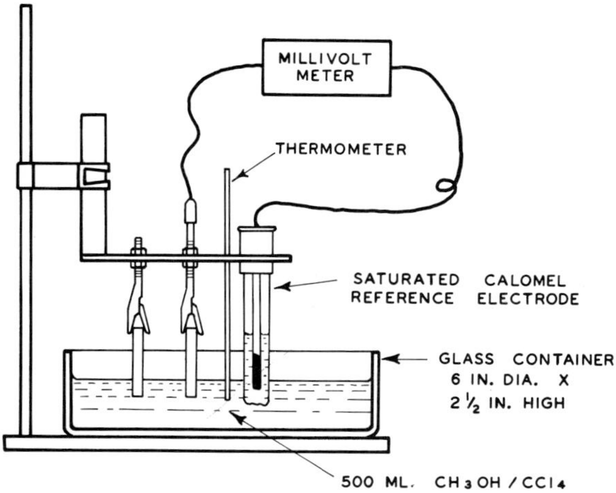

FIG. 12—*Assembled apparatus used in corrosion potential test.*

men size shall be ¼ by ¼ by 2 in., wherever possible, with the 2 in. dimension in either the long transverse or longitudinal direction. With material thicker than ¼ in., the specimen width shall be such that an area close to ½ in.^{2} per specimen is exposed to the solution when the specimen is immersed to the prescribed depth of ½ in. (For example, a ¹⁄₁₆-in.-thick specimen can have a width of ½ in.).

Specimens shall not be taken any closer to an exposed edge (after aging) than the thickness of the material.

Test duration: 35 min.

Suspend specimens so that the bottom edges are in the same horizontal plane.

Heat Treatment Details

NOTE—These treatments were not intended to produce commercial tempers but rather were customized to result in different degrees of resistance to exfoliation.

7178 Plate

Plant production solution heat treat at 875 F, cold-water quenched and stretcher straightened.

Artificially aged in laboratory furnace as follows:

Low resistance—Lot 1—3 h at 250 F plus 4 h at 325 F.
Low resistance—Lot 2—3 h at 250 F plus 8 h at 325 F.
High resistance—3 h at 250 F plus 18 h at 325 F.

(NOTE—All plates from 1 ingot source, all laboratory aging done in a single furnace load with each item removed after the above indicated second step soak).

2024 Plate

Solution heat treated for 30 min at 915 to 920 F metal temperature followed by a boiling water quench (nonstandard quench media used to induce high susceptibility to exfoliation).

7075 Extrusion

Plant production solution heat treat at 870 F, cold-water quenched and aged 24 h at 250 F.

Subsequently artificially aged in laboratory furnace as follows:
Low resistance—Lot 1—8 h at 310 F.
Low resistance—Lot 2—2 h at 310 F.
High resistance—20 h at 325 F.

References

[*1*] Lifka, B. W. and Sprowls, D. O., *Corrosion,* SBIIA, Vol. 22, No. 1, Jan. 1966, p. 7.

[*2*] Romans, H. B., *Materials Research and Standards,* MTRSA, Nov. 1969.

[*3*] Ketcham, S. J. and Shaffer, I. S. in *Proceedings,* 4th International Congress on Metallic Corrosion, 1969 (to be published).

[*4*] Horst, R. L., Jr., and Lifka, B. W., *Corrosion,* SBIIA, Vol. 26, No. 3, March 1970.

[*5*] Zaretski, E. M. and Kireeva, A. F., *Zavodskaya Laboritoriya,* ZVDLA, Vol. 29, 1963, pp. 1098–1101.

E. D. Verink, Jr.,[1] *and R. H. Heidersbach, Jr.*[2]

Evaluation of the Tendency for Dealloying in Metal Systems

REFERENCE: Verink, E. D., Jr., and Heidersbach, R. H., Jr., **"Evaluation of the Tendency for Dealloying in Metal Systems,"** *Localized Corrosion—Cause of Metal Failure, ASTM STP 516,* American Society for Testing and Materials, 1972, pp. 303–322.

ABSTRACT: Literature in the field of dealloying is reviewed. Special emphasis is placed on promising new techniques which appear to have predictive capacity in assessing the likelihood of dealloying. These involve use of electrochemical hysteresis methods to generate experimental potential versus pH diagrams for alloys. Superposition of these experimental diagrams over the theoretical potential versus pH diagrams (Pourbaix diagrams) for the constituent metals of the alloy provides a basis for prediction of the tendency for dealloying as a function of potential and pH. Alpha brass (70Cu-30Zn) in 0.1 *M* sodium chloride is used as an example. The method provides new insights into the explanation of the mechanisms involved in dezincification. For initially copper-free chloride solutions, ranges of potential are indicated in which selective leaching of zinc predominates, where alloy dissolution with replating is expected, and where alloy dissolution without replating occurs.

KEY WORDS: dezincification, electrochemistry, corrosion, tests, evaluation, copper-containing alloys, brasses, potential theory, pH

This paper reviews the current status of dealloying corrosion in metal systems giving particular emphasis to newer methods and techniques of investigation and evaluation which offer special promise.

Dealloying is a corrosion process whereby one constituent of an alloy is removed preferentially from the alloy leaving an altered residual structure [*1,2*].[3]

While dezincification is the most commonly experienced form of dealloying other examples have been reported in practice [*3–13*]. These include loss of nickel [*3,14–16*], aluminum [*10,17–23*] and tin [*24,25*]

[1] Professor and assistant chairman, Department of Metallurgical and Materials Engineering, University of Florida, Gainesville, Fla. 32601.

[2] Department of the Army, Construction Engineering Research Laboratory, Champaign, Ill.

[3] The italic numbers in brackets refer to the list of references appended to this paper.

from copper alloys; iron from cast iron [*26*]; nickel from alloy steels [*27*]; and cobalt from Stellite [*28*].

Since the phenomenon was first reported by Calvert and Johnson [*29*] in 1866, the literature has been filled with reports of research efforts aimed at clarifying the mechanisms of dealloying. Nonetheless, there still is no general agreement as to the detailed mechanism involved. One group contends that the entire alloy is dissolved and that one of its constituents then is replated from solution [*24,28,30–41*]. Another contends that one species is dissolved selectively from the alloy leaving a porous residue of the more noble species [*16,42–48*]. Still others believe that both mechanisms take place [*17,49–55*]. The authors agree with this third group for reasons which will become evident.

The study of dealloying phenomena is fraught with a number of complications. Generally, such reactions are relatively slow, and a lengthy exposure period is required to cause dealloying of sufficient amount to facilitate evaluation. Consequently, there is considerable interest in accelerated tests for evaluation of tendencies of alloys to dealloy. Many techniques have been employed. For example, electrolyte compositions have been adjusted by using more concentrated solutions or solutions having variations in oxidizing power [*53*]. Specific ions have been added to stimulate dealloying, for example, saturated cuprous chloride solutions have been used to accelerate the dezincification of copper-base alloys [*56*]. Electrochemical stimulation also has been used. Unfortunately, all too often the test methods employed can be criticized as having biased the experimental result, and, although specific techniques are now available which can cause dealloying to occur in the laboratory, nevertheless, there still is no firm basis for predicting the likelihood of dealloying in service based on these tests.

Recent Work on Mechanism

Much of the present dilemma regarding the mechanism of dezincification has resulted from physical observations which seemed to support one or the other of the two most accepted theories regarding mechanism.

X-ray diffractometer data from α-brass [*1*] exposed 20 to 30 days in *5 N* hydrochloric acid (HCl) at 50 C show interpeak scattering between the position of the (111) peaks for α-brass and pure copper. This indicates a face-centered-cubic structure of lattice parameter intermediate between the starting material and that of pure copper and is evidence of a selective leaching mechanism. The intensity between the two peaks on the corroded specimens is higher than would be obtained due to overlap between the "tails" of the copper and the brass peaks. However, this intensity probably would go undetected on a standard powder pattern obtained by film methods. Recent X-ray diffraction and electron microprobe studies have produced data which indicate that a selective removal process of dealloy-

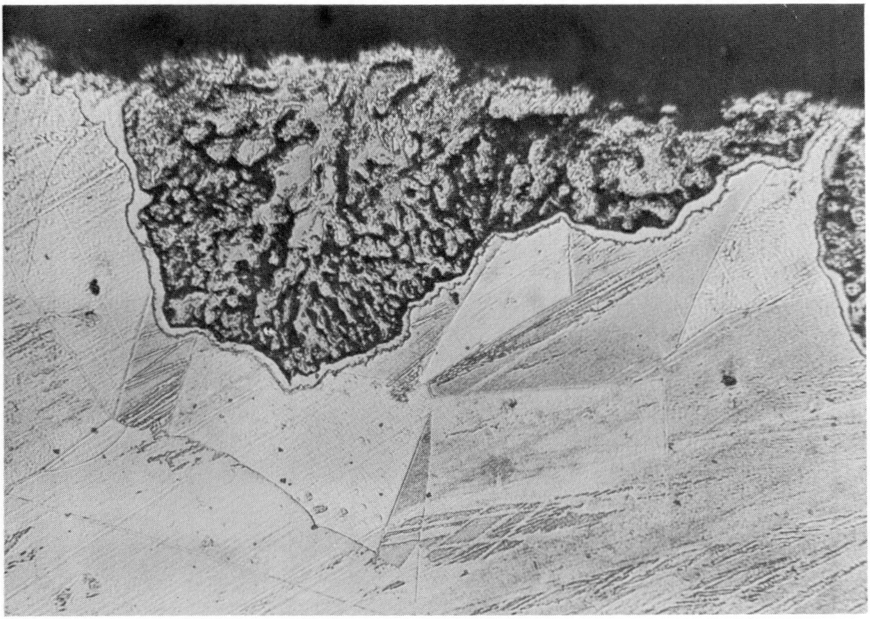

FIG. 1—*Dezincification plug in 70–30 brass exposed for 79 days in 1* N *NaCl at room temperature* (×200).

ing can occur in copper-gold and copper-zinc alloys under certain circumstances [*1,57–59*].

Figure 1 shows a "classical" example of plug-type dezincification in 70Cu-30Zn, α-brass. At the interface between the unattacked alloy and the copper "sponge" there is a dense copper-rich metallic layer; the porous nature of the sponge offers little resistance to the circulation of solution to the reaction interfaces.

Figure 2 shows a dezincified region below the original metal surface. This is one of a group of specimens which showed X-ray evidence of selective leaching of zinc. The perturbations above the original surface are large, dense deposits of copper. Thus, this specimen illustrates that both mechanisms can occur on one specimen.

Abrams [*31*] first showed that copper deposition could occur where flow was restricted. Pits and crevices are practical examples of flow-restricted configurations. Figure 3 is a metallographic cross section of a specimen of Monel 400 showing a deposit of copper which had occurred in a crevice after exposure for 12 months in seawater.

Figure 4*a* is a scanning electron micrograph (SEM) of a copper deposit on an α-brass specimen which had been exposed for 10 days in 5 *N* HCl at 50 C. Figure 4*b* is a nondispersive X-ray pattern obtained using the SEM and shows the formation to be a copper deposit.

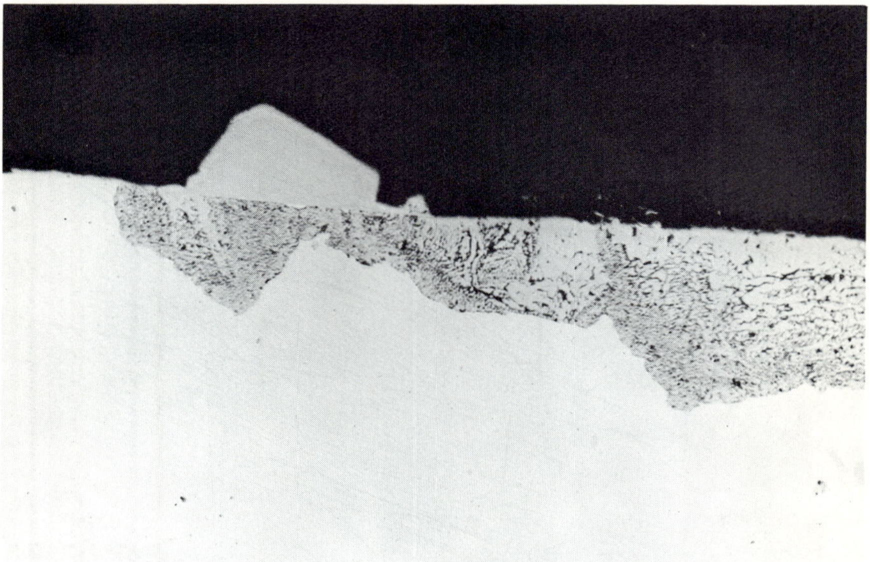

FIG. 2—*Copper deposit on surface and selective leaching of zinc (spongy area) occurred on the same 70–30 brass specimen dezincified for 10 days in 5* N *HCl at 100 C (×121).*

FIG. 3—*Copper deposit on Monel 400 in crevice-corrosion specimen exposed to seawater for 12 months (×268). Specimen, courtesy U. S. Naval Ship Research and Development Center.*

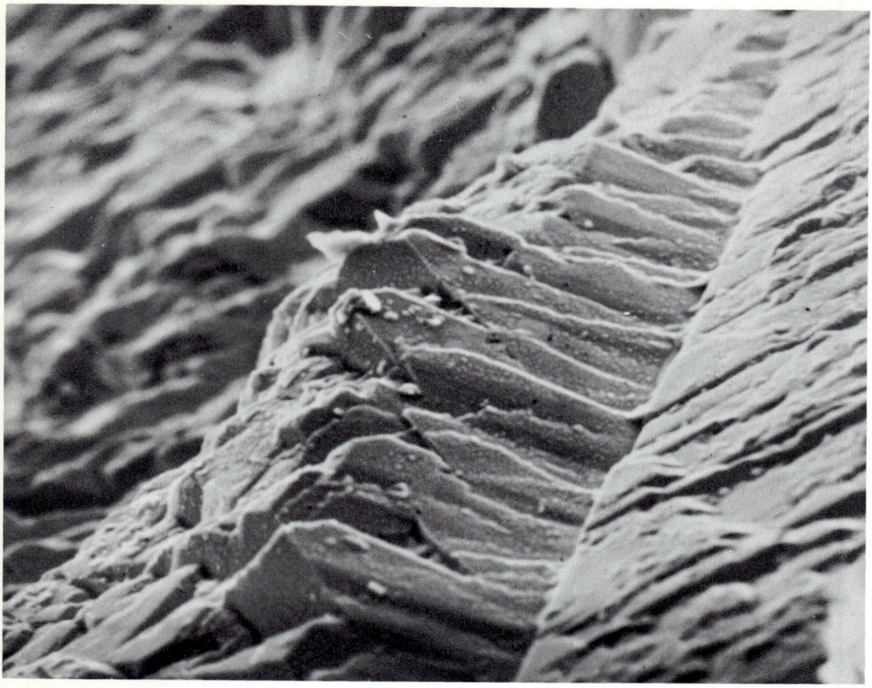

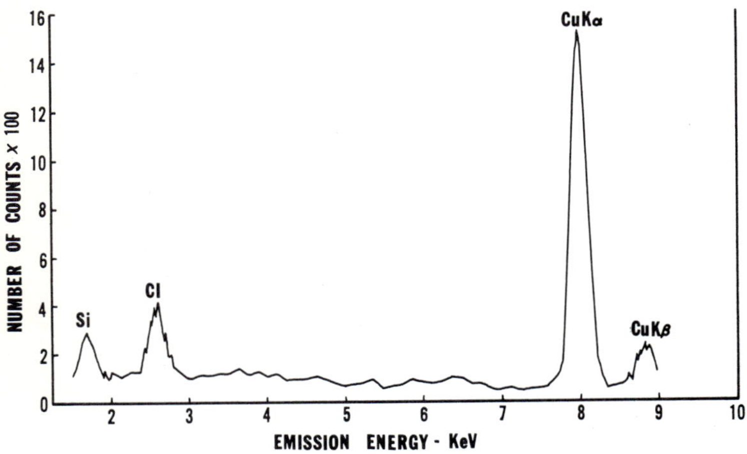

FIG. 4—(a) *Scanning electron micrograph of copper deposit on the surface of a 70–30 brass specimen exposed for 10 days in 5* N *HCl at 50 C* (×*1000*) and (b) *nondispersive X-ray pattern of mineral shown in* (a).

New Techniques

Work now underway at the University of Florida provides encouragement that the tendency for dealloying may be predictable on the basis of experimentally determined potential/pH diagrams constructed from

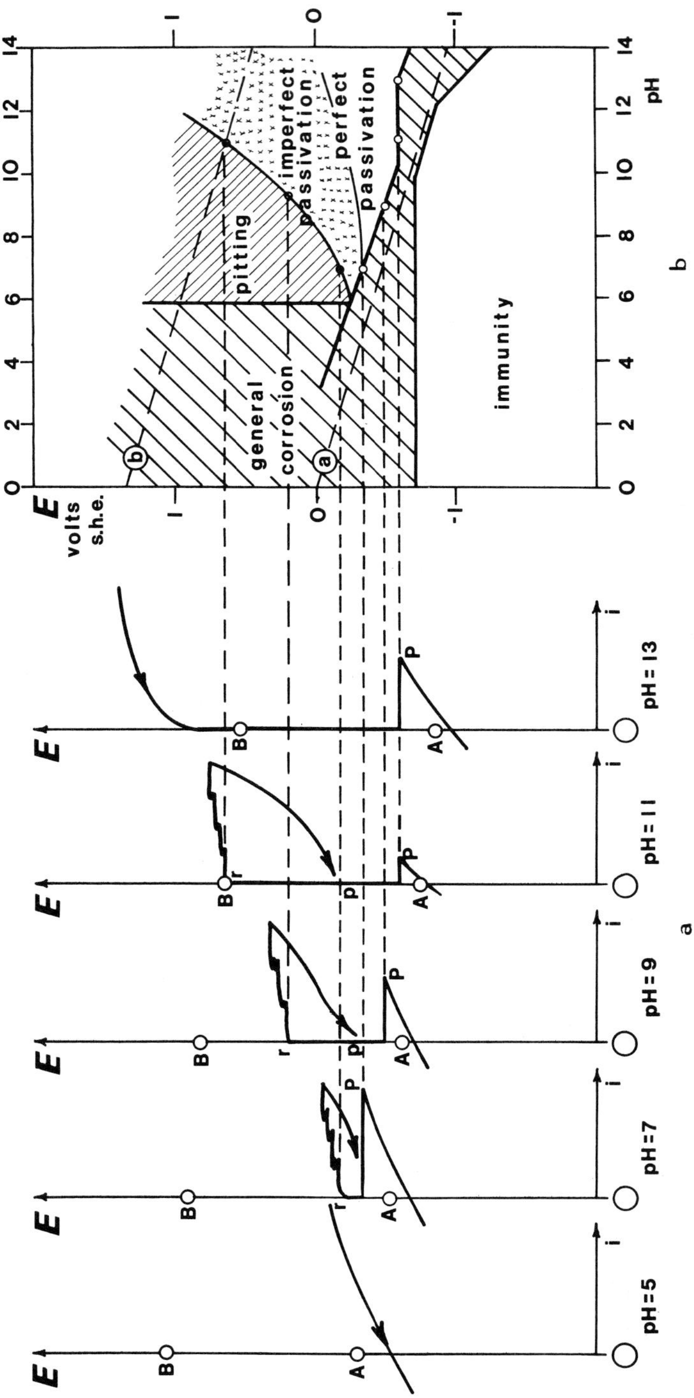

FIG. 5—(a) *Potentiokinetic polarization curves using electrochemical hysteresis method for Armco iron in* 10^{-2} M *chloride solutions of various pH's and* (b) *experimental potential/pH diagram constructed from electrochemical hysteresis data in* (a).

electrochemical information generated by the electrochemical hysteresis method [*60,61*]. In this method, potentiokinetic polarization curves (potential/current density) are run from minus potentials to plus and then returned over a range of interest in solutions of various pH's. Electrochemical features such as "zero current potentials," "passivation potentials," "rupture potentials," and "protection potentials" are noted and plotted on potential/pH coordinates as indicated in Figs. 5*a* and *b*. Connecting the data points delineates regions of "immunity," "general corrosion," and "passive" behavior. Details of the experimental procedure appear elsewhere [*60*]. Figure 6 is the equilibrium potential versus pH diagram drawn by Van Muylder et al [*62*] for the Cu-Cl-H_2O system assuming a chloride ion concentration of 0.1 *M*. Figure 7 is a simplified version of this diagram assuming ionic species to be present in concentrations of 10^{-6} *M*. Figure 8 is a similar simplified potential versus pH equilibrium diagram for the Zn-H_2O systems [*64*]. The equations used in constructing these diagrams are given in Table 1. Figure 9 is the experimental potential versus pH diagram for 70–30 Cu-Zn alloy prepared by Fort.[4] This diagram was constructed using techniques just described. Superposition of the diagrams, Figs. 7–9 gives Fig. 10. In making such a superposition the assumption is made that from an energy standpoint behavior of zinc atoms in the 70-30 brass matrix will be similar to zinc atoms in a zinc matrix to a first approximation. It is evident that the experimentally constructed diagram shares a number of features with the equilibrium (potential versus pH) diagram for copper. For example, in acid solutions the line, #76, separating copper from $CuCl^-_2$ (molarity $= 10^{-6}$) virtually coincides with the zero current potential on the upward scan of the electrochemical hysteresis circuit for α-brass (70Cu-30Zn) in 0.1 *M* chloride.

The boundary between Cu_2O and copper, line #12, corresponds to the sloping line segment of the experimental diagram parallel to it extending from about pH = 7.3 to 8.3. The displacement downward is believed to be related to the zinc content of the alloy. The "jog" in this line at pH $\cong$ 8.3 may be caused by the presence of ZnO on the surface of the alloy above pH = 8.3.

Lines 12 and 12*b* represent the Cu/Cu_2O equilibria for the nonhydrated and hydrated species of Cu_2O, respectively. The corresponding line on the experimental diagram coincides with the position of the calculated line for hydrated Cu_2O (see Fig. 7). The films formed on the specimens were too thin for positive identification by X-ray methods. It is hoped that optical methods will be successful in identifying them.

The line on the experimental diagram above and parallel to the Cu/Cu_2O equilibrium at pH's above 8.3 coincides with the calculated position of the metastable equilibrium line for the coexistence of copper with

[4] Fort, W. C., University of Florida, private communication.

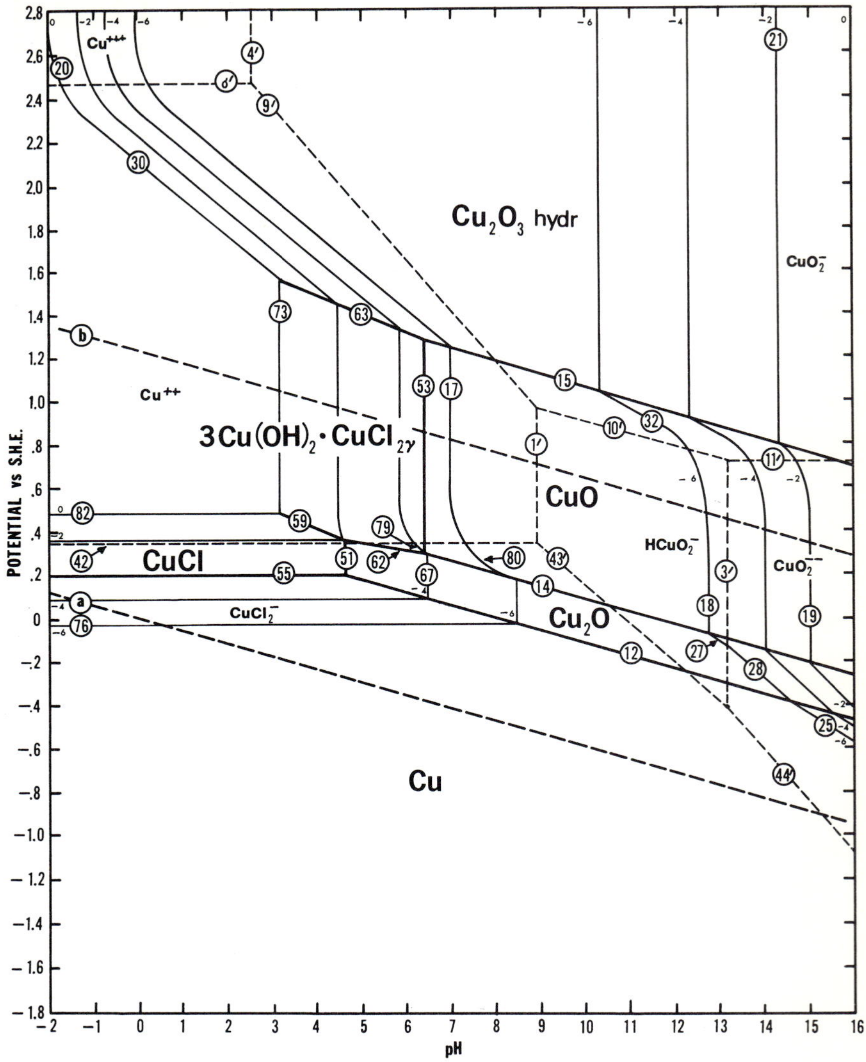

FIG. 6—*Equilibrium potential versus pH diagram for the Cu-Cl-H_2O system at 25 C for solutions containing 0.1* M *chloride ion. Solid species were assumed to be Cu, Cu_2O, CuCl, $3Cu(OH)_2{\cdot}CuCl_{2\gamma}$, CuO, and Cu_2O_3 hydrated* [62].

$3Cu(OH)_2{\cdot}CuCl_{2\gamma}$ in 0.1 *M* chloride (shown as Eq 57 in Table 1 and in Fig. 7). It also is fairly close to the calculated position of line 14*b* (see Table 1) which involves hydrated Cu_2O in equilibrium with CuO. However, the greenish color and the presence of chlorides in the reaction product film strongly suggest that at least some metastable tri-hydroxy-chloride is present. The difference in slope between −0.0443 required for Eq 57

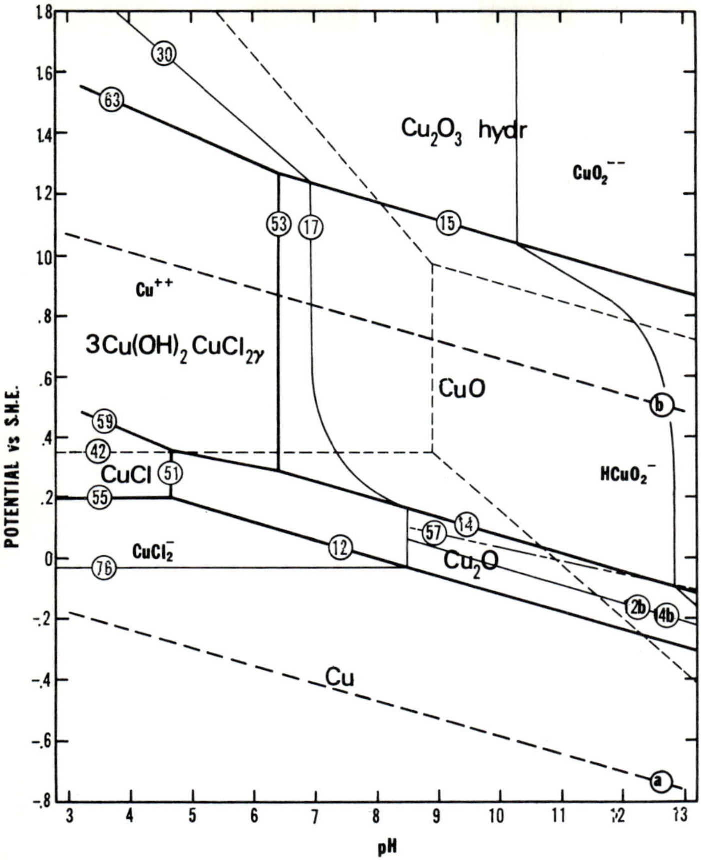

FIG. 7—*Simplified Cu-Cl-H_2O diagram at 25 C for solutions containing 0.1* M *chloride ions and concentrations of ionic species equal 10^{-6}* M. *Redrawn from Ref* 62.

and -0.0591 required by Eq 14 is too subtle to discern as yet considering the scatter in the present data for α-brass (70Cu-30Zn).

On the return scan of the electrochemical hysteresis circuit a zero current potential at 0.200 V_{SHE} (SHE refers to the standard hydrogen electrode) is observed which is independent of pH, at least to pH = 11. This feature is observed with pure copper and with copper-rich alloys of the copper-nickel and of the copper-zinc systems exposed in chloride solutions. These loci coincide with the position of the Cu/CuCl equilibrium potential in 0.1 *M* chloride.

Relevance to Dealloying

Although a vast amount of research remains to be done to verify the suggestions which are presented herein, there is considerable encourage-

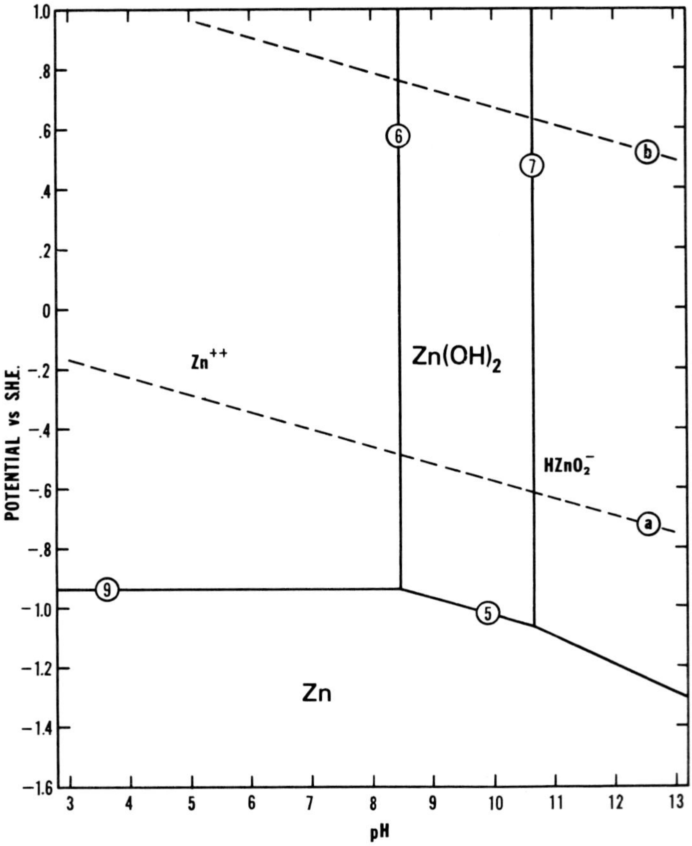

FIG. 8—*Simplified Zn-H₂O diagram for concentration of ionic species equal 10*$^{-6}$ M. *Redrawn from Ref* 63.

TABLE 1—*Equations used in construction of potential versus pH diagrams for the Cu-Cl-H_2O system and the Zn-H_2O system.*

Cu-C*l*-H_2O System [*62*] Eq No.[a]	
12	$2Cu + H_2O = Cu_2O + 2H^+ + 2e$
(*a*)	$E = 0.471 - 0.0591$ pH (no hydrated oxides)
(*b*)	$E = 0.572 - 0.0591$ pH (assumes hydrated Cu_2O)
14	$Cu_2O + H_2O = 2CuO + 2H^+ + 2e$
(*a*)	$E = 0.669 - 0.0591$ pH
(*b*)	$E = 0.568 - 0.0591$ pH (assumes hydrated Cu_2O, but nonhydrated CuO)
16	$2CuO + H_2O = Cu_2O_3 + 2H^+ + 2e$
(*a*)	$E = 1.648 - 0.0591$ pH

TABLE 1 (*continued*)

17 $$Cu^{++} + H_2O = CuO + 2H^+$$

(*a*) $\log(Cu^{++}) = 7.89 - 2\ pH$

42 $$CuCl_2^- = Cu^{++} + 2Cl^- + e$$

(*a*) $E = 0.465 = 0.0591 \log \frac{(Cu^{++})}{(CuCl_2^-)} + 0.1182 \log (Cl^-)$

51 $$2CuCl + H_2O = Cu_2O + 2Cl^- + 2H^+$$

(*a*) $\log(Cl^-) = -5.66 + pH$

53 $$3Cu(OH)_2 \cdot CuCl_{2\gamma} = 4CuCO + 2Cl^- + 2H^+ + 2H_2O$$

(*a*) $\log(Cl^-) = -7.40 + pH$

55 $$Cu + Cl^- = CuCl + e$$

(*a*) $E = 0.137 - 0.0591 \log(Cl^-)$

57 $$4Cu + 6H_2O + 2Cl^- = 3Cu(OH)_2 \cdot CuCl_{2\gamma} + 6H^+ + 8e$$

(*a*) $E = 0.461 - 0.0443\ pH - 0.0148 \log(Cl^-)$

59 $$4CuCl + 6H_2O = 3Cu(OH)_2 \cdot CuCl_{2\gamma} + 2Cl^- + 6H^+ + 4e$$

(*a*) $E = 0.785 - 0.0886\ pH + 0.0295 \log(Cl^-)$

62 $$2Cu_2O + 4H_2O + 2Cl^- = 3Cu(OH)_2 \cdot CuCl_{2\gamma} + 2H^+ + 4e$$

(*a*) $E = 0.451 - 0.0295 \log(Cl^-) - 0.0295\ pH$

67 $$2CuCl_2^- + H_2O = Cu_2O + 4Cl^- + 2H^+$$

(*a*) $\log(CuCl_2^-) = 4.45 - pH + 2 \log(Cl^-)$

76 $$Cu + 2Cl^- = CuCl_2^- + e$$

(*a*) $E = 0.208 + 0.0591 \log(CuCl_2^-) - 0.1182 \log (Cl^-)$

80 $$CuCl_2^- + H_2O = CuO + 2Cl^- + 2H^+ + e$$

(*a*) $E = 0.932 - 0.1182\ pH - 0.059 \log(CuCl_2^-) + 0.1182 \log (Cl^-)$

82 $$CuCl = Cu^{++} + Cl^- + e$$

(*a*) $E = 0.537 + 0.0591 \log(Cu^{++}) + 0.0591 \log(Cl^-)$

$Zn\text{-}H_2O$ System [*63*]
Eq No.

5 $$Zn + H_2O = ZnO + 2H^+ + 2e$$

(*a*) $E = -0.439 - 0.0591\ pH$

6 $$Zn^{++} + H_2O = ZnO + 2H^+$$

(*a*) $\log(Zn^{++}) = 10.96 - 2\ pH$

7 $$ZnO + H_2O = HZnO_2^- + H^+$$

(*a*) $\log(HZnO_2^-) = -16.68 + pH$

9 $$Zn = Zn^{++} + 2e$$

(*a*) $E = -0.763 + 0.0295 \log(Zn^{++})$

[a] Equation numbers refer to circled numbers on diagrams, Figs. 9–13.

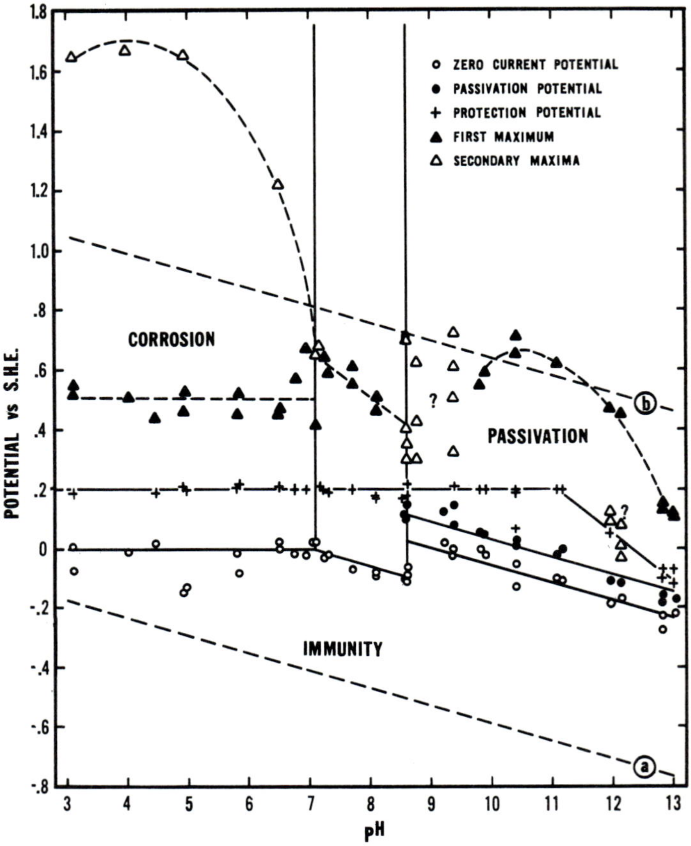

FIG. 9—*Experimental potential versus pH diagram for 70Cu-30Zn in 0.1* M *chloride at 25 C.*[4]

ment to believe that experimental potential versus pH diagrams, interpreted in conjunction with the equilibrium Pourbaix diagrams of the constituent metals, can provide useful information regarding the theoretical tendency for dealloying. Such predictions must be verified by long-term tests to determine reaction kinetics and the compositions and morphologies of reaction products. Such collateral investigations have been facilitated greatly by the development of modern research tools including new types of X-ray analysis, scanning electron microscopy, electron microprobe analysis, ellipsometry, atomic absorption spectroscopy, etc.

For pure copper in deaerated solutions, the zero current potential on the upward potential sweep of the electrochemical hysteresis circuit indicates the position of the so-called "immunity" line at a given pH. As in the case of theoretical Pourbaix diagrams, "immunity" does not imply absolute freedom from corrosion but rather a negligible amount

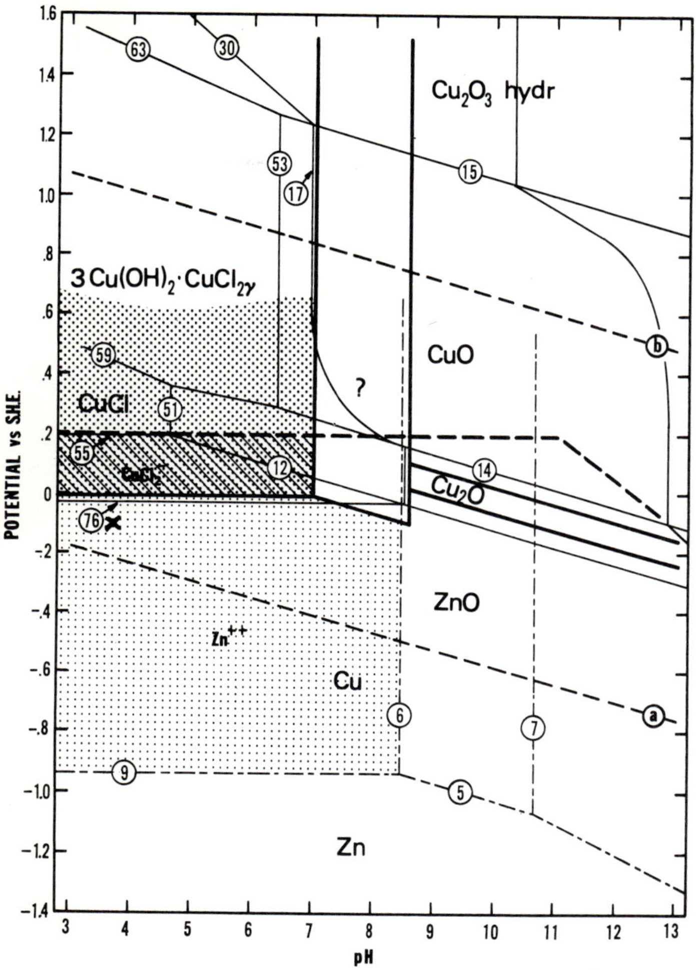

FIG. 10—*70Cu-30Zn alloy in 0.1* M *chloride solutions. Superposition of Figs. 7–9. Small dots indicate the domain in which selective removal of zinc is expected in solutions free of copper ions. Larger dots indicate the domain in which both copper and zinc dissolve. Crosshatching indicates the region in which copper is expected to deposit.*

of corrosion. The potential at which this zero current is observed often is fairly close to the calculated position of the metal/metal ion coexistence potential for a metal ion concentration of 10^{-6} *M*. Thus, the arbitrary choice of 10^{-6} *M,* often chosen as a definition of "noncorrosion," seems appropriate. Alloys which are rich in one component (for example, 70–30 Cu-Zn or 90–10 Cu-Ni) tend to have many features in common with the diagram for the major component (in this case, copper). The "immunity" line, however, appears to have a somewhat different significance for alloys than for pure metals. While the kinetics of alloy dissolution for α-brass are slow below the alloy "immunity" line this does not rule out the possibility of dealloying. Referring to Fig. 10, the immunity line for 70–30 Cu-Zn in 0.1 *M* chloride was about 0.000 V_{SHE}. Between this potential and approximately −0.940 V_{SHE} (line #9) there is a theoretical tendency for the selective removal of zinc from the alloy. Polycrystalline specimens of 70–30 Cu-Zn potentiostated at −0.150 V_{SHE} in 0.1 *M* chloride of pH = 4 (noted by "X" on Fig. 10) showed zinc ion pickup in the solution without the presence of detectable amounts of copper after 107 h at room temperature, based on data obtained by atomic absorption spectroscopy. The limit of detection by this method is less than 0.25 ppm. Pickering and Byrne [57] showed zinc ions but no detectable copper ions from α-brass exposed in sodium sulphate (Na_2SO_4) solutions of pH = 5 after 20 h potentiostated at various potentials below about +0.050 V_{SHE}.

Some researchers add copper ions to accelerate dezincification. Accelerated test methods in which copper ions are added to the solution will result in deposition of copper on the surface of the alloy at any potential more active than about +0.200 V_{SHE}, as shown in Fig. 10. Therefore, one might be led to the erroneous conclusion that only one mechanism exists (dissolution of both components of the alloy followed by deposition of the more noble species). Below about 0.000 V_{SHE}, as shown in Fig. 10, selective leaching can occur in the absence of copper ions.

At potentials more positive than 0.000 V_{SHE}, both constituents of the 70–30 Cu-Zn alloy go into solution. Atomic absorption studies indicate that the ratio of copper/zinc for α-brass in the bulk 0.1 *M* chloride approaches that of the alloy (above 1.5) very quickly above 0.000 V_{SHE}, perhaps within less than 100 mV. Pickering and Byrne [57] report that "two modes of dissolution may occur depending on the potential: preferential dissolution of the more noble metal to simultaneous dissolution of both components." They report preferential dissolution below about +0.200 V_{SHE} and simultaneous dissolution above +0.200 V_{SHE}.

Pickering and Byrne's studies were carried out in Na_2SO_4 solutions at pH = 5. According to those authors, Cu^{++} was believed to predominate. Studies at University of Florida were carried out in chloride solutions. Under these circumstances cuprous ion is favored as the result of the stabilizing effect of the formation of $CuCl^{-}{}_2$ ion [65]. Consequently, for a given amount of current passed, the equivalent weight of copper dissolved would be greater in chloride solutions than in solutions free of chloride ion. This probably accounts for our observation that copper/zinc ratios approaching that of the alloy were achieved very soon after raising the potential above 0.000 V_{SHE}. Observation of the specimens under test in the potential range between 0.000 V_{SHE} and +0.200 V_{SHE} showed no visual evidence of copper enrichment on the specimen surface when solutions were stirred. Copious quantities of copper were deposited on the auxiliary electrode, however. Tests in which no stirring was done also resulted in some deposition on the platinum auxiliary electrode; however, noticeable amounts of copper also were deposited loosely on the surface of the specimens. These observations confirm the conclusions of Abrams [*31*] on the effect of stagnation. Crevices and other occluded cells may be considered extreme cases of stagnation in which concentrations of various species within the occluded cell may vary drastically from those in the bulk solution [*66*].

In summary, at potentials between 0.000 V_{SHE} and +0.200 V_{SHE} both constituents dissolve, but, in the presence of occluded cells or stagnant conditions, copper may deposit on the specimen without evidence of selective leaching. Naturally in this potential range (as in the previously mentioned range below 0.000 V_{SHE}) accelerated tests involving copper ion additions will result in deposition of copper.

Above +0.200 V_{SHE} both constituents of the alloy dissolve in the proportions found in the alloy without evidence of deposition of copper. If the potential has been held above +0.200 V_{SHE} for an appreciable period of time, reducing the potential to +0.200 V_{SHE} or below will result in deposition of copper presumably by the reaction $CuCl + e \rightarrow Cu + Cl^-$.

Summary of Observations (Refer to Fig. 10)

Potential, V_{SHE}	*Observation*
−0.940 to 0.000	1. Preferential dissolution of zinc from 70–30 Cu-Zn alloy essentially without dissolution of copper. Kinetics controlled by "blockage" of the surface by residual copper, and ability of zinc to diffuse to the reaction front.
	2. If copper ions (or other reducible species) are added to the solution, deposition of the added species will result.
0.000 to +0.200	1. Both constituents of the alloy dissolve. The ratio of copper/zinc in solution will be influenced by the valence of the predominant copper ion. In chloride solutions formation of $CuCl^-{}_2$ stabilizes the monovalent copper species and little or no selective leaching of zinc is observed. Some selective leaching of zinc has been reported in nonchloride solutions [*47*].
	2. Copper will deposit from unstirred solutions or in occluded cells.
	3. Copper ions added to the solution will

deposit, particularly in stagnant solutions or in occluded cells.

Above +.200 V_{SHE}

1. Both constituents of the alloy dissolve in the ratio in which they appear in the alloy.
2. No deposition of copper occurs from 0.1 *M* chloride solutions.
3. If potential is dropped to +0.200 or below, deposition of copper will result.

90–10 copper-nickel alloys also undergo dealloying with loss of nickel. The conditions leading to a tendency for dealloying of 90–10 Cu-Ni may be predicted as in the case of α-brass. Figure 11 is a composite of the Cu-Cl-H_2O diagram [*62*] for 0.1 *M* chloride (Fig. 6) plus the equilibrium diagram for nickel [*64*] (with Ni^{++} assumed to be 10^{-6} *M*), plus the experimental diagram developed by Parrish for 90–10 Cu-Ni in 0.1 *M* chloride solutions [*67*].

In the potential range between about 0.000 V_{SHE} and +0.200 V_{SHE}, both constituents of the alloy dissolve. Deposition of copper occurs in crevices [*61*] and pits [*68*] and presumably in unstirred solutions.

Specimens which have been exposed at potentials above 0.200 V_{SHE} will experience deposition of copper upon reducing the potential below 0.200 V_{SHE} in 0.1 *M* chlorides.

Verink and Parrish [*63*] have reported the theoretical possibility of dealloying of 90–10 Cu-Ni in domains where a passive region in the equilibrium diagram for one alloy constituent coincides with a general corrosion region for the other constituents.

Summary

Dealloying phenomena are encountered in a number of metal alloy systems. It is a problem of considerable practical importance in the use of copper-base alloys, particularly those of the copper-zinc system.

For over 50 years there has been disagreement regarding the mechanism of dealloying. It now appears that dealloying may proceed by either a selective leaching process or a dissolution and replating mechanism or both.

The development of new, highly sophisticated investigative tools such as the SEM, microprobe, special X-ray techniques, spectrophotometric analysis, electrochemical techniques, etc., provides great opportunities for characterizing and evaluating dealloying phenomena.

The organization of these observations on an orderly basis such as appears possible by use of potential versus pH diagrams seems to hold great promise in the quest for better understanding. In turn, a clearer understanding of the mechanisms involved will permit development of alloys and procedures with greater engineering reliability.

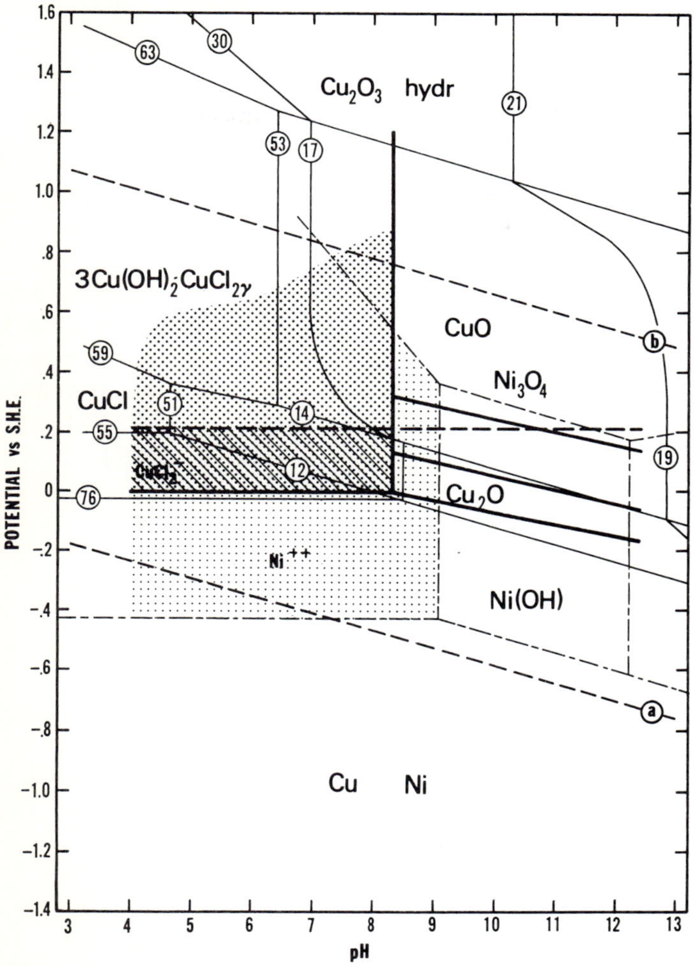

FIG. 11—*90Cu-10Ni alloy in 0.1* M *chloride solution. Superposition of the experimental potential versus pH diagram of Parrish on the Cu-Cl-H_2O diagram (Fig. 7) and the Ni-H_2O diagram for $Ni^{++} = 10^{-6}$* M. *Small dots indicate domain in which selective dissolution of nickel is expected in a copper free solution. Larger dots indicate domain in which both nickel and copper dissolve. Crosshatched area is domain in which copper is expected to deposit.*

Acknowledgments

The electrochemical hysteresis method used to construct experimental potential versus pH diagrams was developed and a portion of the work on 90–10 Cu-Ni alloy was performed as part of the research supported by the Advanced Research Project Agency under Contract No. N-00014-68-A-0173 administered by the Office of Naval Research, Washington, D. C. The balance of the copper-nickel alloy work and part of the copper-zinc alloy work were supported by the Department of Interior, Office of Saline Water, Washington, D. C., under Contract No. 14–30–2600. The National Association of Corrosion Engineers, Houston, Texas, through their Committee on Research, supported the studies of dealloying. To each of these sponsors, the authors express deep appreciation.

Special thanks are due Marcel Pourbaix who served as consultant on the ARPA contract and whose continued association with the University of Florida as co-director of the Center of Applied Thermodynamics and Corrosion serves as an on-going source of stimulation.

Appreciation also is expressed to several graduate research assistants in the Department of Metallurgical and Materials Engineering at the University of Florida; to W. C. Fort, III, for the use of some of his data, to K. K. Starr and T. S. Lee, III, for their helpful review and comments during the preparation of the manuscript.

References

[*1*] Heidersbach, R. H., Jr., *Corrosion,* SBIIA, Vol. 24, No. 2, 1968, pp. 38–44; Vol. 25, No. 7, 1969, pp. 285–286; Vol. 26, No. 5, 1970, p. 217; Vol. 26, No. 8, 1970, p. 256; Vol. 26, No. 10, 1970, pp. 445–447.

[*2*] Tammann, G., *Zeitschrift für Anorganisch Allegemeine Chemie,* ZAACA, Vol. 107, 1919.

[*3*] Bird, D. B. and Moore, K. L., *Materials Protection,* MAPRA, Vol. 1, No. 10, 1962, p. 70.

[*4*] Blumer, A. F., *Corrosion,* SBIIA, Vol. 5, No. 144, 1949.

[*5*] Crenell, J. T. and Sawyer, L. J. E., *Journal of Applied Chemistry,* JACHA, Vol. 12, No. 4, 1962, p. 170.

[*6*] Matsuda, M., *Journal,* Japan Institute of Metals, NIKGA, Vol. 26, No. 1, 1962, p. 124.

[*7*] Niederberger, R. B., *Modern Castings,* MODCA, Vol. 45, No. 3, 1964, p. 115.

[*8*] Polushkin, E. P. and Shuldener, M., *Transactions,* American Institute of Mining, Metallurgical, and Petroleum Engineers, TAIME, Vol. 161, 1945, p. 214.

[*9*] Simmons, A. L., *Metal Progress,* MEPOA, Vol. 57, 1950, p. 496.

[*10*] Tracy, A. W. in *Atmospheric Corrosion of Non-Ferrous Metals, ASTM STP 175,* American Society for Testing and Materials, 1956, p. 67.

[*11*] Van der Baan, Sj., *Corrosion,* SBIIA, Vol. 6, 1950, p. 14.

[*12*] Vernon, W. H. H., *Transactions,* Faraday Society, TFSOA, Vol. 23, 1927, p. 170.

[*13*] Wilkins, R. A., *Mechanical Engineering,* MEENA, Vol. 58, No. 12, 1936, p. 57.

[*14*] Henderson, J. L. and Roadhouse, C. L., *Journal,* Dairy Science, JDSCA, Vol. 23, 1940, p. 215.

[*15*] Stewart, W. C. and LaQue, F. L., *Corrosion,* SBIIA, Vol. 8, 1952, p. 259.

[*16*] Taylor, F. and Wood, J. W., *Engineering,* ENGNA, Vol. 149, 1940, p. 58.

[*17*] Britton, S. C., *Journal,* Institute of Metals, JIMEA, Vol. 67, 1941, p. 119.

[*18*] Gleekman, L. W. and Swandby, R. K., *Corrosion,* SBIIA, Vol. 17, 1961, p. 144t.

[*19*] Hamstead, A. C., *Industrial and Engineering Chemistry,* IECHA, Vol. 50, No. 10, 1958, p. 87A.

[*20*] Hesse, A. H., Myskocoski, E. T., and Loring, B. M., *Transactions,* American Foundrymen's Association, TAFOA, Vol. 51, 1944, p. 821.

[*21*] Peters, B. F., Carson, J. A. H., and Barer, R. D., *Materials Protection,* MAPRA, Vol. 4, No. 5, 1965, p. 24.

[*22*] Schussler, M. and Napolitan, D. S., *Corrosion,* SBIIA, Vol. 12, 1956, p. 107t.

[*23*] Serre, J. and Lawreys, J., *Corrosion Science,* CRRSA, Vol. 2, 1965, p. 135.

[24] Clark, W. D., *Journal,* Institute of Metals, JIMEA, Vol. 73, 1947, p. 263.
[25] Trautzel, P. and Treadwell, W. D., *Helvetica Chimica Acta,* HCACA, Vol. 34, 1951, p. 1723.
[26] Leedon, L. M., *Journal,* American Water Works Association, JAWWA, Vol. 38, 1946, p. 1392.
[27] Dvorak, A., *Strojirentsvi,* STRJA, Vol. 12, No. 1, 1962, p. 39, referenced by *Corrosion Abstracts,* CRNAA, Vol. 2, 1963, p. 11.
[28] Fontana, M. G., *Industrial and Engineering Chemistry,* IECHA, Vol. 39, No. 5, 1947, p. 87A.
[29] Calvert and Johnson, J., *Journal,* Chemical Society, JCSOA, Vol. 19, 1866, p. 436.
[30] Lynes, W., *Proceedings,* American Society for Testing and Materials, ASTEA, Vol. 41, 1941, p. 859.
[31] Abrams, R. B., *Transactions,* American Electrochemical Society, TAESA, Vol. 42, 1922, p. 39.
[32] Nixon, C. F., *Transactions,* American Electrochemical Society, TAESA, Vol. 45, 1924, p. 297.
[33] Hollomon, J. H. and Wulff, J., *Transactions,* American Institute of Mining, Metallurgical, and Petroleum Engineers, TAIME, Vol. 147, 1942, p. 297.
[34] Thompson, D. B., *Australasian Engineer,* AUEGA, Vol. 48, Oct. 1954.
[35] Zender, A. R. and Bulow, C. L., *Heating, Piping, and Air Conditioning,* HPAOA, Vol. 16, 1944, p. 273.
[36] Dixon, E. S., *Bulletin,* American Society for Testing and Materials, ASTBA, No. 102, 1940, p. 21.
[37] Rhodes, F. H. and Carty, J. T., *Industrial and Engineering Chemistry,* IECHA, Vol. 17, 1925, p. 909.
[38] Scully, J. C., *The Fundamentals of Corrosion,* Pergamon Press, London, 1966.
[39] Uhlig, H. H., *The Corrosion Handbook,* Wiley, New York, 1948.
[40] Hashimoto, K., Ogawa, W., and Shimodaira, S., *Journal,* Japan Institute of Metals, NIKGA, Vol. 4, No. 1, 1963, p. 42.
[41] Horton, R. M., *Corrosion,* SBIIA, Vol. 26, No. 7, 1971, p. 160.
[42] Tabor, L. E., *Journal,* American Water Works Association, JAWWA, Vol. 48, No. 3, 1963, p. 239.
[43] Gastev, S. S., *Izvestiia, Akademii Nauk SSR Mettally,* IZNMA, Vol. 3, 1965, referenced in *Corrosion Abstracts,* CRNAA, Vol. 6, 1966, p. 446.
[44] Bialosky, J. M., *Corrosion and Metal Protection,* Vol. 4, 1947, p. 15.
[45] Marshakov, I. K., Bogdanov, V. P., and Aleikina, S. M., *Russian Journal of Physical Chemistry,* RJPCA, Vol. 38, No. 7, 1964, p. 960; Vol. 38, No. 8, 1964, p. 104; Vol. 39, No. 6, 1965, p. 804.
[46] Bassett, W. H., *Chemical and Metallurgical Engineering,* Vol. 27, 1922, p. 340.
[47] Efrid, K. D., M.S. thesis, University of Florida, Gainesville, Fla., 1970.
[48] Rubin, B. T., Ph.D. thesis, University of Pennsylvania, Philadelphia, Pa., 1969.
[49] Piatti, L. and Grauer, R., *Werkstoffe und Korrosion,* WSKRA, Vol. 14, No. 7, 1963, p. 551.
[50] Colegate, G. T., *Metal Industry,* MINYA, Vol. 73, No. 507, 1948, p. 531.
[51] Kenworthy, L. and O'Driscoll, W. G., *Corrosion Technology,* CRRTA, Vol. 2, 1955, p. 247.
[52] Evans, V. R., *The Corrosion and Oxidation of Metals,* E. Arnold, 1960.
[53] Stillwell, C. W. and Turnipseed, E. S., *Industrial and Engineering Chemistry,* IECHA, Vol. 26, 1934, p. 740.
[54] Verink, E. D., Jr., and Parrish, P. A., *Corrosion,* SBIIA, Vol. 26, 1970, p. 5.
[55] Fink, F. W., *Transactions,* Electrochemical Society, TESOA, Vol. 75, 1939, p. 441.
[56] Lucey, V. C., *British Corrosion Journal,* BCRJA, Vol. 1, 1965, p. 7; Vol. 2, 1965, p. 53.
[57] Pickering, H. W. and Byrne, P. J., *Journal,* Electrochemical Society, JESOA, Vol. 116, No. 11, 1969, p. 1492.

[58] Pickering, H. W., *Proceedings,* Conference on Fundamentals of Stress Corrosion Cracking, Columbus, Ohio, 1967, Staehle, R. W., ed., pp. 159–177, published 1969, National Association of Corrosion Engineers.

[59] Pickering, H. W. and Byrne, P. J., *Journal,* Electrochemical Society, JESOA, Vol. 118, No. 2, 1970, p. 209.

[60] Verink, E. D., Jr., and Pourbaix, M., "Use of Electrochemical Hysteresis Techniques in Developing Alloys for Saline Exposures," *Corrosion,* SBIIA, Vol. 27, No. 12, Dec. 1971.

[61] Pourbaix, M., *Corrosion,* SBIIA, Vol. 26, 1970, pp. 431–438.

[62] Van Muylder, J., deZoubov, N., and Pourbaix, M., Report 101, CEBELCOR, Brussels, Belgium, July 1962.

[63] deZoubov, N. and Pourbaix, M., *Atlas of Electrochemical Equilibria in Aqueous Solutions,* Pergamon Press, London, CEBELCOR, Brussels, Belgium, 1966, pp. 406–413.

[64] Pourbaix, M., *Atlas of Electrochemical Equilibria in Aqueous Solutions,* Pergamon Press, London, CEBELCOR, Brussels, Belgium, 1966.

[65] Latimer, W. M., *Oxidation Potentials,* 2nd ed., Prentice-Hall, Englewood Cliffs, N. J., 1961.

[66] Pourbaix, M., Report 127, CEBELCOR, May 1965.

[67] Parrish, P. A., M.S. thesis, University of Florida, Gainesville, Fla., June 1970.

[68] Van Muylder, J., Pourbaix, M., and Van Laer, P., CEBELCOR Report 127, Brussels, Belgium, May 1965.